Boston Studies in the Philosophy and History of Science

Volume 327

More information about this series at http://www.springer.com/series/5710

Johannes Lenhard • Martin Carrier
Editors

Mathematics as a Tool

Tracing New Roles of Mathematics in the Sciences

Springer

Editors
Johannes Lenhard
Department of Philosophy
Bielefeld University
Bielefeld, Germany

Martin Carrier
Department of Philosophy
Bielefeld University
Bielefeld, Germany

ISSN 0068-0346 ISSN 2214-7942 (electronic)
Boston Studies in the Philosophy and History of Science
ISBN 978-3-319-85400-7 ISBN 978-3-319-54469-4 (eBook)
DOI 10.1007/978-3-319-54469-4

Printed on acid-free paper

This Springer imprint is published by Springer Nature
The registered company is Springer International Publishing AG
The registered company address is: Gewerbestrasse 11, 6330 Cham, Switzerland

Dedicated to Ann Johnson (1965–2016)

This volume owes much to Ann. It synthesizes the work of the cooperation group "Mathematics as a Tool" that gathered over several years at the Center for Interdisciplinary Research (ZiF), Bielefeld. The core group consisted of P. Blanchard, M. Carrier, J. Jost, and J. Lenhard. They partly acted as authors; most of the chapters are by scholars who were involved in workshops or other events organized by the ZiF group. Ann, who sadly passed away shortly before the book came out, was more than a brilliant participant in a workshop. She and her work have been an important source of inspiration for the entire group. Our joint work on the history and philosophy of mathematization, including earlier workshops at the ZiF, motivated Martin Carrier and myself to set up the ZiF cooperation group.

My own work owes much to Ann. We enjoyed an unusual collaboration that was based on the stunning experience of thinking on the same wavelength – notwithstanding our different characters and different disciplinary backgrounds. We quickly agreed how history, technology, mathematics, and philosophy could be mixed in fortunate ways. Maybe, we were rather thinking on wavelengths interfering in a way that many crests and no troughs remained. It is a marvellous experience that is, I believe, very rare in a researcher's career.

Ann had the gift to transform persistent effort into uplift for new thought. I hope this volume can achieve something similar. If one or another chapter makes the reader start thinking about what could be fruitfully taken up, changed, and amplified, this is pretty much what Ann would have liked. Let us carry on in her spirit.

Bielefeld, Germany — Johannes Lenhard
Bielefeld, Germany — Martin Carrier

Contents

Contributors

David Aubin Institut de mathématiques de Jussieu-Paris rive gauche (CNRS, Paris-Diderot, UPMC), Sorbonne Universités, Paris, France

Anouk Barberousse Université Paris-Sorbonne, Paris, France

Martin Carrier Department of Philosophy, Bielefeld University, Bielefeld, Germany

Michael Eckert Deutsches Museum, München, Germany

Nicolas Fillion Department of Philosophy, Simon Fraser University, Burnaby, BC, Canada

Hans Hasse Laboratory of Engineering Thermodynamics (LTD), University of Kaiserslautern, Kaiserslautern, Germany

Julie Jebeile IRFU/Service d'Astrophysique, CEA Paris-Saclay, Gif-sur-Yvette, France

Ann Johnson Science and Technology Studies Department, Cornell University, Ithaca, NY, USA

Jürgen Jost Max Planck Institute for Mathematics in the Sciences, Leipzig, Germany

Santa Fe Institute, NM, USA

Tarja Knuuttila University of South Carolina, Columbia, SC, USA

University of Helsinki, Helsinki, Finland

Johannes Lenhard Department of Philosophy, Bielefeld University, Bielefeld, Germany

Andrea Loettgers University of Bern, Bern, Switzerland

University of Geneva, Genève, 4, Switzerland

Miles MacLeod University of Twente, Enschede, The Netherlands

Domenico Napoletani University Honors Program and Institute for Quantum Studies, Chapman University, Orange, CA, USA

Marco Panza CNRS, IHPST (UMR 8590 of CNRS, University of Paris 1 Panthéon-Sorbonne), Paris, France

Chapman University, Orange, CA, USA

Henrik Kragh Sørensen Section for History and Philosophy of Science, Department of Science Education, University of Copenhagen, Copenhagen, Denmark

Daniele C. Struppa Schmid College of Science and Technology, Chapman University, Orange, CA, USA

Ido Yavetz Cohn Institute for History and Philosophy of Science and Ideas, Tel Aviv University, Tel Aviv, Israel

Introduction: Mathematics as a Tool

Johannes Lenhard and Martin Carrier

The role mathematics plays in the sciences has been assessed differently and in conflicting ways. Put very roughly, a strong view holds that mathematically formulated laws of nature refer to or reveal the rational structure of the world. By contrast, a weak view denies that these fundamental laws are of an essentially mathematical character, and rather suggests that mathematics is merely a tool for systematizing observational knowledge summarized in these laws so that one can make additional use of these laws.

In the present volume, we want to put forward a position that combines features of both viewpoints. This position is supposed to characterize the use of mathematics in certain specific areas where mathematical reasoning is employed. It is intended to bring out characteristic features of making practical use of mathematical instruments. The position presents a strong view about how mathematics functions as a tool; "strong" because it assigns an active and even shaping role to mathematics. But at the same it refrains from any claims about the mathematical structure of the universe that is (allegedly) mirrored by mathematical theories in the physical sciences. Employing mathematics as a tool is independent from the possible mathematical structure of the realms of objects under consideration. Hence the tool perspective is *contextual* rather than *ontological*.

When mathematics is used as a tool, it cannot be guided exclusively by internal mathematical reasoning. Instead, what qualifies as adequate tool-use is also determined by the problem at hand and its context. Consequently, tool-use has to respect conditions like suitability, efficacy, optimality, and others. Of course, no tool will provide anything like the unique solution. On the contrary, the notion of tool stresses that there is a spectrum of means that will normally differ in how well they serve particular purposes. This practical outlook demands a new view

J. Lenhard (✉) • M. Carrier
Department of Philosophy, Bielefeld University, Bielefeld, Germany
e-mail: johannes.lenhard@uni-bielefeld.de; martin.carrier@uni-bielefeld.de

J. Lenhard, M. Carrier (eds.), *Mathematics as a Tool*, Boston Studies in the Philosophy and History of Science 327, DOI 10.1007/978-3-319-54469-4_1

on the concept of validity in mathematics. The traditional philosophical stance emphasizes the *permanent* validity of mathematical theorems as a pivotal feature. The tool perspective, in contrast, underlines the inevitably *provisional* validity of mathematics: any tool can be adjusted, improved, or lose its adequacy upon changing practical conditions.

This contextual and malleable nature of mathematical knowledge used as an instrument is a pivotal element of the entire book. We are not so much interested in tools that are ready-made, off-the-shelf products. Rather, we are concerned with the practice of mathematization in which processes such as developing tools, constructing models and designing procedures are intertwined with each other.

The introduction will work out this perspective in more detail, partly in (illuminating) contrast to popular accounts of mathematics and mathematization. We understand our endeavor as a fundamentally interdisciplinary one that needs to be informed from at least three angles: the recent practice of using mathematics, the history, and the philosophy of science. When putting this volume together, we did not attempt to separate these perspectives, but to interlock them in a (hopefully) enlightening way. Before we give a brief overview of the topical structure in Sect. 3, let us provide some philosophical and historical context.

1 Perspectives on Mathematics

1.1 Mathematics as the Language of Nature – Promises and Limitations of Mathematization

The use of mathematical laws for describing and explaining natural phenomena is among the chief epistemic achievements of the Scientific Revolution of the seventeenth century. Medieval scholarship had joined Aristotle in emphasizing the difference between ideal mathematical postulates and real physical phenomena and had considered it impossible, for this reason, to accomplish a mathematical science outside of the ideal realm of celestial bodies. By contrast, the pioneers of the Scientific Revolution, such as Galileo Galilei, René Descartes, and Johannes Kepler, suggested to seek for mathematical laws of nature and conceived physics as a mathematized science. Galileo's law of freely falling bodies, Descartes's (or Snel's) law of refraction and his law of inertia, or Kepler's laws of planetary motion implemented this new idea of mathematical laws of nature. Underlying this approach was the assumption that nature exhibits a mathematical structure. As Galileo put it, the book of nature is written in mathematical language; or in Kepler's words, God used geometrical figures in creating the world. In an influential historical account, Alexandre Koyré featured a Platonic vision of a mathematically structured nature as a methodological key element of the Scientific Revolution (Koyré 1968, 1978). Newton's *Philosophiae Naturalis Principia Mathematica* is often regarded as an early climax of the endeavor to capture the blueprint of the universe in mathematical terms. Michael Mahoney aptly pointed out that granting mathematical structure ontological impact was a key move during the Scientific Revolution that turned nature into a mathematical realm (Mahoney 1998).

A second vision of the Scientific Revolution consisted in connecting understanding and intervention. Francis Bacon and Descartes developed the idea of an applied science in which knowledge about natural processes provides a basis of technology. The best way to take nature into the service of humans is to elucidate her inner workings. This idea has proven highly effective for mathematical science. Mathematical laws of nature make predictions possible under (hypothetically) varying conditions which, in turn, make such laws suited to supporting technical intervention. The outcome of technical procedures needs to be anticipated with high precision, and mathematical laws are apt to meet such requirements.

However, the success of the mathematical sciences is not unlimited. It is true, these sciences have managed to enormously enhance their grip on complicated phenomena in the past 150 years. Even if mathematical treatment was often able to strip off constraints of idealizations and controlled laboratory conditions and to cope with ever more complex systems, the full complexity and intricacy of real-world situations, until the present day, still poses various difficulties and obstacles to their mathematical treatment.

An early exemplar of a complexity problem is the so-called three-body problem that Henri Poincaré demonstrated to be unsolvable around 1900. The difficulty concerns the calculation of the motion of bodies under the influence of their mutual gravitational attraction. This dynamical model proved to be so complex that no analytical solution could be derived. Consequently, in spite of the fact that the underlying mathematical laws are known comprehensively, the long-term motions that result from their rule cannot be foreseen.

A second, similarly famous example, and an early one for the study of "complex systems," goes back to the meteorologist Edward Lorenz. He found in the early 1960s that a given set of differential equations produced different results depending on tiny variations in the initial conditions. Even minute dissimilarities, fluctuations, and uncertainties in these initial conditions led to quite distinct subsequent states. Lorenz's discovery is called "deterministic chaos" today: the time evolution of relevant physical systems depends so sensitively on the precise initial conditions that no long-term prediction is feasible. Even if the nature of the system and the laws governing its time development were known without remainder, its future course could not be anticipated reliably. Speaking more generally, systems attain a high degree of complexity in the relevant sense if individual entities interact on a certain level and their interaction produces emergent phenomena on a higher level, especially when the manifestation of the higher-level phenomena depends upon the details of lower-level interactions.

In other examples of complexity, the means of mathematical modeling are insufficient for deriving valuable solutions. The Navier-Stokes equations are believed to completely capture the behavior of fluids. But these equations cannot be solved in general; solutions can only be given for special cases. Likewise, Schrödinger's equation is taken to account comprehensively for non-relativistic quantum phenomena. Yet already the helium atom can be treated only approximately. The examples show that the concept of complexity has many faces and, more importantly, that the complexity of the conditions pushes the mathematization of nature to its limits—at least regarding predictions and, consequently, targeted interventions in the course of nature.

In a different vein, some fields of science appear much less suited to mathematical treatment—quite independently of their complexity. An important area is the life sciences. Molecular biology is mostly directed at disclosing mechanisms which are governed by spatial structures. Molecules mesh with each other and occupy receptors mostly in virtue of their spatial properties. To be sure, all these mechanisms may be based on quantum laws, and spatial features can be expressed geometrically, i.e., mathematically. Yet, up to these days, mathematical approaches hardly contribute significantly to understanding the relevant interactions. Spatial figures, mechanisms, and feedback loops are the essential concepts. Admittedly, mathematization has led to a couple of statistical rules, like Mendel's laws, and evolutionary models, or has yielded reaction rates of biomolecules. But it is an open question whether mathematics will be able to add important further insights into the life sciences. In short, nature seems not to welcome mathematization indiscriminately.

Thus, the strong view regarding the mathematization of nature seems to be constrained to rather narrow limits. Mathematical laws of nature and the option of putting them to use are restricted to a closely encircled range of sciences and phenomena. However, a look into recent scientific endeavors reveals that mathematical procedures flourish and fulfill various tasks outside this range. Fields such as molecular biology or systems biology make ample use of mathematical methods, without, however, embodying the vision of the strong view, i.e., without assuming that mathematics provides access to the inner workings of the pertinent mechanisms and without building targeted intervention on mathematical insights. This observation motivated us to look at the more modest use of mathematics as a tool.

1.2 Mathematics as a Tool

Mathematical analysis can be helpful in practical respect even if the pertinent fundamental processes cannot be understood or caught productively in mathematical terms. Mathematics deals with structures and such structures can be found at various places, not only in the makeup of bodies and their interactions. It is part of the power of mathematical methods to be able to disclose features and identify patterns in all sorts of data, independently of their nature and origin.

As a result, mathematics is essential in bringing out structures in data and making use of them, as well as in establishing links between theory and data. As to the first item, mathematics is helpful for identifying data patterns and thus makes data-driven research possible. The increase of computing power in the past decades has opened up a new path toward analyzing data, a path that does not lead through constructing theories and building models. As to the second item, mathematics is suited to forging links between theories and phenomena. Measurement procedures can be accounted for mathematically even if the constitution of the pertinent objects and interactions is not governed by mathematical laws. For instance, schematic screening is dependent on mathematical methods. High-throughput procedures or DNA-arrays require mathematical assistance in registering and processing the data. Automated pattern recognition or statistical analysis are indispensable for gaining access to relevant features of the data.

We speak of mathematics as a tool in order to designate the usefulness of mathematics even in areas where the fundamental objects and their interactions seem not to be governed by mathematical laws of nature. Mathematics as a general approach to disclosing and handling general structures irrespective of their physical nature can yield productive insights even into such areas. Using mathematics as a handy vehicle provides new opportunities for taking advantage of mathematical methods. Employing mathematics as an instrument in this sense can break new ground in practical respect even in fields whose basic principles are resistant to mathematization. Thus, mathematical methods are analyzed regarding their instrumental virtues.

It is not customary to look at mathematics from this perspective. The standard account of mathematics is centered on deducibility and provability and associated with coherence and certainty. The tool account is different in that mathematics is analyzed as an instrument, and, consequently, cast for an auxiliary role. If mathematics is used as a tool, the agenda is set by other fields; mathematics helps to solve problems that emerge within these fields. Consequently, the particular internal coherence of mathematics does not by itself guarantee progress. For example, mathematical considerations like deriving new consequences from already accepted model assumptions are only a first step in considering the adequacy of a mathematical model. Further, as it will become clearer when we address particular cases, the instrumental use of mathematics often proceeds in a rough and ready way and seems to be part of endeavors that look tentative and deficient in epistemic respect. Still, it is a virtue rather than a vice of mathematics that it proves to be helpful even under conditions that leave much to be desired in epistemic respect. Using mathematics as a tool is in no way meant in a pejorative sense. It rather emphasizes the productivity of mathematics in coping with structures of arbitrary nature.

Using mathematics as a tool is by no means a recent development. On the contrary, the ancient slogan "saving the phenomena" was popular to describe the instrumental use of mathematics in astronomy. Mathematics was used to demonstrate the compatibility of observations with cosmological and physical principles and with additional auxiliary assumptions. This compatibility was established by reproducing and anticipating astronomical observations on their basis. However, many features of this mathematical account were not taken seriously as a description of nature. The narrow empirical basis of terrestrial observers was assumed to be insufficient for disclosing the true motions of the planets. As a result, in the tradition of saving the phenomena, the success of explaining and predicting celestial phenomena was not attributed to the truthful representation of the universe. Mathematics was used as a tool.

Although the tool perspective is relevant for large parts of mathematics as it occurs in the sciences, it has not yet received much scholarly attention. There is a prominent strand in the philosophy of mathematics that discusses "applicability" in a very principled sense, arguing about the indispensability of mathematics and its ontological status in specific contexts where "the unreasonable effectiveness of mathematics" (Wigner 1960) is a striking issue. A typical question is whether indispensability does or does not grant reality of mathematical entities. A representative study is Colyvan (2001), though the controversy about indispensability is going on, cf. Leng (2002), or Saatsi (2011), among others. One line in this discussion is particularly relevant for our perspective, namely seeing mathematics as a heuristic tool. Formal manipulation of an equation, for instance, might open up new contexts for using this (new) equation. The studies by Steiner (1998) and Bangu (2012) highlight this aspect.

The tool perspective, however, is more related to actual scientific practices than to something as fundamental as indispensability. Given the recent practice turn in the philosophy of science, the hesitant uptake of the tool perspective might still come as a surprise. One reason is that this perspective requires adopting an interdisciplinary approach that looks at mathematics in connection with other disciplines. The recent work on "mathematical practice," however, primarily looks at what mathematicians do rather than focusing on how mathematics functions in a wider context (Mancosu 2008, van Kerkhove et al. 2010).

Among the factors that contributed to eclipsing the instrumental use from the philosophical and (to a lesser extent the) historical view is the traditional distinction between pure and applied mathematics. This distinction suggests a division of labor between the pure branch that creates and validates mathematical knowledge and the applied branch that draws particular conclusions from this body of knowledge so as to solve extra-mathematical problems. In this approach, construction and application of mathematics look like two separate activities, and the applied branch is marked as being completely epistemically reliant on the pure branch. The distinction between pure and applied is sometimes maintained while the hierarchical relation is reversed. Mathematical structures are then claimed to be abstractions from structures in the real world on which pure mathematics would build its theories. Mathematics, accordingly, would be primarily a natural science (Kitcher 1983).

We do not share either view. The counterpoint we will develop abandons this assumption of unidirectional dependence and assumes, on the one hand, that using mathematics as a tool has an impact on the corpus of mathematical knowledge. New mathematical knowledge springs from developing methods to cope with practical challenges. Yet we also believe, on the other hand, that the system of mathematical knowledge shapes the development and instrumental use of mathematical models. Mathematics and the various sciences we consider benefit from proceeding in such reciprocal dependence. Mathematics is no ready-made repository simply to be tapped. On the contrary, problems and the tools to their solution co-evolve. Looking at the invention and implementation of tools thus highlights how dynamical the process of bringing mathematics to bear on the demands of practice actually is.

We want to propose five characteristic aspects of mathematics as a tool that exhibit significant differences to the standard account mentioned. These five partially overlapping topics will be discussed in the following section. Section three will bring together the relevant observations and draw conclusions about the general lessons to be learned from regarding mathematics as a tool. They will challenge the philosophical viewpoint that modern science is characterized by a uniform method and also call into question the role mathematics plays in arguments about science and rational order.

2 Characteristic Features of the Instrumental Use of Mathematics

We present five features that are intended to sketch characteristic modes of using mathematics as a tool and also lead to open research questions. Each of them is rich enough to justify separate treatment, but there is also much overlap among them so that in practice concrete instances usually involve more than one feature.

2.1 Mathematics as a Mediator Between Theory and Observation

Mathematical theories are indispensable for forging links between theory and observation. This applies to measurement, i.e., to connecting theoretical quantities with data as well as to technology. The use of mathematics is essential for registering phenomena and for generating and shaping phenomena. In the former case, mathematics is employed as observation theory. In the latter, it is used as an instrument of intervention in the course of nature.

Mathematical theories of observation have been in use for a long time. Isaac Newton measured the gravitational attraction of the Earth by bringing mathematical laws of mechanics to bear on pendulum oscillation. Measuring electric current intensity by means of a galvanometer relies on Maxwell's equations. Similarly, the use of mathematics for ameliorating technological intervention has a long tradition. In 1824, Sadi Carnot was the first to apply highbrow theoretical principles (caloric theory) to the operation of advanced technical devices (the steam engine). His analysis supported the conclusion that the only way to enhance the efficiency of the steam engine was to increase the relevant temperature differences. We will not dwell on such familiar ground. We rather aim to address observation theories of a hitherto unfamiliar sort, namely, when the pertinent mathematical procedures are not tied to specific physical circumstances or phenomena. Mechanics, electrodynamics, or thermodynamics are physical theories that are employed in registering mechanical, electrodynamic, or thermodynamic quantities. By contrast, the mathematical observation theories we turn to are independent of any particular realm of objects. Using neural networks, statistical procedures, or Bayesian formulas for surveying, exploring and interpreting the data, does not hook up with any properties of the objects under scrutiny. In other words, these mathematical procedures do not follow the causal path from the objects and their interactions to the display of certain values (as in the case of substantive observation theories). Rather, certain patterns in the data are identified irrespective of their physical nature and causal origin.

A paradigm case is represented by DNA-microarrays that assemble thousands of different DNA strands fixed on a tiny grid. A substance is washed over the grid, and the binding patterns the substance exhibits with respect to the DNA strands reveals a lot of information about its ingredients. In this way, thousands of measurements are performed simultaneously, and the information is displayed in the pertinent configurations. The latter are color coded in heat maps and analyzed by sophisticated mathematical techniques. The salient point is that these techniques address the structure of the patterns obtained, not how the substance under scrutiny has produced the patterns. A typical question to be answered is: How similar in some relevant sense is an observed pattern of an unknown substance to a pattern of a known substance (like tissue indicating a certain disease)?

Domenico Napoletani, Marco Panza, and Daniele Struppa (2011) view this example as an instance of a new paradigm, the "microarray paradigm." As they argue, model-based scientific understanding typically proceeds by tying macro-phenomena to their underlying micro-structures. These phenomena are interpreted by tracing them to their micro-causes by appeal to mathematical observation theories. However, the mirco-array paradigm remains at the macro-level and uses mathematics for uncovering structures at this level. Napoletani et al. criticize that this novel use of mathematics renounces causal understanding right from the start. In this vein, they consider this approach to be superficial and call it "agnostic science." Yet it merits attention that such mathematized, computer-assisted methods are capable of tracking patterns in data that would have gone unnoticed otherwise and of identifying completely new kinds of phenomena. Such approaches are fueled by a pragmatic attitude. A useful mathematical procedure does not need to build on the causal processes that govern deep down the unfolding of the relevant phenomena. Rather, it suffices to find some convenient short-cut that may thrive on very particular conditions.

2.2 Data-Driven Research and the Use of Big Data

Generalizing these considerations leads to the phenomenon of "data-driven research" (cf. J. Jost's contribution to this volume). Data-driven research can be contrasted with "model-driven research," in which theoretical expectations or a micro-causal model distinguishes certain patterns in the data as significant.

Data-driven research is purportedly different in that it starts off from the data without an underlying model that decides about their significance. For instance, in pharmaceutical research, large amounts of potential agent substances are scanned automatically (high-throughput screening). The question posed merely is whether a certain substance binds to a certain receptor with certain intensity. Masses of data are produced by such a large-scale screening process, and the result always is that certain patterns are selected.

Data-driven research involves two components: Large amounts of data are produced or scanned and these data are sufficient for suggesting a tentative solution to the problem at hand. That is, data are not gathered for examining a certain theoretical hypothesis or for filling the gaps left by such hypotheses. In the model-driven approach, data are essential, too. Often, models need to be adjusted to the data which is done by measuring pertinent parameter values. By contrast, data-driven research does not begin with such a theoretical model. It rather begins with the data themselves. The goal is to identify or extract data patterns by bringing sophisticated mathematical techniques to bear, i.e., the data are used for suggesting or generating a hypothesis, rather than merely examining it.

Take the challenge of forecasting the future path of a hurricane. A model-based approach draws on fluid dynamics and seeks to anticipate the development of the pertinent storm. However, the performance of this method has proven to be rather poor, and various data-driven alternatives exist. One is to use databanks that store measurements of past hurricanes and sift out those that evolved under similar conditions as the one under investigation. Based on these data, one can make a statistically valid prediction without resorting to physical dynamics. Another alternative is using artificial neural networks. Such a network can be regarded as a mathematical function that transforms certain input strings of data into a certain outcome. The function representing the network can be adjusted by having the network run frequently over similar input strings and correcting the outcome. This can be viewed as a training process. Yet the details of the processing scheme remain obscure. The way in which certain input is transformed into certain output remains opaque in detail. Yet it can safely be assumed that this transformation process is a far cry from the atmospheric processes that make the storm develop the way it does. The relevant causal factors are in no way mirrored in the network. However, training neural networks by using past storm histories provides better forecasts of the path and intensity of a given storm than the best available simulation of the underlying atmospheric processes (Napoletani et al. 2011). It is the interplay between computer-based learning algorithms and available data that determines the value of this instrumental approach.

The crucial question in methodological respect is whether data-driven research is really tantamount to shaking profoundly the heuristic and explanatory strategies in mathematical science. The claim we wish to examine more closely is whether the emergence of "big data" together with the availability of increasingly sophisticated mathematical tools have produced an important alteration in the relevant research strategies. More specifically, does data-driven research merely provide an additional opportunity for developing and judging assumptions in science or have other strategies been sidelined in the wake of its emergence? Does data-driven research push mathematical models and (micro-)causal theorizing out of their central role in science? Does considering data structures replace analyzing micro-causal structures?

2.3 *Tuning Models Mathematically*

Theoretical mathematical models are rarely perfect. Even if a model is very good, it regularly provides a kind of mathematical skeleton that includes a number of parameters whose quantitative assignment has to be read off from empirical data. A case in point is the gravitational constant that does not follow from mathematical arguments though it does follow mathematically that there *is* a gravitational constant. Such parameter evaluations are normal and not specific for using mathematics as a tool. However, it is a different matter when important features of models are determined by the particular problem at hand, using mathematical procedures for enriching models with refinements, adjustments, and correction factors. These procedures are to be distinguished from filling numerical gaps left in theory-based models.

An example is employing highly generic models and having their entire makeup or constitution shaped by recourse to experience. Neural networks again provide a case in point. Such networks embody highly arbitrary functions, and the particular function selected for transforming given input strings into outcome is produced by appeal to a particular set of data. In this case, the influence of the data on the model is much stronger than in filling a gap by recourse to experience. Rather, the theoretical framework is so malleable that arbitrary relations can be reproduced. The framework normally poses merely weak constraints on the model; rather, it is the data that shape the model. The extent to which the architecture matters, i.e., whether the data should decide about everything, is discussed under the heading of shallow vs. deep learning.

There are many options in between in which the mathematical structure neither determines the behavior nor leaves learning or adaptation completely open. Arguably, in the most common cases a theoretical mathematical structure calls for essential additions to tune model behavior. Chemical process engineering provides an example. Molecular structure and thermodynamic properties of substances provide a rough approach to the chemical processes when certain substances encounter under specific combinations of pressure, temperature, and the like. Such top-down models are completed by bottom-up modeling. Mathematical features (tuning knobs) are introduced that are inspired by observation or by trial and error. Such features typically lack a theoretical foundation and sometimes even an empirical interpretation. They help produce correct results, but they do not translate into any coherent picture of the underlying chemical processes. For instance, adding a certain substance to a chemical process may augment the yield of the reaction. Such an influence is then modeled by inserting a corresponding factor into the account. But the factor may not be understood and remain ad hoc (cf. the detailed studies by MacLeod and also Hasse and Lenhard, chapter "Boon and Bane: On the Role of Adjustable Parameters in Simulation Models", this volume).

As a result, the models crafted in this fashion typically contain elements that lack any clear physical interpretation. They fail to represent relevant processes that unfold in the corresponding system. A worry raised in this connection concerns the limits of generalizing such makeshift models. If a model is tuned to specific cases, does it extend to similar cases? If the applicability of tuned models is limited to a narrow range of cases, does this pose a severe difficulty in practice? Or is the limitation compensated by the availability of a greater variety of (specialized) models?

2.4 Using Alternative Routes to Solving Equations

Solving differential equations by a computer simulation proceeds by not literally solving these equations but rather by calculating values of its discretized proxy. Solutions are calculated point by point at a grid and for specific parameter values. As a result, using computational models does not simply mean to enhance the performance of mathematical models but rather changes the ways in which these models are constructed and the modes in which they operate. Digital computers require discrete versions of all relevant objects and operations. This requirement presents a kind of instrumental imperative that drives traditional mathematical modeling into new pathways.

Most notably, traditional mathematics is full of continuous quantities and relationships. Yet they need to be re-modeled as discrete entities so as to become tractable by digital computers. Typically, the discretization produces unwanted effects that have to be neutralized by "artificial" measures. An early example is the "artificial viscosity" that John von Neumann came up with to re-introduce the possibility of very steep wavefronts after discretization of super-sonic waves (Winsberg 2003; for a case in meteorology, see Lenhard 2007).

The discretization of continuous equations is usually not a straightforward process, since it can be accomplished in a number of different ways. Even if these options seem to come down to the same thing when the space-time grid is sufficiently refined and eventually approaching continuity, they can make a big difference in practice. The reason is that any actual computation takes place with finite resolution where the dynamical properties of the discrete versions can differ considerably. One has to ensure computational stability as an extra condition imposed by using the computer as a tool, and this demands guaranteeing that the discrete models behave adequately, as compared to the observed phenomena and the corresponding continuous model.

Parameterization schemes arise from the need to find discrete counterparts to the quantities in continuous models. A telling example is the dynamics of clouds whose (micro-)physical basis is to a large part not yet known. However, cloud dynamics forms an important part of the general circulation models of the atmosphere. They involve the calculation of this dynamics from values defined at grid points. Clouds, however, are sub-grid phenomena (a typical grid may work with horizontal cells of 100 km $\times$ 100 km). Hence the effects of clouds for the whole dynamics need to be expressed as a kind of net effect at the grid points, i.e., they have to be parameterized. What are appropriate parameterization schemes? What are effective parameters that adequately summarize cloud dynamics? How are these parameters estimated reliably? These questions are interdependent and can be solved only by extensive simulation procedures that assume parameter values, observe the model dynamics, re-adapt the values, etc. Hence adequate parameterization has to find a balance between accurate description and effective manipulability.

Parameters are defined and tuned, and become usable by computational strategies of calibration. Typically, successful tuning parameters cannot be ascribed a particular meaning in terms of the pertinent theory such as meteorology. Rather, they acquire their semantic role in the particular context, the particular discretization, the particular embedding in a parameterization scheme, and the particular process of adaptation. In general, one has to be aware of the following threat: The performance achieved after extensive model adaptations does not necessarily reflect the adequacy of the theoretical premises built into the model. The more parameters the model contains, the higher the chance that a certain characteristic of the fitted model arises solely from the instrumental adaptations rather than the underlying theory (see also Sect. 2.3).

Another example is so-called multi-scale methods. To describe the binding behavior of, say, an enzyme, researchers employ a couple of different models that are based on different theoretical approaches. An enzyme is built from amino acids and typically contains about 8000 atoms, but only a small fraction of those atoms, maybe 50, is relevant for the binding behavior. These atoms are modeled with quantum mechanical methods that provide the most detailed view and the most precise results but are computationally too expensive to cover the entire enzyme. Even for these about 50 atoms one has to resort to simplified modeling strategies, like density functional theory. The remaining majority of atoms is modeled via (classic) force fields, i.e., with so-called molecular dynamics methods. This approach is much more coarse-grained and ignores quantum effects, but experience has shown that this limitation does not result in a significant loss in adequacy of model behavior.

The largest errors occur through the ways in which the quantum and the classical regime are coupled together. The coupling cannot be determined theoretically as the two regimes are not compatible. The only solution is to introduce a tentative coupling and to adapt it so that the coupled model attains a good fit to already known cases. Such loops of adaptations, checks, and re-adaptations treat the coupling mechanism as a variable. Without these loops multi-scale methods could hardly be used at all. A similar conclusion holds for the coupling of computationally simpler and more complex methods that work on the same scale, like density functional methods that are amended by more accurate but also computationally much more expensive coupled-cluster methods for particularly sensitive regions. In examples like those discussed, the issue is not to create a correct representation in the language of mathematics, but rather to use the qualities of mathematics as a tool for adjusting coupling mechanisms in a controlled way. In short, mathematics is used as a tool, a good-enough, or even optimal, compromise between accuracy and tractability.

The importance of computational modeling is underlined by an additional example from theoretical physics. In quantum mechanics, so-called 'Feynman diagrams' provide a key to numerical algorithms. Principled formal mathematical theory has to be transformed into methods that allow one to effectively squeeze out the relevant numbers. Historian of science David Kaiser has pointed out that Feynman diagrams are among the most important calculation tools of theoretical physicists (2005, 156). Both aspects, being theoretical and being a tool, are not contradictory, but complement each other. Feynman diagrams have a formal basis in path integrals and at the same time entail highly useful combinatorial features. These features are used to break down overly complex computations into manageable parts. The results of these simplification procedures are empirically adequate but their theoretical justification remains somewhat shaky. Mathematically, the "integration-by-parts" identities are the pivotal components. They were established in the early 1980s, but only since around 2000 have computer-assisted methods been developed that offer automated recursion procedures for fitting practical problems (of particle physics) to the mathematical identities. The point is that these formulas allow certain adaptations to fit concrete cases. They do not cover the relevant problems exactly; hence a perfect fit cannot be achieved. The task to find an optimal fit is tackled in a computational way; namely, the integration-by-parts formulae are adapted to precision measurements by an automated procedure. These precision measurements have become available only since around 2000, and thus measurement and computational modeling are strongly interlinked here.

The performance of computational models can even run counter to theory-based expectations. Linear programming, a method of numerical discrete optimization, provides an illustration. The great popularity of linear programming methods did not emerge along with mainframe digital computers, but only much later when smaller machines had become easily available. The mathematical theory of linear programming is well known, but the performance of a model in a concrete situation depends heavily on particular conditions. The point is that linear programming includes an essential part of trial-and-error, because the performance of algorithms often hinges on tricks (or ad-hoc adaptations) that lack any theoretical justification. Only since around 1990 has the exploratory use of such methods become feasible, due to the cheap availability of computational tools (see also Johnson and Lenhard 2011). For instance, the dual simplex algorithm works much better than the primary one, as has been observed in many cases, but this difference has no known theoretical reason. Although it is arguably true that without theory one would not have invented the dual algorithm, it is also true that one has "no idea about performance from theory" (Martin Grötschel, personal communication). Hence one is dependent on the (cheap) computer as a device that is able to work in a successful way with linear programming as a mathematical tool.

The common ground among all these examples is the replacement of the mathematically distinguished and correct approach to solving the relevant equations with an alternative route. The reason for doing this is that the correct approach is unfeasible under the conditions at hand. Yet the drawback is that the methods used in its stead are not demonstrably adequate. They yield only approximate solutions whose adequacy cannot be, as the case may be, estimated reliably. (On the intricate relation between solution and approximation in the context of computational methods, cf. Fillion, chapter "The Vindication of Computer Simulations", this volume).

To what extent is using mathematics as a tool in this sense different from the sense sketched in Sect. 2.3? This time the *ersatz* procedures pursued hook up with the presumed causal dynamics. Neural networks admittedly fail to track the underlying processes whereas discretized equations, Feynman diagrams, or the use of density functionals reproduce to a certain extent the conceptual backbone and causal image of the pertinent full-scale theories. Following the theoretical lead on an alternative route is distinct from abandoning this theoretical lead altogether. But is this difference a matter of degree or of principle? Are the two poles of the spectrum linked to different modes of using mathematics as a tool?

2.5 Non-representational Idealizations

Idealizations are intimately connected to the role of mathematics in the sciences. All mathematical operations inevitably deal with objects or models that are in some sense idealized. The point is what sense of idealization is the relevant one in our context. "According to a straightforward view, we can think of idealization as a departure from complete, veridical representation of real-world phenomena" (Weisberg 2013, 98). All mathematical models suit this description. They can be adequate without having to be completely faithful representations. The interesting point rather is in what ways they can deviate from the representational ideal.

We want to introduce a type of idealization that cuts across the various suggestions to capture different sorts of idealized models. All these suggestions deal with object-related idealizations. Idealizations create a simpler version of the relevant objects and their relationships so that mathematical models of them become more tractable. In principle, these simplifications could be removed by de-idealizing the model step by step and thereby making the models more and more complex. However, from our perspective, the most important distinction is a different one, namely, between object-related and tool-related idealizations. The latter result from the properties of the tool and make sure that the tool can be used in the first place. Tool-related idealizations exhibit no clear relation to issues of representation. The question of how adequate the tool works in a certain situation needs to be tackled independently.

Philosopher of science Robert Batterman (2010) has pointed out that asymptotic limits or singularities regularly destroy the representation relation, because the objects covered by the model simply do not exist. So why should scientists be interested in these objects? The main reason lies in the mathematical properties of the objects. A key notion in this context is that of tractability. A large number of efforts in mathematization are intended to enlarge the realm of tractability. Paul Humphreys, for instance, has characterized computer simulation by the way it extends the realm of mathematical tractability (Humphreys 2004). We want to stress that there also is another dynamics in play, namely, modeling systems in an idealized way *so that* they can be treated mathematically.

Of course, object-related idealizations can be seen in this way, too. In thermodynamics, for instance, assuming a heat bath serves to keep temperature constant and thereby simplifies the pertinent equations in a crucial way. But there are also tool-related idealizations that do not relate to the object-side in any clear fashion. Examples are typically of a more technical nature, though they are not rare at all. Quantum chemistry, for example, has developed strategies to cope with the computationally extremely complex Schrödinger equation. One of the groundbreaking strategies is to use a set of "linear basis functions" for approximation. Importantly, these functions have not been introduced for some (quantum) theoretical reason, nor do they idealize away certain features of the complex interaction of electrons. Rather, these functions have mathematical properties that allowed computation in the first place. Consequently, the justification of this modeling step had to be checked by comparing the computed results to empirical data (cf. also Lenhard 2014).

Another illustration, and one with media coverage at that, is financial economics. Stochastic analysis here works as a tool for the pricing of options and other derivatives and for appropriately calculating hedging strategies. To be tractable, most models have to ignore transaction costs. It is not clear, however, whether these costs are indeed not relevant for a first approximation. Rather, their neglect is necessary to be able to find solutions for the mathematically posed problems of pricing. That is, this idealization is tool-related. Sociologist of science Donald MacKenzie (2008) expressed a similar train of thought when he gave his book about the role of the computer in financial economics the programmatic title that the computer is *An Engine, Not a Camera*. That means the tool suggests modeling steps that change the framework and the objects of the field. Rather than represented, the objects are transformed. The application of instruments does not simply record objects and their relationships, but changes the corresponding realm. In the case of financial economics, many modeling assumptions come from the motive to access the realm tractable for stochastic analysis. In this sense, the model idealizations are tool-related. Which types of idealizations in scientific practice are of this sort? What are the differences between recent computer-based methods and older approaches to using mathematics?

2.6 Synthesis

We have discussed five modes of using mathematics as a tool which partly overlap and are partly heterogeneous in kind. The instrumental use of mathematics is not governed by a single scheme. Rather, the heterogeneity of tools and of tool usages teaches lessons about mathematization, its dynamics and its limits. The examples also suggest that the modes of instrumental use are manifold and contrasting. The emerging picture of the relevant approaches resembles a patchwork of tools. This picture can be regarded as a contribution to the philosophy of mathematics, a picture arguably in line with Mark Wilson's metaphor of "façades" (Wilson 2006). He argues that even classical mechanics does not form a single coherent realm, but is rather grouped into different patches that are merely "glued" together so that they create the impression of a (homogeneous) façade.

A look behind this façade reveals a dynamic and non-homogeneous picture which stands in contrast to accounts in the philosophy of science that concentrate on success stories of a very peculiar kind, like Newtonian mechanics. In the eighteenth century, rational mechanics seemed to be the royal road leading toward unified science. Eminent figures in the early nineteenth century, like Adrien-Marie Legendre, even held that scientific and mathematical knowledge were close to their completion. This expectation proved to be mistaken. One hundred years later, in the early twentieth century, the sciences had assumed an unprecedented dynamics and brought forth an unparalleled variety of mathematical tools. Another century later, in recent times, mathematical models and methods, especially computer-related ones, are widespread in nearly every branch of the sciences. However, we perceive a rather heterogeneous picture that looks like a far cry from the older expectation that mathematization would lead to unification. There are great examples of mathematically formulated theories that promise to govern large parts of the world, such as those of Newton and of Schrödinger, but there are also great exceptions – in the twofold sense of being admirable and rare.

We would like to highlight two findings that seem to be characteristic of the instrumental use of mathematics: *Control replaces explanation,* and *validation is accomplished by use*. Control often is a practical goal, for instance, when a satellite has to be navigated into a stable orbit. The clincher of using mathematics as a tool lies in providing knowledge about how to control processes or systems. In the traditional understanding of the role of mathematics in the sciences, its usefulness in formulating comprehensive, unifying accounts of the phenomena is stressed. Mathematics provides the deductive links that connect the first principles with their empirical consequences and thus produces a pyramid-shaped body of knowledge. The rule of first principles over the phenomena is established by mathematical derivation. In the present context of mathematics as a tool, however, the deductive links are often tenuous, restricted to a narrow realm of conditions, forged in an ad-hoc manner, or intransparent. As a result, the explanatory power conferred on the principles by applying mathematics is considerably weakened (though there are particular modes of explanation linked to the tool-perspective). Yet the power of prediction need not be reduced at the same time. On the contrary, accurate predictions are an essential criterion for judging the quality of mathematical tools. One of the conclusions to be drawn from such considerations is that predictive power may not be completely dependent on insights into the inner workings of nature. Instead, predictive power is often established by using mathematics as an instrument. Consequently, the validation of such tools is accomplished by examining their practical achievements. The pivot of validation is practical use.

3 The Chapters

The contributions to this volume cover a wide array in terms of history, of philosophical approaches, and also in terms of examples studied. They address a broad spectrum of interdisciplinary perspectives. If some readers find one or another passage too historical, philosophical, or mathematical for their personal taste, we would like to reassure them. All texts allow readers to skim such passages, if they prefer to do so. We agree with all contributors to the book – and hopefully with all our readers – that this kind of interdisciplinary variety (and ambition) is what the topic of mathematics as a tool requires, and that the price of difficult accessibility of some parts is worth paying.

The chapters are organized into three major parts, each one exemplifying a particular perspective on mathematics as a tool, though this clustering does not exhaust the topical richness of the chapters. The first part addresses the organization of science and explores how tool development and use, on the one hand, and the social and conceptual organization of science, on the other, mutually influence each other. Ann Johnson focuses on the role of mathematics in engineering over a wide historical time span and juxtaposes a rational with an empirical culture of prediction. She assesses the value of mathematics as a tool in several dimensions, in particular regarding prediction and explanation, and explains how these goals exhibit a tension with each other. Tarja Knuuttila and Andrea Loettgers deal with contemporary synthetic biology. They show how practices of mathematical modeling are imported from other disciplines and how they are influential in organizing the interdisciplinary work in this new field. This work is reconstructed as being geared toward design principles rather than concrete regulatory networks. Ido Yavetz takes us back to ancient times. He presents a dense investigation of the *Almagest* and elucidates the mathematical methods used by Ptolemaios. Yavetz argues that two different methods were available at the time, namely, trigonometry and numerical computation, and that Ptolemaios chose numerical computation for practical, tool-related reasons, rather than for conceptual ones. Henrik Sørensen completes the first part. He picks up the notion of "cultures of mathematization" and characterizes a culture by the ways tools are adjusted to phenomena. He addresses the discovery of quasi-crystals, where mathematical models were pivotal to conceptualize – not merely to represent – new phenomena.

Part II gathers contributions that re-evaluate established concepts of scientific methodology and bring to light how mathematical tools transform these concepts. Hans Hasse and Johannes Lenhard investigate the role of adjustable parameters, concentrating on the use of thermodynamics in engineering. They argue that simulation methodology fundamentally alters the significance of adjustable parameters. They are transformed from a subordinate detail into a key factor of (predictive) success. Miles MacLeod follows a similar line in investigating the newly developing field of systems biology. MacLeod underlines the crucial role of parameter adjustments for constructing predictively accurate models. They rely, however, on measurements that are often difficult to conduct, if they are feasible at all. Quite in line with Knuuttila and Loettgers (chapter Mathematization in Synthetic

Biology: Analogies, Templates, and Fictions), he stresses the heuristic advantages of instrumental mathematical models and the problems emerging with respect to their validation. Nicolas Fillion's chapter tackles numerical methodology. He capitalizes on a thorough examination of the approximate nature of numerical solutions and elaborates the large practical impact of the latter. His conclusion is a warning against conceiving the numerical part of simulation as too simple a tool. Anouk Barberousse rounds off the second part with her study on what is called "Empirical Bayes," a new and widely used statistical strategy. Her main point is that computational methods have not only made this strategy possible and have thereby modified statistical practice, but that these tools have also transformed the basic concepts and hence challenge the principles of the Bayesian approach.

The third part reflects on the tool character of mathematics from various perspectives. In his study on mathematical tools in astronomical observatories, David Aubin analyzes the evolving and discrepant conceptions of tool and instrument and how mathematics changes from a particular tool into a powerful instrument. What turns a tool into an instrument, Aubin argues, is mainly its active role in shaping a research field. Michael Eckert examines the notion of idealization in the context of hydrodynamics and hydraulics. He highlights how idealizing away fluid resistance provided the key to understanding the phenomenon of fluidity. This shows that tool-related practice and development of theory interact. In this case, dropping fluid resistance is the first step of developing ways for dealing with it mathematically. What happens when established mathematical tools are subjected to mathematical analysis, i.e., when such tools are themselves conceived as objects for mathematical analysis and improvement? Domenico Napoletani, Marco Panza, and Daniele Struppa investigate such processes in new approaches to optimization that are "forced" and not related to the structure of the problem at hand. They point out that benchmark problems and comparisons are crucial for the methodology and also point out the problems raised by social effects of black box methods, i.e., methods for problem-solving that do not require a thorough understanding of the problem. Jürgen Jost, finally, reflects on how computing technology, having both tremendously increased the rate of data acquisition and enabled new methods of analyzing them, makes us re-think the role of mathematics. He discusses a fine sample of recent developments in mathematics and, like Napoletani et al. above, discerns a new wave of mathematization that now has mathematical tools as objects. Mathematics thus is developing from a tool for science into a science of tools.

The chapters tightly interlock not by chance, but because they arose out of a three-year cooperation group at the Center for Interdisciplinary Studies (ZiF), Bielefeld University, with Philippe Blanchard, Martin Carrier, Jürgen Jost, Johannes Lenhard, and Michael Röckner as permanent members. This group's work comprised historical and philosophical workshops as well as extensive small-group discussion-interviews with practitioners in a variety of scientific disciplines that all use mathematical tools in one way or another. We want to thank them for devoting their sparse time to our project, thus giving us the opportunity to widening our horizon and adapting our position more than once.

The authors gratefully acknowledge support by the DFG SPP 1689.

References

Bangu, S. (2012). *The applicability of mathematics in science: Indispensability and ontology*. Basingstoke: Palgrave MacMillan.

Batterman, R. W. (2010). On the explanatory role of mathematics in empirical science. *British Journal for the Philosophy of Science, 61*, 1–25.

Colyvan, M. (2001). The indispensability of mathematics. *Philosophia Mathematica, 3*(6), 39–62.

Humphreys, P. W. (2004). *Extending ourselves. Computational science, empiricism, and scientific method*. New York: Oxford University Press.

Johnson, A., & Lenhard, J. (2011). Toward a new culture of prediction: Computational modeling in the era of desktop computing. In A. Nordmann, H. Radder, & G. Schiemann (Eds.), *Science transformed? Debating claims of an epochal break* (pp. 189–199). Pittsburgh: University of Pittsburgh Press.

Kaiser, D. (2005). Physics and Feynman's diagrams. *American Scientist, 93*, 156–165.

Kitcher, P. (1983). *The nature of mathematical knowledge*. New York: Oxford University Press.

Koyré, A. (1968). *Newtonian studies*. Chicago: University of Chicago Press.

Koyré, A. (1978). *Galileo studies*. Hassocks: Harvester Press.

Leng, M. (2002). What's wrong with indispensability? *Synthese, 131*(3), 395–417.

Lenhard, J. (2007). Computer simulation: The cooperation between experimenting and modeling. *Philosophy of Science, 74*, 176–194.

Lenhard, J. (2014). Disciplines, models, and computers: The path to computational quantum chemistry. *Studies in History and Philosophy of Science Part A, 48*, 89–96.

MacKenzie, D. (2008). *An engine, not a camera. How financial models shape markets*. Cambridge, MA/London: MIT Press.

Mahoney, M. S. (1998). The mathematical realm of nature. In D. E. Garber et al. (Eds.), *Cambridge history of seventeenth-century philosophy* (Vol. I, pp. 702–755). Cambridge: Cambridge University Press.

Mancosu, P. (Ed.). (2008). *The philosophy of mathematical practice*. Oxford: Oxford University Press.

Napoletani, D., Panza, M., & Struppa, D. C. (2011). Agnostic science. Towards a philosophy of data analysis. *Foundations of Science, 16*, 1–20.

Saatsi, J. (2011). The enhanced indispensability argument: Representational versus explanatory role of mathematics in science. *British Journal for the Philosophy of Science, 62*(1), 143–154.

Steiner, M. (1998). *The applicability of mathematics as a philosophical problem*. Cambridge, MA: Harvard University Press.

Van Kerkhove, B., de Vuyst, J., & van Bendegem, J. P. (Eds.). (2010). *Philosophical perspectives on mathematical practice*. London: College Publications.

Weisberg, M. (2013). *Simulation and similarity: Using models to understand the world*. Oxford: Oxford University Press.

Wigner, E. P. (1960). The unreasonable effectiveness of mathematics in the natural sciences. *Communications on Pure and Applied Mathematics, 13*, 1–14.

Wilson, M. (2006). *Wandering significance. An essay on conceptual behavior*. Oxford: Clarendon Press.

Winsberg, E. (2003). Simulated experiments: Methodology for a virtual world. *Philosophy of Science, 70*, 105–125.

Part I
Organizing Science

Rational and Empirical Cultures of Prediction

Ann Johnson

Mathematics plays a self-evidently important role in making scientific predictions. The rise of science as an epistemically superior mode of knowledge production over the past four centuries has depended on making accurate predictions; the apparent certainty of scientific knowledge has often been borne out by accurate predictions. Mathematics has been unarguably 'effective' in this sense. The question I want to explore here is how predictions have improved, that is how mathematics has become *more* effective, if effectiveness is measured in terms of producing accurate predictions.

The easy answer is that predictions improve when the underlying mathematics change. Some mathematical approaches yield better predictions for certain systems than do other approaches—there is a question of 'fit' with regard to mathematical models. However, changing mathematical approaches constitutes something other than simple positivism, i.e., a consistent march of progress, yielding better and better predictions. Changes in underlying mathematical techniques can also be destabilizing; apparently better predictions can generate profound questions about why they predict more accurately. Asking why predictions are improving often calls into question what appears to be known about the systems being modeled. Therefore even as predictions improve, mathematical ontologies can become less secure. Modelers often fear that there is no underlying mathematical explanation—are accurate predictions anything more than elaborate coincidences? Specifically, do good predictions explain anything if they are merely extrapolations of empirical data? Does the mathematics need to offer an ontological statement about the truth of the model or is a good match good enough?

In this paper I look at engineering mathematics (also referred to as engineering science or simply 'theory') in the late eighteenth and early nineteenth centuries. This

A. Johnson (✉)
Science and Technology Studies Department, Cornell University, Ithaca, NY, 14850, USA
e-mail: aj532@cornell.edu

J. Lenhard, M. Carrier (eds.), *Mathematics as a Tool*, Boston Studies in the Philosophy and History of Science 327, DOI 10.1007/978-3-319-54469-4_2

was a period in which competing mathematical approaches were epistemologically destabilizing. Engineering science was changing in the hunt to produce better predictions, but in doing so engineers were questioning the certainty (and ontology, thought they would never have thought of it as such) of their knowledge. Rational mechanics (also called mixed mathematics) dominated engineering mathematics (and natural philosophy) in the eighteenth century. More and more sophisticated, often calculus-based, models aimed to provide better predictions of real world phenomena, like ballistics, kinematics, and statics and the strength of materials. While models got more complicated and harder to solve, and were therefore often considered more sophisticated mathematically, in many cases they failed to produce accurate predictions at all.

Engineers had two general responses to this failure. In one set of responses they focused on making the real world more predictable, for example, by producing materials that behaved in more predictable ways. There was an effort to categorize and standardize materials and machines. They sought to design technologies that acted more rationally—i.e., that produced a narrower array of phenomena. This meant materials that had a smaller set of responses, e.g., cannon that shot projectiles more consistently to the same distance. In other words, they focused on making the real world more precise; accurate predictions depended on predictable behavior. Measuring precision and predictability was part of an experimental problem; it represented engineers taking an empirical tack. There was also a mathematical dimension to the question of making the technologies act the "same" way each time. What did "the same" actually mean? How much deviation in the responses was too much? At the end of the eighteenth century, questions about the regularity of responses were not yet statistical for engineers; they were ontological and deeply practical.

In dealing with real materials and phenomena, engineers tried different mathematical approaches in addition to their efforts to make the material world more predictable. By firing cannon repeatedly they could create charts of the responses, distances, trajectories, and measureable external factors like wind direction and speed. No mechanics equations in the eighteenth century produced better predictions than these data arrays. The same went for strength of materials. They tested the responses of material after material, found out where materials yielded, deformed, or broke, and then charted those results and used simple arithmetic and algebraic formulae to predict the responses of structures. As much as they wanted to design sophisticated mathematical equations, the accuracy of their data tables was impossible to duplicate. But data wasn't unproblematic to manage. Imagine testing the bending strength of a beam. Some beams might have knots in the timber and break easily, while others might be perfectly clear and hold much larger than expected loads. Which data outliers should be discarded? Too many weak beams would mean an inefficient use of timber; too many strong ones would mean beams breaking. Neither was going to improve the reputation of the engineer. The data used in the tables needed to be representative, but engineers at the turn of the nineteenth century had no mathematical methods for eliminating error; Gauss, Laplace, Legendre and others were only beginning to develop these techniques for astronomy and geodesy. Least squares and other methods wouldn't be commonplace in engineering for generations. Still, in spite of their accuracy, the approach of data tables raised real questions about the ontology and epistemology of their work—was empiricism a legitimate route for mathematization, especially in the decades/century before the development of robust statistical methods? The development of statistics in the early nineteenth century extended this empirical path and started to stabilize some of the underlying epistemological concerns.

The case primarily explored here is statics and the strength of materials – strength of materials being one of Galileo's "Two New Sciences." They were chosen here to show some of the conflicts between what I will term two different cultures of prediction: the rational and the empirical. In this paper I will compare and contrast two different paths used by engineers to predict the behavior of the material world. Rational approaches derived from the rational mechanics of the seventeenth century. Rational mechanics was a seductive idea; figuring out equations, derived from Newtonian laws of motion and force promised to elevate engineering questions to real philosophical status. Finding the right equation that would predict the flight of a projectile or the displacement of a beam seemed an engineer's dream. The problem was that these equations, as they were constructed in the eighteenth century, produced universally poor predictions. Engineers knew that equations needed more variables to be able to take known phenomena like wind resistance into account; analysis provided the mathematical framework, but engineers struggled to model phenomena beyond Newton's variables of mass, force, velocity, and acceleration. Engineers of the eighteenth and nineteenth century usually knew why rational mechanics produced inaccurate predictions, but they were rarely positioned to solve the problems.

As a result, engineers were more comfortable making predictions based upon empirical testing. Often this meant that the mathematics were unsophisticated, based on long tables of raw data, and that standardizing testing methods was important. In addition, it pushed engineers into a taxonomic vein, where different technologies and materials had to be characterized and classified as similar in order to produce useful tables. Still, empirical predictions offered a number of ancillary benefits. Empirical predictions were not only being more accurate, they were also easier to communicate and teach. The growing importance of cheap paper and printing made table-filled handbooks much more accessible than they would have been a century earlier. While engineers might have liked to produce Newtonian equations that were practically useful, in the end they needed to produce practically useful predictions more than they needed to participate in the culture of rationality.

1 Mathematics in Engineering

The role of mathematics in engineering is neither an inevitable nor natural development, although it often seems so. From the perspective of the twentieth and twenty-first centuries the use of mathematics in engineering is now so pervasive that it is hard to see it as contingent at all. Historians have gone back to many episodes in the history of engineering to uncover the use of mathematics, in say, the building of the cathedrals, the pyramids, and Roman aqueducts. In most of these cases some underlying mathematically principles have been found, even some useful calculations. However, in looking for the precursors of mathematical thinking, scholars are telling, perhaps unintentionally, a normative story, a story of how it could have been, had mathematics been privileged as a way of knowing in the past as it is in the present. It is too hard to imagine an epistemology in which structures as grand and complex as, say, the Pantheon or the Duomo were built without mathematical predictions using experiential methods. As a result seekers of mathematical justification find them, in principles, just not in the archival evidence, such as it is, of large, pre-modern building projects.

Resisting the temptation to understand the use of mathematics after the scientific revolution as contingent is equally difficult. The story of the mathematization of mechanics is paradigmatic in the history of science and follows a well-known arc from observational science to awkward mathematization (Copernicus) to more elegant mathematics (Kepler) to the generalized axioms of Newton's Principia, all the way to Lagrange's sophisticated extension of celestial mechanics. Practitioners in the seventeenth through nineteenth century also saw and approved of this story. The story seems to tell us what we want to believe about the way mathematization should be. It becomes a normative story and other episodes are fit into its pattern, a pattern that explains nothing less than the emergence of physics as paradigmatic mathematical science. Engineering practitioners also believed that the mathematization of their work would and has followed a rational and progressive pattern. As a result of the widespread (aspirational) belief in the natural course of mathematization in mechanics and engineering, some scholars have imposed this story. (Heyman 1998; Narayayan and Beeby 2001)

Claiming that mathematization was contingent is not a claim that engineers in the eighteenth century weren't trying to make engineering a more mathematical endeavor. A number of historians have explored especially French efforts in the eighteenth century to impose mathematics on engineering practice. France is the center for these efforts because the French state build the social and proto-professional infrastructure needed to be able to teach aspiring engineers how to use mathematics. Ken Alder argues persuasively that the French state also used mathematical knowledge and skills to qualify students for admission to the engineering academies. (Alder 1997, 1999) Janis Langins's history of fortifications engineering in France in the eighteenth century shows the deep-seated mathematical beliefs that repeatedly shaped what engineers thought they were doing. Jean Errard, a fortifications engineer from Lorraine, wrote a treatise in 1600 titled, "Fortification Reduced to an Art and Demonstration." In this treatise he claims mathematical rigor and contempt for mere "empirics." But there is no real rigor in his treatise. He cannot calculate any of the design principles he lays out. Langins argues, "it is not always clear what engineers mean when they talk of mathematics." (Langins 2004) Later, in describing the non-use of mathematics in one of Augusto Ramelli's works, Langins writes, "mathematics seems nothing more than a rational approach to nature and problem solving." (Langins 2004) Langins' argument is engineers wanted recourse to the mathematical (geometric) virtues of demonstrability and certainty and felt that a rational science of fortress designed based on geometry could posses those same virtues, as mechanics did. The fact that mechanics was still only an "in principle" science that did not produce good coherence with the behavior of real objects and materials did not worry these engineers.

Mathematics offered engineers what Peter Galison and Lorraine Daston term epistemic virtues, and engineers wanted to claim them for their own work, so they did. (Galison and Daston 2007) In *Objectivity*, Galison and Daston focus on representations, mostly in scientific atlases as the kinds of objects to consider when constructing epistemic virtues, but do different epistemic virtues attach to different schemes of mathematization? Are the virtues of geometry different from those of the calculus or of statistics? More to the point, are there different virtues in advancing rational mechanics versus testing materials to produce more accurate and useful strength of materials tables? I will argue that these modes do represent different epistemic virtues and follow Galison and Daston's claims that "science dedicated above all to certainty is done differently—not worse but differently—from a science that takes truth-to-nature as its highest desideratum." (Galison and Daston 2007) The certainty/truth-to-nature dichotomy doesn't exactly map onto rationalism/empiricism, but it offers important parallels. The different epistemic goals of rationalism and empiricism in the hands of engineers present different epistemic virtues and different mathematical tools are methodologically appropriate for achieving these ends.

2 Navier, Ponts des Invalides, and the Challenge to the Analytical Ideal

Eda Kranakis makes an even bolder set of claims for the rhetorical value of mathematics in engineering in her comparative study of French and American bridge designers. She writes, "In France, the prevailing view was that mathematical theory should precede and guide both experimental research and empirical practice... In the US the prevailing view was that experimental research should guide design efforts and that they should emerge from experimental and empirical work." (Kranakis 1997) French engineers wanted to deduce technological solutions from mathematical theory. (Kranakis 1997) Engineers like Navier believed that constructive practice could be fully deduced from mathematical theory. Antoine Picon calls this the "analytical ideal." (Picon 1987–1988) When Navier's models failed to predict the behavior of real materials, he blamed the engineers who needed to "give more careful attention to how necessarily idealized theories could be made to yield useful information with minimum distortion and error." (Kranakis 1997) Navier put his own mathematical approach to the test with the Pont des Invalides suspension bridge in 1826. When the anchors of the bridge failed (following some flooding caused by a broken water main), Navier's mathematical model failed to predict the problem (strain in the buttresses' anchors that led to cracking). Kranakis writes, "He evidently felt confident that he could determine the resultant force accurately enough to position a comparatively slim buttress at just the right point to provide the necessary resistance. Yet he gave no indication of how to ensure that in practice such buttresses would be placed exactly where theory said they should be." (Kranakis 1997) Kranakis rightly points out that Navier's error was not inevitable—the shear and lift forces on the buttress that failed were predictable, but only through empirical means. Navier dismissed the need to test his design, and did not compensate for his lack of empirical knowledge by strengthening the buttress. He trusted the analysis, yet after the fact he admitted that his theory could not predict the behavior of the anchorage. Navier's superiors had also made the anchorage much stronger and larger than Navier had specified.

Navier's unquestioning belief in his mathematics was the problem. Not all French contemporaries felt sympathy for Navier. In *Le Curé de Village* Balzac wrote critically of both Navier and his Ponts et Chaussées apologists, "All France knew of the disaster which happened in the heart of Paris to the first suspension bridge built by an engineer, a member of the Academy of Sciences; a melancholy collapse caused by blunders such as none of the ancient engineers—the man who cut the canal at Briare in Henri IV's time, or the monk who built the Pont Royal—would have made; but our administration consoled its engineer for his blunder by making him a member of the Council-general." (de Balzac 1899; Kranakis 1997)

3 American Responses and the Empirical Turn

The experience of Navier's bridge came quickly to the US. As Kranakis shows, most American engineers were brutally empirical, experiential even, and once they found a design that worked they often used it everywhere, hence the attraction of patented bridge truss designs. Materials might change based on local availability of materials, and some engineers would do tests on the new materials, but by the 1840s there were dozens of material property handbooks that provided arithmetic calculations of material strength for use in the field by engineers with limited mathematical training. However, there was a group of engineers concerned with Navier's failure, and who shared his belief in the certainty and deductive utility of mathematical models. West Point (US Military Academy) engineers were trained in primarily French methods from 1802 through the 1850s. Claudius Crozet, a polytechnicien, first brought his textbooks to West Point in the first decade of its existence. After a period of some instability, Sylvanus Thayer took over the Academy in 1817, fresh from an 18 month course at the École Polytechnique.[1] Upon assuming the superintendent's position he

[1]It is worth noting that while Thayer used the École Polytechnique as a golden engineering credential, the École des ponts et chaussées and École du génie de mézières probably offered more relevant, if less elite, training.

re-dedicated the Academy to French rational mechanics. Then in the 1830s French-trained Dennis Hart Mahan took over the cadets' instruction in statics and wrote his own treatise, following French standards. Mahan first introduced calculus-based theory.

There was a full generation of West Pointers working on American infrastructure when Navier's bridge collapsed in 1826. They wrote extensively about the accident in their preferred publishing venue, *The Journal of the Franklin Institute* (hereafter, JFI). Alexander Dallas Bache had recently taken over the Franklin Institute and reformulated the Journal. The JFI then in turn shaped the kinds of things engineers knew or were expected to know. Bache—a West Point trained engineer, probably best known for his efforts to secure government patronage for American science, and the later director of the U.S. Coast Survey– immediately raised the technical literacy of the papers published in the Journal. Bruce Sinclair, in his history of the Franklin Institute, discusses Bache's aspiration toward publishing in this way:

> To Bache research always implied publishing. Publishing was infused with the same values the same ambitions as original investigation in science. Publication was the way to reputation for men with career aspirations. But it was also the yardstick of a man's talents, because it revealed whether or not he had the ability to frame important questions and provide conclusive results. Publication was the most positive means of erecting standards for the form and content of science. (Sinclair 1974)

Prior to Bache taking over the JFI, under editor Thomas Jones, the JFI had been characterized by its efforts to render science "useful in plain language." (Sinclair 1974) But Bache wanted to reach a different audience, so when he took control of the Journal he sought out different authors, usually outside the Franklin Institute. The kinds of articles they produced were markedly different from those Jones published. For Bache, interested in making science and engineering professional activities, professionalism meant, at least in part, specialization. Articles were increasingly technical, increasingly mathematical, and written for an increasingly narrow audience with particular technical knowledge. No longer were there wide ranging, but mainly descriptive articles like Jones' own "Observations on the connection of mechanical skill with the highest attainments in science: an introductory course on mechanics" or English engineer Thomas Treadwell's empirical tract on strength of materials, "Observations on the use of Cast Iron &c." The series of articles which best characterizes the changes Bache made was a long series on the experiments undertaken at the Franklin Institute on the strength of boilers. Running for six consecutive issues and taking over 100 pages these articles laid out a significant research program in the strength of materials producing results which were useful in application to mechanical construction with wrought iron. This series of articles was written for the specialist—one had to know quite a bit about working with iron as well as quite a bit of mathematics in order to find the boiler experiments series useful.

One of Bache's favorite JFI contributors was Joseph Totten, a member of the US Army Corps of Engineers and one-time mentor of Bache's. Totten had been trained at West Point in the first decade of the nineteenth century and was stationed at Fort Adams in Newport, Rhode Island in the 1820s and 1830s. When Totten failed to receive the funds he needed to actually build and arm the fort, he took the opportunity to create a kind of post-graduate school for West Pointers. He would get the highest ranked graduates for a year or two, and they would come to Fort Adams to help him undertake experiments to determine the properties of American stone, mortar, timber and other materials. Given their sophisticated mathematical training at West Point, this experimental experience was the corrective that Navier himself lacked. This educational system, combining the rational and empirical, was created in the 1820s in a joint effort between Totten and Thayer, as they became increasingly aware that West Pointers weren't ready to undertake the kinds of projects that they were needed for. Certainly, they knew about Navier's accident in 1826 because of its coverage in the JFI.

The research Totten undertook during his 15 years at Fort Adams ultimately proved to be just as important for Bache's career as it did for the dozens of other West Pointers who passed through the Fort. The partnership between Totten and Bache was mutually beneficial to the two men, especially through their record of publication. Prior to Bache taking the helm at the JFI, Totten had published little of the important Strength of Materials work he oversaw at Ft. Adams. There were a couple of articles published in the JFI prior to Bache's editorship and two articles were published in the journal edited by Yale's Benjamin Silliman, the *American Journal of Science*. However, Totten and his group produced dozens of articles for the JFI. Totten's own articles contained not only his research; he also translated several French works with important notations referring to the differences one might find in native American materials.

Totten's efforts reveal the shortcomings of the engineering theories West Pointers were learning as well as a turn toward the empirical as a solution. When French equations produced poor predictions, Totten and his engineers focused on generating new data about materials. Working with Lieutenant Thomas Brown, Totten undertook a series of experiments on pine and spruce timbers. The equations textbooks contained might work but the tables they contained derived from French timbers were inaccurate for American materials. Totten took this, in part, as a patriotic act to show the superiority of American materials, but primarily he was interested in making better predictions of beams' and columns' behavior. Using very simply equations like

$$\frac{bd^3}{aL^3} = W$$

or

$$\frac{bd^2c}{L} = W$$

Brown and Totten created simple, arithmetic formulae into which their new data could be inserted and results calculated. In this equation, b and d represent the dimensions of the timber (its breadth and depth dimensions), l represents the beam's length and W is the weight applied (in a point load). The first equation calculates how much weight the beam can hold before unacceptable deformation (which Brown and Totten do not clearly define as a ratio of deformation to span, as would have been common then and now; I think they assume the reader knows this, and it is a standard ratio). The second equation calculates the load or length at which the beam will break—W or L will be much larger in this equation than in the deformation calculation. In the article, Totten and Brown offered tested values for a, a "constant" of stiffness and c, a "constant" of strength. In the equations, the empirical values are called "constants" because they are dimensionless. It is important to note that neither of these equations analyze dimensionally into the correct units; such formalities were ignored by Brown and Totten.[2] To mathematicians and mechanicians, this could be nothing other than sloppy mathematics. But dimensional analysis could be solved by giving the empirical values the right units. This simply wasn't a concern of Brown and Totten; to them the equations generated acceptably accurate predictions of the load capacity of timber beams. West Pointers definitely had the algebraic skills to manipulate the equations to solve for whichever variable was unknown, whether solving for load or for length of the member. (Brown 1831)

American engineers did not seem to worry about dimensional analysis in the antebellum period. What this means is that formulae could be combined in very unique ways—but purists would object. Such combinations were anathema for polytechniciens. Combining formulae meant that American engineers had far more tools at fingertips than others did. What these kinds of formulae did was to produce outcomes that have been empirically determined to be relatively accurate predictions. In some (many) cases, they were simply curve-fitting exercises with some physical rationale messily tacked on. But they worked for the purposes of prediction, which was valorized. Rather than marking American engineering practices as immature or unsophisticated, such efforts showed that flexibility and modification in the field was more important than fidelity to principle.

Totten believed his statics was an engineering theory that was useful and useable by not only West Point trained engineers, but also non-university trained engineers, who by the 1803s and the railroad boom would begin to dominate the profession, at least in America. These empirically-based approaches would fortify American engineers against the Navier errors. Simple algebraic expressions, plus well-tested empirical values represented the epistemic virtue of early nineteenth century American engineers, and they can be contrasted against the epistemic virtues of Navier and the polytechnicians. Totten valued prediction and representing the materials at hand accurately. This entailed a taxonomic project to classify building materials of different kinds and origins. Navier valued a mode of engineering calculation

[2]Dimensional analysis is a common enough engineering technique—it means making sure that the dimensions of a problem will cancel out to leave the answer in the desired units, whether pounds, kilos, inches, or newtons.

that represented theoretical mechanics of the highest order and was expressed in sophisticated mathematical forms, including the calculus and analysis.[3] (Heyman 1972; Gilmour 1971) The epistemic virtues of the two approaches contrast clearly.

4 Social Causes and Consequences of Mathematization

Denaturalizing mathematics in engineering highlights the question of why engineers mathematized at all and why they mathematized in different ways in different places. Perhaps engineers presented their work as theoretical and scientific in order to impress their audiences; mathematization was characteristic of a grab for professional status. While this makes sense to a twenty-first century audience, there is little historical evidence to support it in the eighteenth and nineteenth centuries, even in France. Frederic Graber's article on the Ourcq canal actually points to the liability of presenting sophisticated mathematics to audiences who lack the skills to assess the veracity of the mathematics. (Graber 2008) Graber argues that the École and Corps des Ponts et Chaussées were far less mathematically sophisticated than the military engineers educated at Mézières or the graduates of the École Polytechnique. Governed by the Ponts et Chaussées Assembly, a group of older engineers, they often found proposals stacked with cutting edge mathematics They were unswayed by mathematics that was unfamiliar to them. The analytic ideal apparently did not extend to them; their epistemic virtues were likely closer to the empiricism of the American engineers like Totten.

Ken Alder also argues that even in eighteenth century French military engineering the social status of sophisticated mathematics was not the motivation for implementing them in French engineering schools. (Alder 1997) As a result, simpler, empirical methods, when they produced more accurate predictions, earned the confidence of engineers and their patrons, and did not have inherently lower status among their users.

Useful rational mechanics created demands that could not be met in the American (or colonial) world. They demanded an extensive system of mathematical education, which itself required the capacity to sort potential students. Ancien Regime France was seeking these to build the capacities in order to strengthen the military by making it less dependent on the aristocracy and more open to other social classes. (Alder 1999) In the US, labor sorting was also an interest of the state and of nascent engineering institutions, but the problem was not breaking the lockhold of the upper classes, but rather determining ways to expand the number of engineers, especially during the years of the early railroad building boom. These differing social class and national priorities clearly pushed in different directions on systems designed to produce engineering and expert knowledge. Epistemic virtues were

[3]Coulomb had introduced the calculus into statics in the 1770s. Military and civil engineering academies taught these methods in the period following the French Revolution.

measured not only by practicing engineers, but also by the states that valued these experts. Methods like Totten's that were relatively easily taught were attractive in the US; whereas methods that appeared to offer openness and meritocracy, as the analytic ideal did, were politically useful in France, both before and after the Revolution.

What the Ponts de Invalides collapse and other episodes with poor predictions showed was that rational mechanics could fail to produce exactly the predictions that would be most interesting and useful to know. In some obvious sense poor predictions weren't the math's 'fault;' rather, they were produced by models that lacked salient details. But those details *were* available to empirical engineers, who built models and discovered material behaviors they did not expect. In terms of producing accurate predictions, the rational could not compete with the empirical. The project of combining the rational and the empirical would dominate further developments in the statics and strength of materials of the nineteenth and twentieth centuries. The development of statistical methods also played a key role in reducing empirical data to fit into rational, analytical frameworks.

5 Engineers' Ways of Knowing: Numeracies

By the late eighteenth century it is clear that engineers in different settings had different preferences, with different epistemic virtues, regarding what types of mathematical models to use. American Army Corps of Engineers men were seduced as West Point cadets by French mathematical sophistication, but found out that on the job their predictions were not as good as they expected. They moved to supplement, and eventually throw over, rational mechanics with empirical testing of materials and models. Others, lacking the introduction to French methods, though empirically derived formulae were more than sufficient. Engineers who chose patented truss bridges from catalogs went one step further and let someone else do the calculating altogether. What is clear is that engineers had (and have) what John Pickstone would call multiple 'ways of knowing' and multiple ways of knowing mathematically. (Pickstone 2000) Pickstone uses the notion of 'ways of knowing' to describe are four-fold: natural historical; analytical; experimental; and technoscientific. Engineers have lots of ways of knowing: mind versus hand; shop versus school; Rational versus empirical. But there are also multiple ways within Pickstone's 'analytical' category, including arithmetic, geometrical, analytical, computational. These ways all have different advantages and disadvantages. Some engineers were omnivorous; they would readily combine different ways of knowing. Others were true believers and purists in understanding different mathematical approaches in epistemically hierarchical ways, usually with the calculus at the top. We can see polytechnicians like Navier as a purists about rational, analytical methods. On the other hand, handbook authors were true believers in plug-and-chug methods, empirically derived and effective in the field and useable by nearly anyone with a grade-school education.

Rather than refer to these different approaches as mathematical ways of knowing, I prefer to think of them as numeracies.[4] Literacy as a concept only makes sense in the context of particular language. I would not refer to myself in the United States as illiterate because I cannot read Russian—the concept of literacy assumes the subject knows a language for communication purposes but is unable to read or write it. Numeracy can and should be developed as a parallel concept. Mathematics educators do address these ideas when they worry about ordering the secondary mathematics curriculum. Students may learn calculus but doing so will not necessarily help them understand statistics.[5] (Hacker 2016; Phillips 2014) Like language literacy, numeracies can also be unproblematically plural – one can use both geometry and calculus to solve equations, but only if one possess numeracy in each mathematical mode. Numeracies must also be seen in context; at times some numercies are higher status than others, but status is dynamic and changing. Like language literacies, numeracies shape and are shaped by the way the world looks. To a world in which land was the most powerful form of wealth, geometrical numeracy was important. Surveyors' skills and the accuracy of their work depended their command of geometry. To statistician (or, say, insurance actuary), aggregate behavior defines a world outlook. Numeracies themselves espouse and offer different, and not necessarily hierarchical, epistemic virtues. Still, in the context of education and institution building, hierarchices are often claimed, as in claims for the superiority of analysis and calculus versus arithmetic and simple algebra.

6 Conclusion

Since the eighteenth century, engineering culture has held two epistemic virtues in productive tension.[6] (Akera 2008) One value is the rationalist dream of calculating the behavior of the physical world through first principles. This dream rests on the classic definition of a scientific explanation as a mathematical description that can produce an observable verification. Prediction is a subordinate to explanation in this rationalist dream. Engineers in France and Francophile engineering schools bought into these virtues—they *believed* them. On the other hand, working engineers recognized that their status and livelihoods depended on quick, accurate predictions of (limited) real world phenomena.[7] They did not need to hold fidelity to first

[4]Here the British tradition of referring to "maths" makes this easier to explain than the American word "math."

[5]This is a recent controversy in the United States.

[6]Here I highlight two epistemic virtues, but I do no mean to imply there were *only* two epistemic virtues in play in this tension.

[7]Limited, because they didn't need to predict all possibilities. This wasn't Hume's problem of induction. They knew the bridge wasn't going to lift off the ground spontaneously, turn into an apple, or vanish.

principles in high regard. Theoretical robustness was a low priority because it didn't produce what was most valuable. If testing produced the best results, the only question was whether empirical data could be extended to non-local situations, situations where the data wasn't a perfect replica, such as scaled models (but did the resultant phenomena scale?). The tension between these two values was productive in that throughout the nineteenth and twentieth centuries better and better methods were developed for improving the accurate or grounded rational models through the use of empirically-determined data.

References

Akera, A. (2008). *Calculating natural world: Scientists, engineers, and computers during the rise of U.S. cold war research.* Cambridge: MIT Press.

Alder, K. (1997). *Engineering the revolution: Arms and enlightenment in France, 1763–1815.* Princeton: Princeton University Press.

Alder, K. (1999). French engineers become professionals; or, how meritocracy make knowledge objective. In W. Clark, J. Golinski, & S. Shaffer (Eds.), *The sciences in enlightened Europe* (pp. 94–125). Chicago: University of Chicago Press.

Brown, L. T. S. (1831). On the strength of pine and spruce timbers. *Journal of the Franklin Institute 7*(April), 238.

de Balzac, H. (1899). *Scenes from country life: La Comédie Humaine* (K. Wormeley, Trans.). Boston: Little Brown.

Galison, P., & Daston, L. (2007). *Objectivity.* New York: Zone Books.

Gilmour, C. S. (1971). *Coulomb and the evolution of physics and engineering in eighteenth-century France.* Princeton: Princeton University Press.

Graber, F. (2008). Purity and theory: Theoretical tools at the Ponts et Chaussées circa 1800. *Technology and Culture, 49*(4), 860–883.

Hacker, A. (2016). *The math myth and other STEM delusions.* New York: New Press.

Heyman, J. (1972). *Coulomb's Memoir on statics: An essay in the history of civil engineering* Cambridge: Cambridge University Press.

Heyman, J. (1998). *Structural analysis: A historical approach.* Cambridge: Cambridge University Press.

Kranakis, E. (1997). *Constructing a bridge: An exploration of engineering culture, design and research in nineteenth-century France and America.* Cambridge: MIT Press.

Langins, J. (2004). *Conserving the enlightenment: French military engineering from Vauban to the Revolution.* Cambridge: MIT Press.

Narayayan, R. S., & Beeby, A. W. (2001). *Introduction to design for civil engineers.* London: Spon Press.

Phillips, C. J. (2014). *New math: A political history.* Chicago: University of Chicago Press.

Pickstone, J. (2000). *Ways of knowing.* Chicago: University of Chicago Press.

Picon, A. (1987–1988). Les ingénieurs et l'idéal analytique à la fin du XVIIIe siècle. *Sciences et techniques en perspective 13*, 70–108.

Sinclair, B. (1974). *Philadelphia's philosopher mechanics: A history of the Franklin Institute, 1824–1865.* Baltimore: Johns Hopkins University Press.

Mathematization in Synthetic Biology: Analogies, Templates, and Fictions

Tarja Knuuttila and Andrea Loettgers

1 Introduction

In his famous article "The Unreasonable Effectiveness of Mathematics in the Natural Sciences" Eugen Wigner argues for a unique tie between mathematics and physics, invoking even religious language: "The miracle of the appropriateness of the language of mathematics for the formulation of the laws of physics is a wonderful gift which we neither understand nor deserve" (Wigner 1960: 1). The possible existence of such a unique match between mathematics and physics has been extensively discussed by philosophers and historians of mathematics (Bangu 2012; Colyvan 2001; Humphreys 2004; Pincock 2012; Putman 1975; Steiner 1998). Whatever the merits of this claim are, a further question can be posed with regard to mathematization in science more generally: What happens when we leave the area of theories and laws of physics and move over to the realm of mathematical modeling in interdisciplinary contexts? Namely, in modeling the phenomena specific to biology or economics, for instance, scientists often use methods that have their origin in physics. How is this kind of mathematical modeling justified?

In the following we will shed light on these questions by focusing on the interdisciplinary research practice of synthetic biology. Synthetic biology is a

T. Knuuttila (✉)
University of South Carolina, 901 Sumter St., Byrnes Suite, Columbia, SC, 29208, USA

University of Helsinki, Helsinki, Finland
e-mail: knuuttil@mailbox.sc.edu; tarja.knuuttila@helsinki.fi

A. Loettgers
University of Bern, Bern, Switzerland

University of Geneva, Rue de Candolle 2, CH-1211, Genève 4, Switzerland
e-mail: andrea.loettgers@unige.ch; andrea.loettgers@csh.unibe.ch

J. Lenhard, M. Carrier (eds.), *Mathematics as a Tool*, Boston Studies in the Philosophy and History of Science 327, DOI 10.1007/978-3-319-54469-4_3

relatively novel field of research located at the interface of physics, biology, engineering, and computer science. Being situated in this complex disciplinary environment makes model building in synthetic biology a highly interdisciplinary task: Methods, techniques, strategies, and concepts from various, even distant fields enter into and get intertwined in the modeling practice of synthetic biology. One unique characteristic of this practice is due to how synthetic biologists combine various kinds of models: model organisms, mathematical models and synthetic models. The latter ones comprise a novel type of models that are constructed from biological components such as genes and proteins on the basis of mathematical modeling. To understand the rationale of this combinational modeling approach one needs to take a closer look at the strategies of mathematization in synthetic biology.

We will discuss two interrelated means through which synthetic biologists study models of gene regulatory networks: analogies and mathematical templates. Synthetic biologists, we argue, proceed to mathematize gene regulatory networks by compound analogies that draw inspiration from engineered artifacts on the one hand, and model systems with non-linear dynamics on the other hand. Engineered artifacts provide material analogs for biological systems, whereas the theory of complex systems offers formal analogs in the form of various mathematical templates for analyzing oscillatory phenomena.[1] A kind of patchwork model results from such compound analogies that, as we will discuss, consists of elements that may even draw into opposite directions. These mathematical models nevertheless allow synthetic biologists to conceptualize biological regulation in terms of positive and negative feedback loops side-by-side with mathematical templates and methods that have been applied in various contexts dealing with rhythmic/cyclic behavior resulting from non-linear dynamics (e.g., physics, chemical kinetics, ecology, economics). In the modeling process the general templates for describing various forms of interaction are adjusted to the subject matter in question, but they remain nevertheless rather abstract, lacking many known empirical details. This contributes to one typical problem constraining the use of template-based mathematical modeling: such models are usually underdetermined by data.

Interestingly, this does not worry synthetic biologists too much. One reason may be that they do not consider their models to be representations of any specific naturally occurring gene regulatory networks. Instead, they consider themselves to be in the business of studying general *design principles* or *network motifs* of gene regulatory systems (i.e., genetic circuits). Being very aware of the fact that the conceptual and mathematical means they use are often transferred from other disciplines, they consider their models to depict only possible mechanisms underlying biological regulation. Such principles could have evolved in natural systems but biological systems might have implemented different kinds of regulatory mechanisms. As a consequence the design principles studied mathematically are best conceived of as fictions and their very fictionality has led synthetic biologists

[1]For a discussion on material and formal analogies, see Hesse (1966), and Knuuttila and Loettgers (2014).

to construct *synthetic models* on the basis of mathematical models. Built from biological material, synthetic models can be considered as experimental objects constructed to study the assumptions and credibility of mathematical models. Yet, as we will show, the relationship between mathematical modeling and synthetic modeling is anything but direct—and synthetic models themselves can also be regarded as fictions, albeit concrete ones. In what follows we will first briefly introduce the field of synthetic biology and then go over to the discussion of the means and process of mathematization in this particular field.

2 Synthetic Biology: A Nascent Interdisciplinary Field

Synthetic biology focuses on the design and construction of novel biological functions and systems. It is often understood in terms of the pursuit for well–characterized biological parts to create synthetic wholes, and as such it has typically been understood as a kind of engineering science in which engineering principles are applied to biology (Church 2005). This view is shared by the public understanding of synthetic biology as well as the practitioners themselves. According to Jim Collins, who introduced one of the first synthetic networks, a toggle-switch, in 2000: "[...] synthetic biology was born with the broad goal of engineering or 'wiring' biological circuitry – be it genetic, protein, viral, pathway or genomic – for manifesting logical forms of cellular control" (Khalil and Collins 2010).

However, a more basic science oriented branch of synthetic biology has developed alongside the more engineering and application oriented approaches. This basic science oriented branch of synthetic biology targets our understanding of biological organization by probing the basic design principles of life by various strategies of modeling (see above). The design and exploration of synthetic models, i.e. engineered genetic circuits constructed from biological material and implemented in natural cell environment, provides the most recent strategy of this kind of approach (Sprinzak and Elowitz 2005).

The two branches of synthetic biology are not isolated but overlap and interact in several important ways. First, both make use of compound analogies to engineered artifacts and abstract model systems showing rhythmic/cyclic behavior. Second, the scientists in both branches employ largely the same theoretical tools and techniques. Third, the results gained in the basic science approach are utilized by the engineering oriented branch and the other way around. The main differences between the two branches thus lie in the primary aims of the scientists working in them, that is, whether they probe design principles in order to learn about the mechanisms operating in biological organisms or in search of design principles, which could be used in the engineering of *novel* biological parts and systems. These differences in aims are largely rooted in the different scientific backgrounds of the scientists. For example, the majority of scientists belonging to the first group probing the basic design principles of biological organization come from physics, whereas most of the scientists belonging to the second, more application oriented group have a background in engineering. Moreover, synthetic genetic circuits are so far largely in their proof of principle phase, and the actual applications of synthetic biology, like the synthetic malaria drug artemisinin, have in contrast resulted from laborious tinkering processes in the lab.

2.1 *Modeling Biological Mechanisms*

Biological systems have an inherent complexity given by the number of the different components and their interactions embodied by them. Metabolic and gene regulatory networks provide examples of biological systems that are extensively studied in synthetic biology (e.g., Bujara et al. 2011; Zhang et al. 2011; Elowitz and Leibler 2000; Nandagopal and Elowitz 2011). In what follows, we will focus on gene regulatory networks. Such networks consist of interacting genes and proteins. Genes and proteins interact via transcription and translation processes. Figure 1 shows a simplified picture of the main steps of these translation and transcription processes. Following the central dogma of molecular biology the DNA (deoxyribonucleic acid) carries all the genetic instructions necessary for the development, reproduction and functioning of an organism. The information stored in DNA is transcribed in the process of RNA (ribonucleic acid) synthesis into individual transportable cassettes, the so-called messenger RNA (mRNA). The individual cassettes carrying the blueprint of a protein as sequences of amino-acids, leave the nucleus and enter a complex protein machinery, the ribosome. In this machinery the transcribed information is translated and used in the formation of the protein.

The transcription process is activated or inhibited by so-called transcription factors. These are proteins binding to the promoter site of the gene. Figure 2 shows an example of such an activator binding to the promoter site of a gene. In the upper part of the picture the transcription factors binding to the promoter site inhibit RNA polymerase. In the lower part of the picture the transcription factor gets released by proteins moving into the cell and the transcription process starts.

Many important biological functions are based on gene regulatory networks. A prominent example is the circadian clock, which regulates day and night rhythm in biological organisms. The early modelers of biological organization had suggested already in the 1960s that the rhythmic behavior observed in the circadian clock is controlled by a molecular feedback mechanism (Goodwin 1963; Winfree 1967). For example, Colin Pittendrigh, who studied circadian rhythms on *Drosophila* wrote that the: "[...] commonest device in installing regulators—from the control of heartbeat to that of protein synthesis—is negative feedback. And one of the innate tendencies of such feedback systems is to oscillate" (Pittendrigh 1961: 122). Yet it has remained an open question as to whether gene regulatory networks in biological systems implement control in the same way as human engineered systems.

Synthetic biologists have followed the tradition of modeling the organization of biological systems in terms of feedback systems, although the relationship of

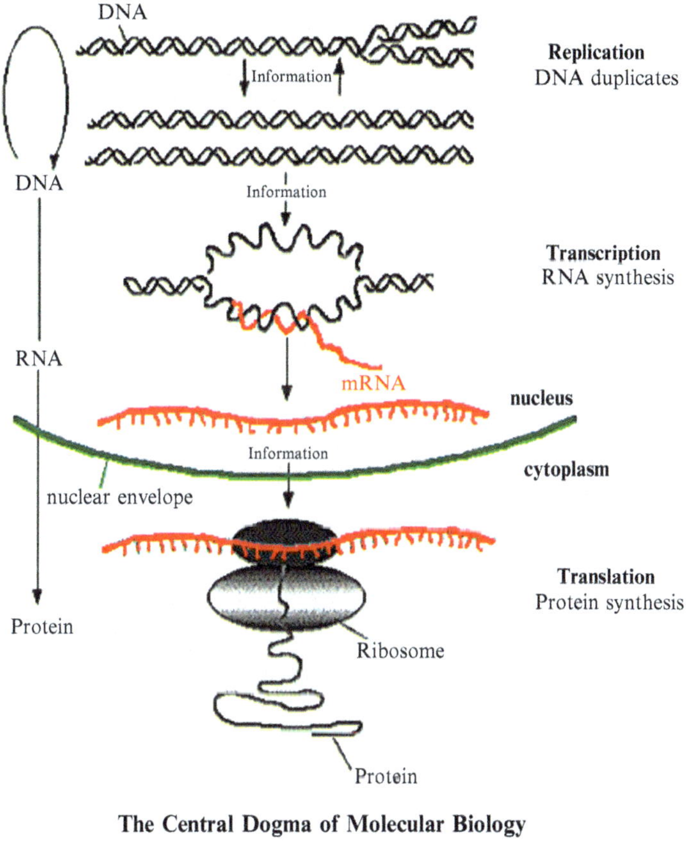

Fig. 1 The diagram shows the main elements of the transcription and translation processes according to what is called 'the central dogma of molecular biology' (https://nanohub.org/resources/17701/watch?resid=17812) The central dogma was introduced by Francis Crick in 1958. The dogma states that genetic information, which is transcribed from DNA into RNA and used in the production of proteins, cannot flow in the reverse direction

this modeling paradigm to experimental results is far from straightforward. The exploration of regulatory networks in model organisms is very complicated even in the case of such "simple" organisms as the bacteria *Escherichia coli*. It requires a lot of experience and skill to determine the constituent elements (genes, proteins) of the network, its structure and the interaction between the elements. Although the results of experimentation with model organisms are interpreted in terms of design principles adapted from engineering, recent results in synthetic biology show that gene regulatory networks can function in rather counter-intuitive ways. Nature seems to make use of different kinds of principles than human engineers, exploiting, for example, stochastic fluctuations (i.e., noise) in a functional way (cf. Çağatay, Turcotte, Elowitz, Garcia-Ojalvo, and Suel 2009). Engineers typically try to eliminate noise from their systems (see below). Such results as these are bound

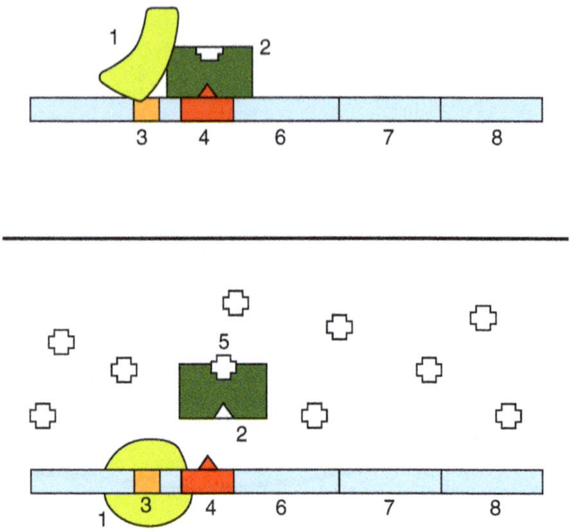

Fig. 2 A transcription factor binding at the promoter site and inhibiting RNA polymerase. ((https://en.wikipedia.org/wiki/Promoter_(genetics)) The numbers represent the following entities: *1*: RNA Polymerase, *2*: Repressor, *3*: Promoter, *4*: Operator, *5*: Lactose, *6*: lacZ, *7*: lacY, *8*: lacA) Only when the transcription factor is released from the promoter site does the transcription process start

to question the basic concepts and assumptions made by mathematical modeling of genetic circuits. This friction between the work on mathematical models vis-à-vis model organisms has led synthetic biologists to introduce a novel, additional model type, a synthetic model, which is located between mathematical models and model organisms. In the next sections we will study how the mathematical models of gene regulatory systems are constructed and the way these mathematical models are related to synthetic models.

3 Analogical Reasoning and the Use of Templates

As discussed above, scientists have assumed for some time that negative and/or positive feedback mechanisms play an important role in controlling biological functions (Jacob and Monod 1961; Goodwin 1963; Winfree 2001). This assumption was to a large part based on drawing analogies to engineered systems and it also formed the basic idea on which the mathematical models of biological regulation were built. The starting point of such mathematical model consists often of what synthetic biologists call a *toy model*. It is a model of a stylized abstract mechanism, such as a simple negative feedback mechanism, which is then being extended and refined taking into consideration some subject specific empirical knowledge. Below we will discuss some essential steps of designing a mathematical model from an initial toy model.

Negative feedback loops provide a control mechanism of a very general character: models of negative feedback can be found from many different contexts such as engineering, biochemistry, and physics. In a negative feedback loop the output feeds back into the input of the system repressing the further output and by

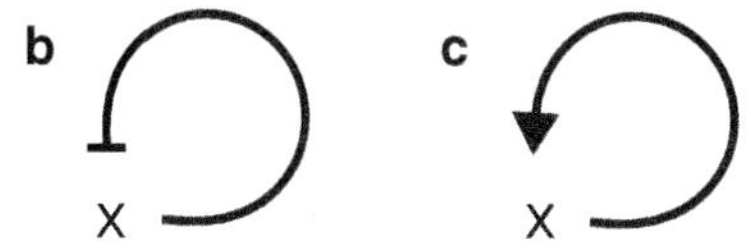

Fig. 3 Two network motifs: *b* depicts a negative feedback loop and *c* a positive feedback loop (Alon 2006: 451)

doing so stabilizes the system. In designing mathematical models of gene regulatory networks an oft-used motif is autorepression. In the case of autorepression, the gene product of a gene A suppresses its own function by binding to its (transcription) site.

The process of autorepression in gene regulatory networks is shown on the left hand side of the picture and the right hand side shows a sketch of a positive feedback loop. The simple diagram (Fig. 3) omits all biochemical details, the structure of genes and proteins as well as such essential parts of the mechanism as the binding of the activator to the promoter site etc. As discussed in the last section, the gene regulatory mechanism is comprised of a transcription and a translation part. During transcription the protein functions as a transcription factor binding to the transcription site of the gene. In the first step one observes an initial rise in the production of the gene product. But when the concentration reaches the repression threshold, which means that the transcription of the gene product becomes repressed, the production rate decreases and the system locks into a steady-state level. This locking into a steady state can be accompanied with oscillations in the protein level. Finding the conditions for sustained oscillations is one of the aims of mathematical modeling because many biological phenomena are periodic/rhythmic, and oscillations are thought to underlie the organization of such important gene regulatory systems as the circadian clock.

However, such oscillations also mark an important difference between feedback mechanisms in engineering and biology. Whereas oscillations in protein levels are essential for controlling biological rhythms, in engineered artifacts oscillations are typically regarded as unwanted and the systems are designed in such a way that oscillations are suppressed. For example, all kinds of electronic control systems typically have to avoid such oscillations in order to function reliably. Familiar examples of such devices are thermostats and cruise controls. This shows that the dynamic features of the regulation mechanisms are different in the case of biological systems despite the initial analogy to engineering. Brian Goodwin described this point in his influential book *Temporal Organization in Cells* (1963) in the following way: "The appearance of such oscillations is very common in feedback control systems. Engineers call them parasitic oscillations because they use up a lot of energy. They are usually regarded as undesirable and the control system is nearly always designed, if possible, to eliminate them" (Goodwin 1963: 5).

Once the simple sketch of a feedback mechanism has been designed, it has to be translated into a mathematical model. Such mathematical model typically consists of a set of differential equations, one modeling the production and degradation of a protein and a second one modeling the mRNA synthesis and degradation.

The model then gets adjusted to the particularities of the biological system under study. This modeling approach is very common in kinetic theory and the differential equations are essentially kinetic equations. The challenge is how to choose the relevant biochemical parameters and to determine their values. These limitations are of both of a practical nature (i.e., how to measure the values of the biochemical parameters, which are part of dynamical processes), as well as theoretical, regarding the lack of knowledge and theoretical insight that would guide the search for the most relevant parameters.

Clearly, this process of model construction disregards most biochemical details as well as the rich structures of genes and proteins. Goodwin discussed this abstract character of mathematization accordingly: "[...] in the study of the dynamic properties of a class of biological phenomena such as we are attempting, it is necessary to extract a manageable number of variables from the very large array which occurs in biological system. Is it ever possible to make such an extraction or simplification without doing violence to the very basis of biological organization, its inherent complexity? There is certainly no *a priori* answer to this question, and the only procedure is to try to find some set of variables which appear to constitute a reasonably self-contained system[2] and see if one can get meaningful and useful results relating to its behavior." (Goodwin 1963: 9). The work by Goodwin on temporal organization in cells has been fundamental in modeling cyclic processes in biological systems such as the circadian clock. He provided the elementary mathematical model that functioned as a basic template for the construction of such synthetic models as the *Repressilator*, nearly four decades later.

Let us finally note how the quote by Goodwin mediates the lingering sense of not knowing much of the details. And even after the 1980s when experimental data on genes and proteins involved in circadian clocks in various model organisms started to accumulate, the situation has not changed too much. The limits to what the scientists know about the components, organization and biochemical details of biological systems such as the circadian clock are still pressing. Because of this the already established conceptual frameworks from other areas, such as negative and positive feedback loops, provide at least a starting point for the first modeling attempts. And a corresponding mathematical framework is provided by the computational templates and methods that are used in modeling non-linear dynamic systems. These initial mathematical models for representing and studying various kinds of abstract feedback systems need mathematical articulation and adjustment in view of the systems at hand, yet modelers have to simultaneously take into account the mathematical constraints on how much detailed information can be expressed and studied by these models.

[2]This notion of a "reasonably self-contained system" bears an interesting link to the theme of fiction discussed below in Sect. 5.

3.1 *Mathematical Templates for the Study of Gene Regulation*

The general equations used by many systems and synthetic biologists (e.g., Goodwin 1963; Elowitz and Leibler 2000; Alon 2006) to describe the processes of transcription and translation are of the following form:

$$\begin{aligned} \frac{dm}{dt} &= \alpha_m - \beta_m \cdot m \\ \frac{dp}{dt} &= \alpha_p - \beta_p \cdot p \end{aligned} \tag{1}$$

where m is the concentration of RNA and p the concentration of protein, α_m and α_p the production, and β_m and β_p the degradation rates of RNA and the protein. This set of differential equations is called rate equations. It is used in chemical kinetics to describe the rate of reactions. These equations provide an example of what Paul Humphreys (2004) calls a computational template. With the concept of a computational template Humphreys refers to genuinely cross-disciplinary computational devices, such as functions, sets of equations, and computational methods, which can be applied to different problems in various domains. An example of such a template is the Lotka-Volterra model, which provides one of the simplest templates for modeling non-linear dynamics. In fact, the rate equations are at a formal level close to Lotka-Volterra equations.[3] The equations are of such a general character that without knowing that they are describing transcription and translation in a genetic network one could as well take them to describe something else, for instance some chemical reaction. In other words, these differential equations are able to describe the general dynamic behavior of various kinds of systems independently from many particularities of these systems. In addition to generality, Humphreys explains the cross-disciplinary usability of computational templates by their tractability. This is one important reason for the introduction of the rate equations from chemical kinetics to the study of genetic regulation: one can easily calculate the steady states of the system.

The steady states are calculated in the following way:

$$\frac{dm}{dt} = 0$$

$$\frac{dp}{dt} = 0$$

[3]For Lotka-Volterra equations as computational templates, see Knuuttila and Loettgers (2011, 2016).

the condition for the steady state is fulfilled by:

$$0 = \alpha_m - \beta_m \cdot m$$
$$0 = \alpha_p - \beta_p \cdot p$$

with:

$$m = \frac{\alpha_m}{\beta_m}$$
$$p = \frac{\alpha_p}{\beta_p}$$

On the basis of these general equations it is possible to specify in more detail the kind of regulation process one is going to study. For example, if the protein p in the set of general equations (1) functions as a repressor one has the case of negative autoregulation/ negative feedback loop. In this case the protein p inhibits the transcription process and therefore its own production. This will lead to oscillations in the protein level.

A first possible step in the adjustment of the differential equations consists in making the assumption that RNAp (RNA polymerase) binds fast to the transcription site being represented by the promoter activity function. This simplifies the problem in such a way that one does not need to take explicitly into consideration the binding of RNAp.[4] The differential equations for the process of autorepression are then of the following form:

$$\frac{dm}{dt} = \alpha_m \cdot g_R(r) - \beta_m \cdot m$$
$$\frac{dp}{dt} = \alpha_p \cdot m - \beta_p \cdot r$$

with $g_R(r)$ as the promoter activity function and r the number of repressors. The differential equations are non-linear and coupled. The change in the number of m (mRNA) depends on the number of the repressors r, and the other way around, the number of repressors on the number of mRNA. The resulting set of non-linear coupled differential equations cannot be solved analytically.

In sum, in mathematizing biological circuit systems synthetic biologists typically start from the analogies drawn to electric circuits and render the network motifs that describe various kinds of feedback loops into equations by using the toolbox of modeling complex systems, especially the non-linear differential equations. However, as a result the models arrived at are abstract in that they lack a lot of details, and furthermore, there is the problem that the formalisms have not typically

[4]Other scientists such as Brian Goodwin take the binding of the RNAp into account. This makes the differential equations more difficult by adding a further variable.

been developed with biological systems in mind, although they have been adjusted to take into account some features of the subject matter in question. This abstract, hypothetical and interdisciplinary nature of the mathematical models of genetic circuits has led synthetic biologists to develop a novel modeling method, synthetic modeling. Synthetic models probe to what extent it is legitimate to suppose that gene regulatory networks function on the basis of feedback mechanisms of some kind. Synthetic models are biological networks that are engineered from genes and proteins on the basis of mathematical models. In that sense they can be considered epistemic tools that are constructed to study the design principles depicted by the mathematical models.[5] In a sense this strategy can be seen as a way to materially embody and recontextualize the template-based, sparse and "foreign" mathematical models into the actual biological conditions where the dynamic, mechanism or function under study is located.

4 Synthetic Modeling – The *Repressilator*

The *Repressilator* is a simple engineered gene regulatory network. It is one of the first and most famous synthetic models, introduced in 2000 by Michael Elowitz and Stanislas Leibler (Elowitz and Leibler 2000). The *Repressilator* consists of three interacting genes connected via a negative feedback loop creating oscillations in the protein level. In gene regulatory systems, as we have seen, oscillating proteins are the essential part of the control. The basic network design is taken from electronics: *The Repressilator* is a biological version of a ring oscillator. Before the *Repressilator* was built, Elowitz and Leibler designed a mathematical model of it utilizing mathematical tools that had been developed to study the biological feedback systems (discussed in the previous section).[6] One particular book was of special importance for the design of the *Repressilator*: *Biological Feedback* by Thomas and D'Ari (1990), which presents a formal methodology for analyzing the dynamic behavior of complex systems.[7] In this book feedback systems are analyzed and described in a very general way—that is, it provides computational templates for analyzing different kinds of feedback systems.

[5]On the notion of an epistemic tool, see Knuuttila (2011).

[6]For example, the properties and dynamic features of network motifs describing recurrent structures in genetic networks (e.g. feedforward and feedback loops) can be analyzed by making use of the Michaelis-Menten equations (Berg et al. 2002).

[7]Personal communication by Michael Elowitz.

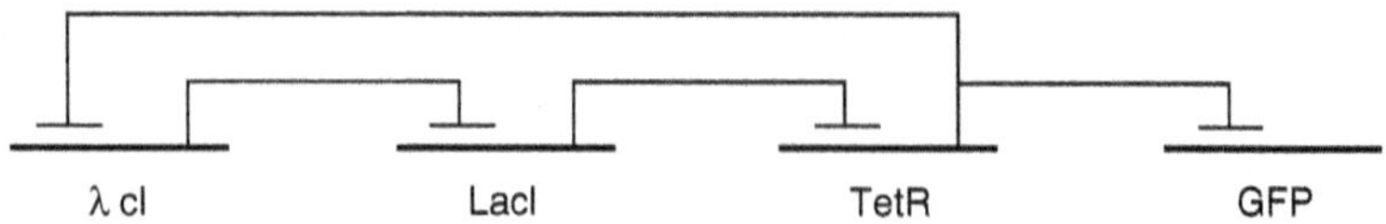

Fig. 4 The main structure of the *Repressilator*

4.1 The Mathematical Model of the Repressilator

Already the seemingly simple set of differential equations presented in the Sect. 3 leads to complex dynamics. More complicated models of gene regulation can be built on this basic template. The mathematical model underlying the *Repressilator* provides an example of such a model (Elowitz and Leibler 2000). The *Repressilator* consist of three genes, *TetR*, *LacI* and *λcl*, which are arranged in such a way that they inhibit each others' activity (see Fig. 4). The fourth gene used in the construction of the *Repressilator* is a Green Fluorescent Protein (GFP). The GFP gene is not part of the differential equations, as it does not contribute to the dynamic of the system (as discussed below). The dynamic of the *Repressilator* results from the following mechanism: The protein related to each gene represses the protein production of its neighboring gene. This leads to oscillations in the protein levels of the respective genes. The mathematical model Leibler and Elowitz constructed was based on the two differential equations for autorepression. In the case of the *Repressilator*, instead of one gene and its protein, one has 3 genes and proteins—and therefore 6 coupled differential equations of the following form:

$$\frac{dm_i}{dt} = -m_i + \frac{a}{\left(1 + p_j{}^n\right)} + \alpha_0$$

$$\frac{dp_i}{dt} = -\beta\left(p_i - m_i\right)$$

$$\text{with}\begin{pmatrix} i = lacl, tetR, cl \\ j = cl, lacl, tetR \end{pmatrix}.$$

The three proteins *lacl, tetR, cl* are produced by the genes of the *Repressilator*. The set of differential equations is basically of the same form as the one discussed above. It consists of a production and a degradation term. As before, p_i denotes the number of proteins and m_i the number of mRNA. In the case of a saturating number of repressors, the number of proteins is given by α_0 because of some leakiness at the binding side. In the case of no repressors, the number of proteins is given by $\alpha + \alpha_0 \cdot \beta$, which denotes the ratio of the protein over the mRNA decay rate. The Hill coefficient denoted by n describes the binding strength of the proteins to the transcription site. Thus the differential equations take into account specific biomolecular properties such as leakiness and binding strength.[8] However, those

[8]Even if all the active sites of the proteins are occupied by repressors one observes some production of proteins, which is expressed by α_0.. This is what is meant by leakiness.

parameters are usually not known and have to be estimated by computer simulations. In those computer simulations a stability diagram is produced marking regions of stable and unstable solutions of the differential equations for different values of α, β and n. Only when the studied state becomes unstable, sustained oscillations may occur. Since Elowitz and Leibler were interested in regulation by oscillations they focused on the latter case. Only sustained limit-cycle oscillations could provide the rhythm for controlling day and night rhythms in biological organisms.

The computer simulations performed by Leibler and Elowitz gave them more insight into the biochemical conditions of sustained oscillations: "We found that oscillations are favoured by strong promoters coupled to efficient ribosome-binding sites, tight transcriptional repression (low 'leakiness'), cooperative repression characteristics, and comparable protein and mRNA decay rates" (Elowitz and Leibler 2000: 336).

To sum up, the preceding discussion on mathematical modeling in synthetic biology shows how a mathematical model of gene regulation is constructed: by introducing rate equations from chemical kinetics and combining them with a special control mechanism adopted from electrical engineering one can arrive at a general form of coupled differential equations. These differential equations need then to be adapted to the subject matter under investigation by specifying parameters such as binding strength and by exploring different possible dynamics related to the parameters. All these modeling activities can be best described as developing and exploring a blueprint for the construction of the subsequent synthetic model.

4.2 *The Repressilator*

The synthetic model, the *Repressilator*, was constructed on the basis of the mathematical model and consists of two parts (Fig. 5). In the diagram the synthetic genetic regulatory network, the *Repressilator*, is shown on the left hand side. The outer part is an illustration of the plasmid constructed by Elowitz and Leibler. The plasmid is an extra-chromosomal DNA molecule integrating the three genes of the *Repressilator*. Plasmids occur naturally in bacteria. In the state of competence, bacteria are able to take up extra chromosomal DNA from the environment. In the case of the *Repressilator*, this property allowed the integration of the specifically designed plasmid into *E.coli* bacteria. The inner part of the illustration represents the feedback loop between the three genes, *TetR*, *LacI* nd λ *cl*, whose dynamics was studied in advance by the mathematical model. The left-hand side of the diagram shows the *Reporter* consisting of a gene expressing a green fluorescent protein (GFP), which is fused to one of the three genes of the *Repressilator*.

The construction of the *Repressilator* critically depended on the development of new methods and technologies, such as the construction of plasmids, Polymerase Chain Reactions (PCR) and Green Fluorescent Proteins (GFP). GFP became available in the mid-1990s (Chalfie et al. 1994) and very soon also fluorescent proteins with yellow (YFP) and red channels (RFP) were introduced (Elowitz et al. 1997). By fusing GFPs into a gene regulatory network, implemented within for

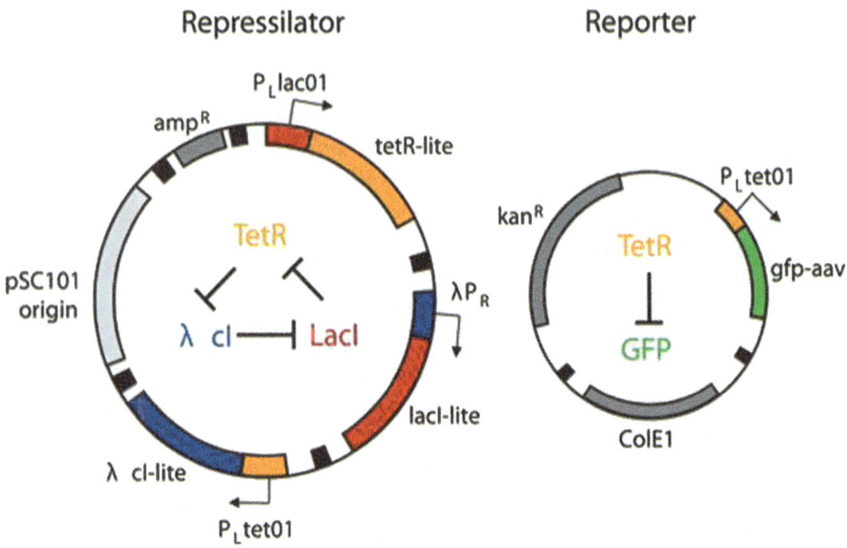

Fig. 5 The main components of the *Repressilator* (left hand side) and the *Reporter* (right hand side) (Elowitz and Leibler 2000: 336)

Fig. 6 The picture shows E.coli bacteria into which next to GFP's also yellow and red fluorescent proteins have been introduced (Elowitz et al. 2000 1184)

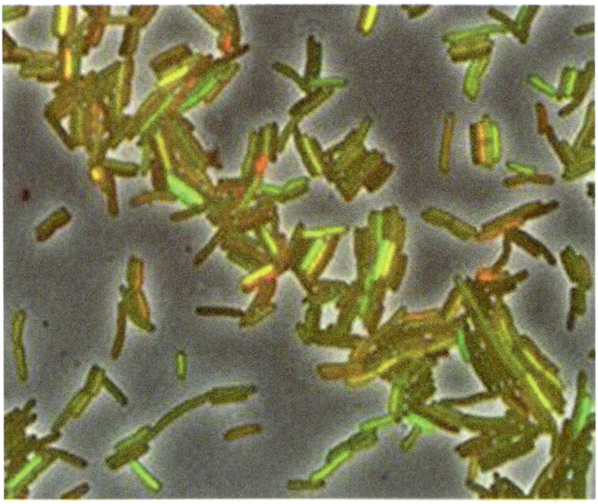

example E. coli, the expression of genes becomes visible and can be analyzed. Figure 6 shows a picture of "blinking bacteria" from the work of Michael Elowitz (Elowitz et al. 2000). In analyzing the intensity of the light emitted by the GFP, YFP, and RFP of the *E.coli*, synthetic biologists like Elowitz and his co-workers, try to get insight into the dynamic of such networks and how they give rise to specific biological functions. This kind of analysis comes with several challenges and difficulties. For example, the measurements may indicate that two genes interact, but this does not necessarily mean that one can assign in a straightforward fashion a mechanism underlying that interaction. Moreover, even if the two genes interacted, this does not indicate that this interaction would play any functional role in the biological system.

The GFP oscillations in the protein level of the *Repressilator* made visible the molecular behavior of transformed cells, i.e. the cells in which the *Repressilator* was implanted. It turned out that the *Repressilator* was indeed able to produce oscillations at the protein level but these oscillations showed irregularities. Interestingly, to find out what was causing such noisy behavior Elowitz and Leibler reverted back to mathematical modeling. In designing the *Repressilator*, Elowitz and Leibler had used a deterministic model. A deterministic model does not take into account stochastic effects such as stochastic fluctuations in gene expression. Performing computer simulations on a stochastic version of the original mathematical model, Elowitz and Leibler were able to reproduce similar variations in the oscillations as observed in the synthetic model. This led researchers to the conclusion that stochastic effects may play a role in gene regulation—which gave rise to a new research program attempting to identify sources of noise in biological systems and the effect of noise on the dynamics of the system (e.g., Swain et al. 2002). This research program makes extensive use of combinational modeling: the role of noise in biological systems was not only studied and explored by making use of mathematical and synthetic modeling but also by comparing the network architectures in model organisms such as *B. subtilis* and in synthetic systems (e.g., Süel et al. 2007). Model organisms have become an increasingly important part of the modeling practice of synthetic biology laboratories.

5 Fictions: Abstract and Concrete

Above we have described the complex interplay of mathematical modeling and synthetic modeling in synthetic biology. Due to the way mathematical models are constructed they remain abstract and describe only possibilities. However, this is also an advantage of mathematical modeling as the abstract general templates make it possible to study several possible scenarios by adjusting them accordingly. This gives modelers a handle on how things could be and what reasons might underpin why these things might be organized in this or that way. The case of the *Repressilator* showed how synthetic modeling can probe the biological realisticness or implementability of the possible mechanisms depicted by mathematical models.

It is already evident how synthetic modeling has affected synthetic biology: Biology in all its complexity has occupied the central stage. Important engineering notions on which synthetic biology has been grounded, such as noise and modularity, have been reinterpreted and some analogies drawn to engineering have been questioned (see Knuuttila and Loettgers 2013, 2014).

Yet, in order to study the new questions raised by synthetic modeling, researchers typically revert back to mathematical modeling. A good example of this is provided by a recent study by Tatiana T. Marguéz-Lago and Jörg Stelling (2010) who, by employing a series of what they call "minimal models," studied some counter-intuitive behaviors of genetic circuits with negative feedback. As discussed above, the *Repressilator* and related studies made synthetic biologists seriously consider how noise could have a functional role in biological organization (cf. Loettgers 2009). Marguéz-Lago and Stelling have further analyzed the implications of stochastic fluctuations (i.e., noise) by mathematical modeling. They write: "It has often been taken for granted that negative feedback loops in gene regulation work as homeostatic control mechanisms. If one increases the regulation strength a less noisy signal is to be expected. However, recent theoretical studies have reported the exact contrary, counter-intuitive observation, which has left a question mark over the relationship between negative feedback loops and noise" (Marguéz-Lago and Stelling 2010: 1743). Marguéz-Lago and Stelling's article is a telling example of how mathematical models are used to explore different possible explanations for such unexpected behaviors. Starting out from a simple *toy model,* one that cannot represent realistically any biological system, the scientists explore the conditions for different observed behaviors. They create different possible design principles, which could occur but do not necessarily exist in any natural systems. Thus the way mathematical models are designed and used in synthetic biology serves to highlight their fictional character.

This exploration of possible natural design principles resonates interestingly with the recent philosophical discussion on the fictional nature of modeling (cf. Suárez 2009). For instance, Peter Godfrey-Smith approaches the contemporary model-based theoretical strategy in terms of *imagined non-actual objects* which are investigated and explored in order to learn something about real-world objects. An important property of these imagined non-actual objects is that they could be concrete if real. Or in the words of Godfrey-Smith: "[. . .] what I see model-builders are after is trying to describe and understand systems that are only imaginary, but which would be *concrete if real*" (Godfrey-Smith 2009: 101). Synthetic biologists proceed in this way, taking this process even a step further by constructing concrete fictions. The mark of fiction is thus not in its imaginary non-concrete nature but its being a self-contained system that can be manipulated and explored (Knuuttila 2009; Rouse 2009). By engineering gene regulatory networks from biological components synthetic biologists design concrete fictions, which can be tested by and compared with mathematical models—or even transferred into an engineered object fulfilling a specific task.

It is not difficult to uncover the fictional features of a synthetic model such as the *Repressilator* although it is a biological construct functioning in a living cell: Its components (and their number and arrangement) had to be chosen in view of what would be optimal for the behavior under study. The genes used in the *Repressilator* do not occur in such a combination in any known biological system but are chosen and tuned on the basis of the simulations of the underlying mathematical model and other background knowledge[9]—in such a way that the resulting mechanism would

[9]This draws synthetic modeling close to simulation modeling, which brings to mathematical modeling exploratory and experimental features (e.g., Lenhard 2007).

allow for sustained oscillations. These technical constraints imply a constraint on what can be explored by such synthetic models: they also study possible design principles in biological systems. In that synthetic models are like mathematical models, they still only provide "how-possibly" explanations. This emphasis is clear from the writings of synthetic biologists.

Leibler and Elowitz did not claim that their synthetic system corresponds to any actual mechanism at work in biological systems. On the contrary, they were very much aware of the limitations of their procedure of drawing analogies to mechanisms, which have been proven to work in engineering but not necessarily in biology. Elowitz and Leibler described their expectations concerning the outcome of the *Repressilator*: "We did not set out to describe precisely the behaviour of the system, as not enough is known about the molecular interactions inside the cell to make such a description realistic. Instead, we hope to identify possible classes of dynamic behaviour and determine which experimental parameters should be adjusted to obtain sustained oscillations" (Elowitz and Leibler 2000: 337). Sprinzak and Elowitz in turn write in the introduction of their review article on synthetic biology: "They [synthetic models] fail to operate as reliably, but they provide a proof of principle for a synthetic approach to understanding genetic circuits" (Elowitz and Sprinzak 2005: 443). Accordingly, synthetic models could provide a proof of principle for the possibility that such a mechanism as negative feedback could function as a control mechanism in biological systems. This is due to the fact that, despite their fictional character, synthetic models are closer to the actual biological organisms, in so far as they are expected to function under the same material constraints as biological systems. This feature draws synthetic models closer to experimentation and because of this they can be seen as partly bridging the gap between experimentation in model organisms and mathematical modeling. But such a proof is of course far from definite, which is precisely the reason synthetic biologists make use of the combinational approach.

Finally, the fictional nature of synthetic models shows also what goes unrecognized if one takes too literally the idea of mathematical models as *blueprints* for the design and construction of synthetic models. Namely, when talking about mathematical models, synthetic biologists often refer to them as blueprints. Yet the notion of a *blueprint* gives the impression of a ready-made, fixed thing that would function in a more definite manner, like architectural plans for a house. To describe the mathematical model underlying a synthetic model as a blueprint partly misses the explorative role of mathematical models. They provide tools for studying possible realizations or scenarios, or what synthetic biologists call design principles or motifs, emulating engineering scientists.

6 Concluding Remarks

Above we have studied the ways in which synthetic biologists make use of compound analogies by invoking engineering notions such as feedback system, and utilizing computational templates from the study of complex systems. We have argued that because the mechanisms underlying biological functions such as the circadian clock are largely not known, scientists probe them by using control mechanisms, which have been proven to work in other scientific contexts. This makes mathematical modeling, we suggest, inherently fictional (cf. Weisberg 2007)—but it simultaneously enables scientists to make use of cross-disciplinary computational templates and modeling methods. Indeed, the tools and templates that have been developed over the last decades by the study of complex systems provide an important reason why synthetic biologists make use of feedback mechanisms in describing and designing mathematical models of gene regulatory networks. Here also the advancement of computer technologies and the possibility of simulating the non-linear dynamics of feedback systems played a prominent role. Only with the availability of computer technologies and simulations could the space of possible dynamic behaviors of mathematical models be explored, and made use of in the construction of synthetic models.

But of course this analogical procedure of transporting concepts and tools from other fields of study is bound to introduce some uncertainties in the new terrain of application. As we have seen, such engineering-inspired control mechanisms may not resemble those that have evolved in natural processes. They are, indeed, merely possible design principles. This then comes close to the present philosophical discussion on the fictional nature of modeling—moreover by providing a rationale for it, something that the philosophical discussion on fictions largely lacks. This fictional character is also affirmed by synthetic biologists themselves who envision that, as a result of the synthetic approach, the entire field of biology might undergo an important change "from a discipline that focuses on natural organisms to one that includes potential organisms" (Elowitz and Lim 2010: 889).

In this paper we have concentrated on the basic science-oriented branch of synthetic biology that seeks to understand the general design principles of biological organization on the level of their minimal, or sufficient, components and features. Let us note, however, the double meaning of the quest for design principles in synthetic biology. On the one hand, as we have discussed, synthetic biologists consciously create fictional systems in order to try out various design principles. In electrical engineering, for example, these design principles have well-understood properties and the challenge is to find out whether, and to what extent, they apply in the context of biology. On the other hand, the study of possible design principles aims for engineering *novel* biological parts or systems. Even if such design principles may not have evolved, they could be constructed and used for various purposes, for example for vaccines (e.g., the work of Jay Kiesling).

Last but not least, the testing of whether a design principle can operate in natural systems requires a laborious combinational use of mathematical models, experimentation on model organisms, and synthetic models. This approach has already led to a change in our understanding of how biological systems function and served to underline their differences vis-à-vis engineered artifacts. It will be interesting to see what kind of impact this will have on the mathematical methods and techniques used in modeling biological organization.

References

Alon, U. (2006). *An introduction to systems biology*. London: Chapman & Hall/CRC Mathematical and Computational Biology.

Bangu, S. (2012). *The applicability of mathematics in science: Indispensability and ontology*. Basingstoke: Palgrave Macmillan.

Berg, J. M., Tymoczko, J. L., & Stryer, L. (2002). *Biochemistry*. New York: W. H. Freeman.

Bujara, M., Schümperli, M., Pellaux, R., Heinemann, M., & Panke, S. (2011). Optimization of a blueprint for in vitro glycolysis by metabolic real-time analysis. *Nature Chemical Biology, 7*, 271–277.

Çağatay, T., Turcotte, M., Elowitz, M. B., Garcia-Ojalvo, J., & Suel, G. M. (2009). Architecture-dependent noise discriminates functionally analogous differentiation circuits. *Cell, 139*(3), 1–11.

Chalfie, M., Yuan, T., Euskirchen, G., Ward, W. W., & Prasher, D. C. (1994). Green fluorescent protein as a marker for gene expression. *Science, 263*(5148), 802–805.

Church, G. M. (2005). From systems biology to synthetic biology. *Molecular Systems Biology, 1*.

Colyvan, M. (2001). *The indispensability of mathematics*. New York: Oxford University Press.

Elowitz, M. B., & Leibler, S. (2000). A synthetic oscillatory network of transcriptional regulators. *Nature, 403*(6767), 335–338.

Elowitz, M. B., & Lim, W. A. (2010). Build life to understand it. *Nature, 468*(7326), 889–890.

Elowitz, M. B., Surette, M. G., Wolf, P.-E., Stock, J., & Leibler, S. (1997). Photoactivation turns green fluorescent protein red. *Current Biology, 7*(10), 809–812.

Elowitz, M. B., Levine, A. J., Siggia, E. D., & Swain, P. S. (2000). Stochastic gene expression in a single cell. *Science, 297*(5584), 1183–1186.

Godfrey-Smith, P. (2009). Models and fictions in science, *Philosophical Studies, 143*(1), 101–116.

Goodwin, B. (1963). *Temporal organization in cells*. London, New York: Academic Press.

Hesse, M. B. (1966). *Models and analogies in science*. Notre Dame: Notre Dame University Press.

Humphreys, P. (2004). *Extending ourselves: Computational science, empiricism, and scientific method*. Oxford: Oxford University Press.

Jacob, F., & Monod, J. (1961). Genetic regulatory mechanisms in the synthesis of proteins. *Journal of Molecular Biology, 3*(3), 318–356.

Khalil, A. S., & Collins, J. J. (2010). Synthetic biology: application come to age. *Nature Reviews Genetics, 11*(5), 367–379.

Knuuttila, T. (2009). Representation, idealization, and fiction in economics: From the assumptions issue to the epistemology of modelling. In M. Suárez (Ed.), *Fictions in science: Philosophical essays on modeling and idealization* (pp. 205–231). New York/London: Routledge.

Knuuttila, T. (2011). Modeling and representing: An artefactual approach. *Studies in History and Philosophy of Science, 42*(2), 262–271.

Knuuttila, T., & Loettgers, A. (2011). The productive tension: Mechanisms vs. templates in modeling the phenomena. In P. Humphreys & C. Imbert (Eds.), *Representations, models, and simulations* (pp. 3–24). New York: Routledge.

Knuuttila, T., & Loettgers, A. (2013). Basic science through engineering: Synthetic modeling and the idea of biology-inspired engineering. *Studies in History and Philosophy of Biological and Biomedical Sciences, 44*(2), 158–169.

Knuuttila, T., & Loettgers, A. (2014). Varieties of noise: Analogical reasoning in synthetic biology. *Studies in History and Philosophy of Science Part A, 48*, 76–88.

Knuuttila, T., & Loettgers, A. (2016). Modelling as indirect representation? The Lotka-Volterra model revisited. *British Journal for the Philosophy of Science*. doi:10.1093/bjps/axv055.

Lenhard, J. (2007). Computer simulation: The cooperation between experimenting and modeling. *Philosophy of Science, 74*(2), 176–194.

Loettgers, A. (2009). Synthetic biology and the emergence of a dual meaning of noise. *Biological Theory, 4*(4), 340–349.

Marguéz-Lago, T., & Stelling, J. (2010). Counter-intuitive stochastic behavior of simple gene circuits with negative feedback. *Biophysical Journal, 98*(9), 1742–1750.
Nandagopal, N., & Elowitz, M. B. (2011). Synthetic biology: Integrated gene circuits. *Science, 333*(6047), 1244.
Pincock, C. (2012). *Mathematics and scientific representation*. Oxford and New York: Oxford University Press.
Pittendrigh, C. S. (1961). On temporal organization in living systems. *The Harvey Lectures, 59*, 63–125.
Putman, H. (1975). What is mathematical truth? *Historia Mathematica, 2*(4), 529–533.
Rouse, J. (2009). Laboratory fictions. In M. Suárez (Ed.), *Fictions in science: Philosophical essays on modeling and idealization* (pp. 37–55). New York/London: Routledge.
Sprinzak, D., & Elowitz, M. B. (2005). Reconstruction of genetic circuits. *Nature, 438*(7067), 443–448.
Steiner, M. (1998). *The applicability of mathematics as a philosophical problem*. Cambridge, MA: Harvard University Press.
Suárez, M. (2009). *Fictions in science: Philosophical essays on modeling and idealization*. New York: Routledge.
Süel, G. M., Kulkarni, R. P., Dworkin, J., Garcia-Ojalvo, J., & Elowitz, M. B. (2007). Tunability and noise dependence in differentiation dynamics. *Science, 315*(5819), 1716–1719.
Swain, P. S., Elowitz, M., & Siggia, E. D. (2002). Intrinsic and extrinsic contributions to stochasticity in gene expression. *Proceedings of the National Academy Sciences, 99*(20), 12795–12800.
Thomas, R., & D'Ari, R. (1990). *Biological feedback*. Boca Raton: CRC Press.
Weisberg, M. (2007). Three kinds of idealization. *The Journal of Philosophy, 104*(12), 639–659.
Wigner, E. P. (1960). The unreasonable effectiveness of mathematics in the natural sciences. *Communication on Pure and Applied Mathematics, 13*(1), 1–14.
Winfree, A. (1967). Biological rhythms and the behavior of populations of coupled oscillators. *Journal of Theoretical Biology, 16*(1), 15–42.
Winfree, A. (2001). *The geometry of biological time*. Heidelberg/New York: Springer.
Zhang, F., Rodriquez, S., & Keasling, J. D. (2011). Metabolic engineering of microbial pathways for advanced biofuels production. *Current Opinion in Biotechnology, 22*(6), 775–783.

Trigonometry, Construction by Straightedge and Compass, and the Applied Mathematics of the *Almagest*

Ido Yavetz

1 Introduction

The earliest surviving trigonometric tables used to compute numerical values for geometrical magnitudes occur in Ptolemy's *Almagest* (composed in the 2nd century AD). Current historical evidence cannot fix with precision the exact origins of such trigonometric tables. However, it seems likely that the Greek astronomer Hipparchos of Nicaea (2nd Century BC) was among the first, if not the very first to compute the ratio of chord to radius for a series of central angles in a circle, and to set the example of their use in astronomy for Ptolemy's later work. By comparison, geometrical methods for the determination of magnitudes are considerably older, and have become highly formalized no later than the end of the 4th century BC, in Euclid's *Elements*. This raises questions with regard to the comparative advantages of trigonometry over the older geometrical methods, and the particular emphasis that they received in the context of Greek mathematical astronomy.

From the point of view of repeatability and communication, numerical solutions of geometrical problems possess distinct practical advantages over the equivalent Euclidean constructions by straightedge and compass. The division of a given length by straightedge and compass, for example, identifies a single point as the required solution with the full confidence and precision of a rigorously justified procedure. In practice, an individual is not very likely to obtain the exact same division of a given length twice in a row, let alone two different individuals following the procedure with different tools for the same given length. The mechanical and material properties of straightedges, compasses, and drafting media together with the "personal equation" of different individuals set practical limits to repeatability,

I. Yavetz (✉)
Cohn Institute for History and Philosophy of Science and Ideas, Tel Aviv University, 2 Fichman St, Tel Aviv, Israel
e-mail: yavetz@tau.ac.il

J. Lenhard, M. Carrier (eds.), *Mathematics as a Tool*, Boston Studies in the Philosophy and History of Science 327, DOI 10.1007/978-3-319-54469-4_4

and make it difficult to set a precise value for the procedural error. On the other hand, barring computational mistakes, for a given quantity the procedure of long division by two would yield the same practical result exactly, whether repeated by the same individual or by different individuals. This alone would recommend Ptolemy's preference of trigonometric tables and numerical evaluation of parameters over the Euclidean procedures of construction by straightedge and compass. However, in the case of Greek mathematical astronomy, of which the *Almagest* is the culminating achievement, some inherent special features add further practical incentives to abandon the straightedge and the compass in favor of trigonometric tables and numerical calculation. To obtain values for the parameters of his planetary models in the *Almagest*, Ptolemy needed to solve a set of essential problems, which he managed by numerical procedures based on a precalculated trigonometric table of chords. This paper examines one of these essential problems in order to show that in principle it has a straight-forward Euclidean solution in the form of geometrical construction by straightedge and compass with no recourse to trigonometry. In practice, however, the geometrical solution encounters severe difficulties, revealing the practical limits of a Euclidean approach to the mathematical astronomy that Ptolemy sought to establish.

Figure 1 illustrates the basic model that serves the three exterior planets (Mars, Jupiter, and Saturn) in Ptolemy's planetary system. The coordinate system that

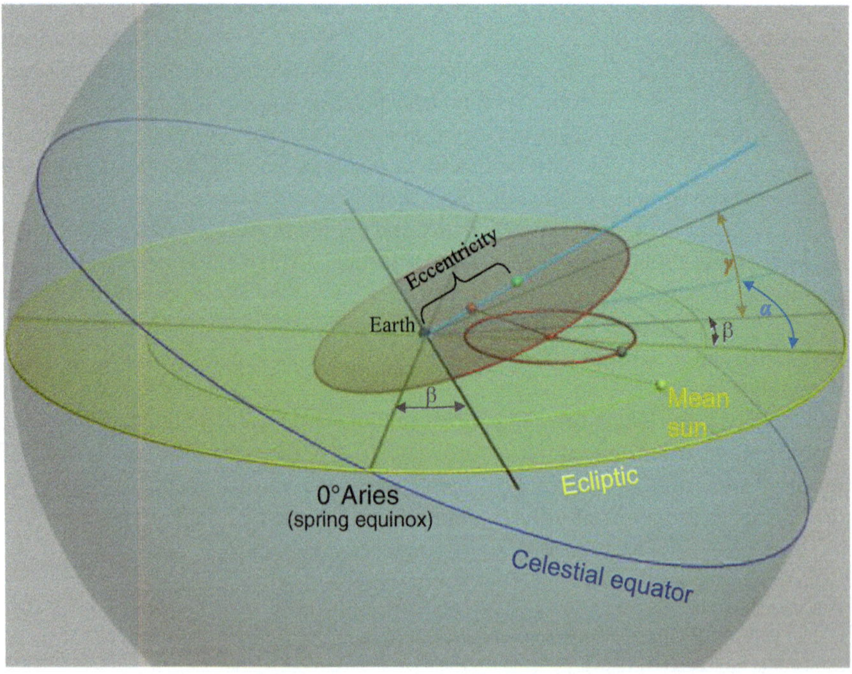

Fig. 1 The basic structure of a Ptolemaic model for the planets Mars, Jupiter, and Saturn. The epicycle plane is parallel to the ecliptic, as instructed in *Planetary Hypotheses*

Ptolemy uses is based on the relative orientations of the celestial equator and the ecliptic – two planes that pass through the centrally located earth and cut the celestial sphere into equal domes. The equatorial plane is perpendicular to the axis around which the stellar sphere turns in one sidereal day, while the sun's motion relative to the stellar background defines the ecliptic plane. The line along which the two planes cut each other also passes through the earth, and its ends mark the spring and fall equinoxes (only when the sun is at one of these points are days and nights equal everywhere on the earth). Still in the ecliptic plane, the end points of the line through the earth perpendicular to the equinoctial line mark the summer and winter solstices. These are respectively the sun's position on the longest and shortest days of the year in the northern hemisphere, and the greatest departures of the sun north and south of the celestial equator.

The basic model consists of an eccentric deferent, whose plane is tilted relative to the ecliptic, and an epicycle whose center the deferent carries on its circumference. The center of the deferent bisects the eccentricity, which is the distance between the earth and the equant point, from which the center of the epicycle appears to move at constant angular speed. As for the plane of the epicycle, Ptolemy has two versions: in the *Almagest*, the plane of the epicycle oscillates about an axis that remains always parallel to the plane of the ecliptic while it is tangent to the deferent at the epicycle's center. Book 13 of the *Almagest* describes the oscillation principle in detail. In the *Planetary Hypotheses*, Ptolemy abandoned the oscillating epicycle in favor of an epicycle whose plane is always parallel to the plane of the ecliptic, and this is the situation depicted in Fig. 1. The mean sun is a theoretical point that serves all of Ptolemy's planetary models in one way or another. It moves at constant angular speed in a circle centered on the earth, with a period of 365.25 solar days. For the exterior planets, the mean sun determines the revolution of the planet on the epicycle's circumference, under the condition that the line of sight from the epicycle's center to the planet is always parallel to the line of sight from the earth to the mean sun. The position of an exterior planet relative to the fixed stars requires the specification of two angles – a longitude and a latitude – and since angles do not depend on the length of their rays, the absolute dimensions of the deferent and epicycle are not required. Recognizing this, Ptolemy sets the size of each deferent to 60 for ease of computation with his sexagecimal system, and then needs only to fix the size of the epicycle in the same units. Altogether, each model requires the specification of seven independent parameters in order to be fully defined. They are as follows (see Fig. 1):

1. The rotational period of the deferent.
2. An initial position of the epicycle center at a particular point in time. (The rotational speed of the epicycle is not an independent parameter, being fully determined by the requirement that at all times the line of sight from the epicycle center to the planet must remain parallel to the line of sight from the earth to the mean sun).
3. The size of the epicycle (assuming a standard size of 60 for the deferent).
4. The size of the eccentricity (the distance from the earth to the equant point).

5. The angle α that defines the orientation of the eccentricity.
6. The angle β that defines the orientation of an axis that lies in the plane of the ecliptic and passes through the earth, about which the plane of the deferent is tilted.
7. Finally, the angle γ defines the tilt of the deferent relative to the ecliptic. Because this angle is very small (never more than a few degrees), no further corrections are indicated for parameters 2-5, which Ptolemy computes as if the deferent lies in the plane of the ecliptic.

Ptolemy's masterly way of showing how to break down into stages the computation of these parameters is the key to the *Almagest*'s dominant influence over the field of mathematical astronomy for about 1300 years. In the framework of a medium sized treatise, he teaches how to plan the minimal number of observations required for each stage, and proceeds to use the results of such observations in detailed computation of each and every parameter. To render the discussion completely self-contained, the *Almagest* includes a derivation of the trigonometric relations that it uses, and a full set of trigonometric tables for the practical work.

In the following pages we look only at parameters 4 and 5 – the size and orientation of the eccentricity – in order to show how they may be obtained by construction with straightedge and compass, and why the construction is not likely to satisfy the *Almagest*'s practical ends. Given the rotational period of the deferent, and an initial position of the epicycle's center at a given point in time, the size and orientation of the deferent's eccentricity determine all the subsequent positions of the epicycle's center. The two parameters must be evaluated together, and Ptolemy solves the problem trigonometrically by three successive approximations. Each approximation takes off from the previous one and approaches more closely the exact, but directly incomputable solution. This is one of several iterative solutions in the *Almagest*, which provides the earliest documented instances of such methods in the history of mathematical physics. Ptolemy's iterative procedure for the size and orientation of the eccentricity of the outer planets has been studied extensively in the modern literature,[1] so there is no need to repeat it here. For the present purposes, we set aside Ptolemy's trigonometric procedures, and show how to employ the same observational data that he specifies in order to evaluate the two parameters by successive approximations using purely Euclidean constructions with straightedge and compass.

[1] Toomer (1998, H321–H347). Pedersen (1974), discusses the procedure as applied to Saturn, for which two iterations prove sufficient. Besides one more computational cycle, the procedure for Mars is essentially the same.

2 Astronomical Observations and Prerequisite Theoretical Considerations

Following Ptolemy's procedure, let the deferent lie in the plane of the ecliptic, represented in Fig. 2 by the plane of the paper. T marks the earth, at rest in the center of the universe. C, at a distance e from the earth, is the center of the deferent. E, at a distance e from C, is the equant from which the epicycle's center appears to move at uniform angular speed on the deferent's circumference. Assuming that the deferent's radius is 60 units, the length TE $= 2e$ and its angular separation from the direction of the spring equinox (marked by 0° in the figure) need to be calculated.

The model's motion always keeps the line of sight from the epicycle's center to the planet parallel to the line of sight from earth to the mean sun. Therefore, when the planet's observed position is exactly opposed to the mean sun's calculated position, the extended line of sight from the earth to the planet must pass through the invisible center of the epicycle, and the measurable longitude of the planet is also the longitude of the epicycle's invisible center. Ptolemy provides data for three oppositions, as follows[2]:

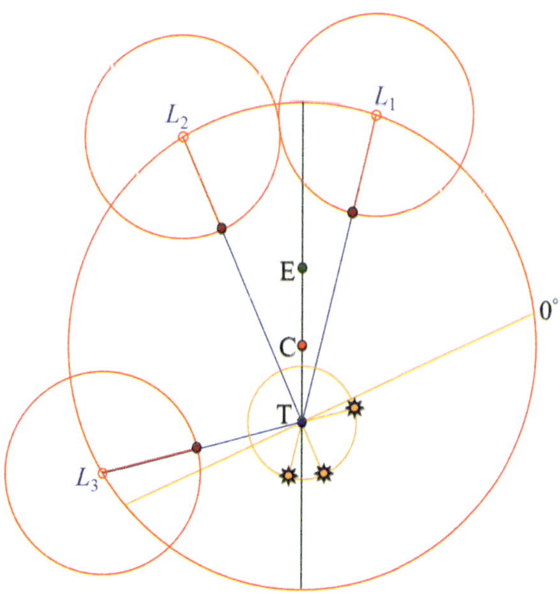

Fig. 2 Three observations of Mars at mid retrogression provide the data needed to calculate the direction and magnitude of the deferent's eccentricity

[2]*Ibid.*, p. 484 [H322]. Ptolemy gives the dates relative to the ruling years of the Roman emperor at the time. Thus he gives T_1 as "fifteenth year of Hadrian, Tybi 26/27 in the Egyptian calendar, 1 equinoctial hour after midnight, at about Gemini 21°." The rendition into modern dates is included in Toomer's translation.

1. $T_1 = 01{:}00$, night of Dec. 14/15, 130 AD; $L_1 = 21°$ in the sign of Gemini.
2. $T_2 = 21{:}00$, Night of Feb. 21/22, 135 AD; $L_2 = 28°50'$ in the sign of Leo.
3. $T_3 = 22{:}00$, Night of May 27/28, 139 AD; $L_3 = 2°34'$ in the sign of Sagittarius.

From these, he extracts two timed angular intervals, (t_1, α) and (t_2, β), (one Egyptian year = 365 solar days):

$t_1 = T_2 - T_1 = 4$ Egyptian years, 69 days, 20 hours (1529.833 mean solar days).
$\alpha = 67°50'$ beyond complete revolutions.
$t_2 = T_3 - T_2 = 4$ Egyptian years, 96 days, 1 hour (1556.042 mean solar days).
$\beta = 93°44'$ beyond complete revolutions.

The angular intervals α and β are measured from the earth, where the rotational speed of the deferent appears irregular. From the equant, however, the deferent's rotation appears uniform. Therefore, given the deferent's period in addition to the times t_1, t_2 in which α and β are covered, one can calculate the angles γ and δ that subtend from the equant the chords that α and β subtend from the earth. For the periods of both the deferent and the epicycle, Ptolemy uses the following observation:

> For Mars, 37 returns in anomaly correspond to 79 solar years (as defined by us) plus about 3;13 days, and to 42 revolutions of the planet from a solstice back to the same solstice, plus $3\frac{1}{6}°$.

This requires some terminological and conceptual clarifications.

1. The "solar year" in this context refers to the tropical year, defined as the time between successive arrivals of the sun to the spring equinox. Ptolemy uses a value of 365;14,48 mean solar days in the sexagecimal standard, namely, $365+\frac{14}{60}+\frac{48}{60^2}$, which is about 365.2467 days (compared to the modern value of 365.2422 days).
2. By "motion in anomaly" Ptolemy refers to the motion of the planet around its epicycle.
3. For the motion of the epicycle's center around the deferent, he uses the term "motion in longitude."

Therefore, the period of motion in longitude, which is the rotational period of the deferent, is the time between successive returns of the epicycle's center to the same ecliptic longitude, e.g., the spring equinox. For the period of return in anomaly, namely the period of the epicycle's rotation, Ptolemy takes the time for the epicycle to complete one revolution relative to the line of sight from the equant to the epicycle's center. The basic planetary model requires that the line of sight from the epicycle's center to the planet must remain parallel to the line of sight from the earth to the mean sun. Therefore, the period of return in anomaly is equivalent to the planet's mean synodic period, e.g. the average period between successive oppositions to the mean sun relative to the terrestrial observer, which is the time between any two successive oppositions to the mean sun relative to an observer at the equant. Had the deferent been stationary, the mean synodic period would be exactly equal to the mean solar year. However, since the deferent turns in the same direction as the epicycle, the following relationship must hold between the mean solar year, y_s, the epicycle's period, p_e, and the deferent's period, p_d:

$$\frac{1}{y_s} = \frac{1}{p_e} + \frac{1}{p_d}.$$

Equivalently, as Ptolemy puts it, during the time it takes for the sun and either Mars, Jupiter or Saturn to return to the exact same ecliptic coordinates,

> ...the number of revolutions of the sun during the period of return is always, for each of them, the sum of the number of revolutions in longitude and the number of returns in anomaly of the planet...

With this in mind, if it takes 79 mean solar years plus 3;13 days for the epicycle to complete exactly 37 rotations, the synodic period rounds off to 779.94 mean solar days (modern value: 779.94). The mandatory relationship between the mean solar year, the epicycle's period and the deferent's period rounds off to $p_d = 686.94$ mean solar days (compared to a modern value of 686.97 for the tropical period of the orbit http://nssdc.gsfc.nasa.gov/planetary/factsheet/marsfact.html). Using this value, both t_1 and t_2 contain more than two and less than three complete revolutions of the deferent. Therefore, the rotation past complete revolutions for either of them can be calculated according to:

$$\frac{t_i - 2p_d}{p_d} = \frac{a_i}{360},$$

yielding $a_1 = \gamma = 81°44'$, and $a_2 = \delta = 95°28'$ which are the values that Ptolemy gives in the *Almagest*.

3 Constructing the Solution: First Iteration

Ptolemy's first iteration assumes that the center of the deferent coincides with the equant. Rather than following him, we begin by assuming that the deferent's center coincides with the center of the earth (although the construction shown here could also begin with Ptolemy's first approximation, with no appreciable difference after two further iterations). Draw a circle, and let its radius represent 60 units of length. Draw a horizontal diameter through the circle, and mark its right end as pointing in the direction of the spring equinox (see Fig. 3 below, laid out with the drafting utilities of 3ds MAX instead of mechanical tools to produce Ptolemy's angular data).

From the spring equinox, mark the arcs L_1, L_2, and L_3 to the three oppositions cited by Ptolemy. Draw the chord L_1L_2, and at L_1 and L_2, construct equal angles of $90 - \gamma/2$ degrees, that meet at an angle γ, to complete a sum of 180° for the resulting isosceles triangle. Construct the circle that circumscribes the triangle, and from any point on the arc of this circle inside the circle of radius 60, points L_1 and

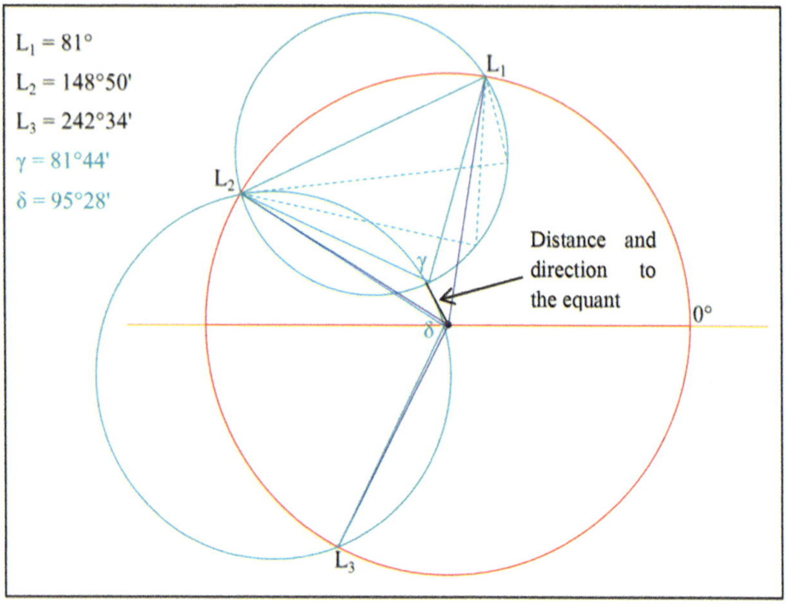

Fig. 3 First iteration – direction and eccentricity of the equant constructed geometrically on the initial assumption that the deferent's center coincides with the earth

L_2 will be seen at angle γ (because all peripheral angles that subtend the same chord are equal). For points L_2 and L_3, construct an isosceles triangle with apex angle δ, and draw the circle that circumscribes it. The only point from which L_1 and L_2 are subtended by γ while L_2 and L_3 are subtended by δ is where the two circles intersect inside the circle of radius 60. This is the only point that can serve as the equant. Its distance from the center and its orientation relative to the spring equinox can be directly measured on the drawing. However, the center of the deferent should not coincide with the earth, but rather bisect the distance from the earth to the equant. Therefore:

4 Constructing the Solution: Second Iteration

Keeping the angles L_1, L_2, and L_3 relative to the horizontal line through the earth to the spring equinox, reproduce the circle of radius 60 not around the earth as before, but around the midpoint between the earth and the equant found in the first iteration (marked by a dashed line in Fig. 4).

Use the points where the original lines to L_1, L_2 and L_3 intersect the new eccentric deferent as the vertices from which to construct the isosceles triangles with apex angles γ and δ, and as before, circumscribe them by circles. Figure 5 shows the result of the second iteration, with the deferent bisecting the new eccentricity.

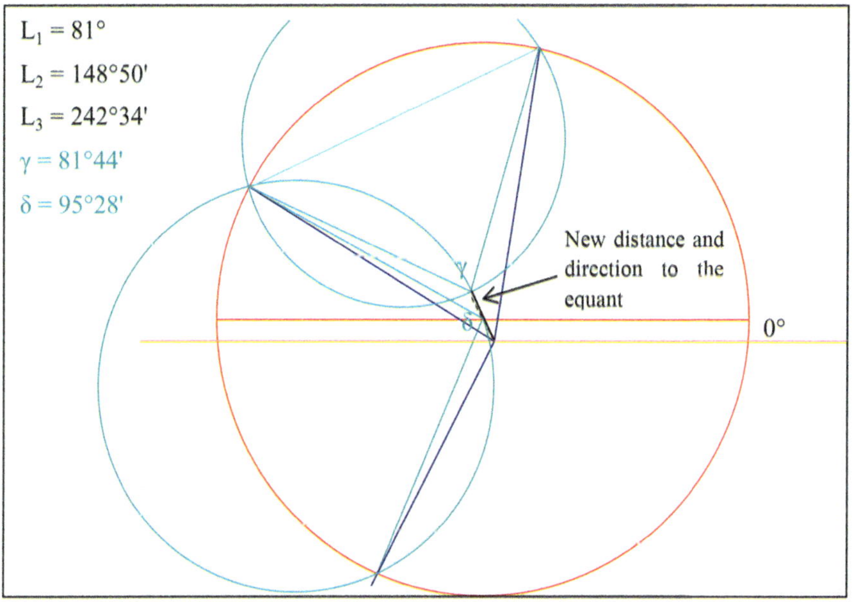

Fig. 4 Second iteration – direction and eccentricity of the equant reconstructed geometrically after the deferent's center has been moved to the midpoint between the earth and the equant in the first iteration

The size and orientation of the equant can be measured directly on the diagram, and the table below compares the measurements to Ptolemy's final result in the *Almagest*:

Graphic measure	Ptolemy
$e = 5.97$	$e \approx 6$
$\theta = 115°47'$	
$\theta - L_1 = 34°47'$	$\theta - L_1 = 34°30'$

Numerical computation using Ptolemy's data yields 6 for e, and $34°29'31''$ for $\theta - L_1$, so Ptolemy's result for the angle is better than the graphically evaluated one, but he also performed three iterations, as compared to only two here. In principle, a third graphical iteration should close the gap to Ptolemy's result. However, a third constructive iteration would prove impractical. The constructed orientation of the equant differs from Ptolemy's by 17 minutes of arc. An angle of 17 minutes of arc is drawn across the bottom of Fig. 5, but the actual length of line relative to which this small angle must be drawn is $2e = 12$, where the radius of the deferent is 60. For any useful drawing of 17 minutes of arc, the diagram's linear dimensions must be at least 15-20 times larger, making the deferent's radius about two meters long. Hand held compasses are out of the question at this scale, and one would probably be better

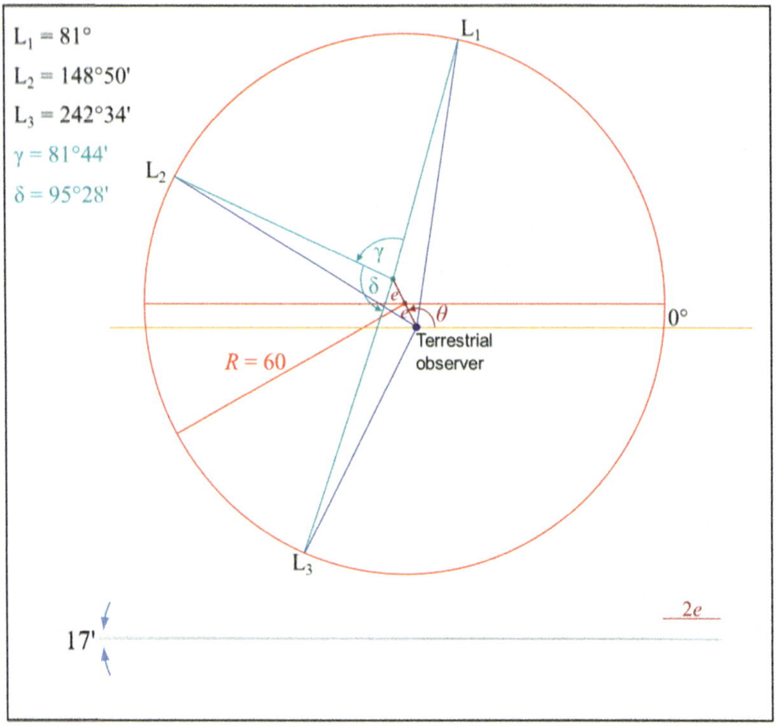

Fig. 5 The second iteration comes to within 17 minutes of arc of Ptolemy's third iteration for the eccentricity's direction, making a third iteration with straight-edge and compass impractical

served drawing the required straight lines with a carpenter's chalk line (indicated in Homer's *Illiad*, book 15:410–415) rather than a straightedge. Ptolemy's numerical procedure, on the other hand, suffers from no such limitations. As observational data becomes more precise, a fourth and fifth iteration may be added, requiring only more numerical calculations and additional references to the trigonometric table.

The practical difficulty that emerges in light of this exercise is not unique to the case of Mars, and many other calculations in the *Almagest* run up against similar practical barriers. Therefore, in addition to the advantages in communication and repeatability that numerical computations enjoy over equivalent geometrical constructions, it appears that Greek astronomy posed problems that overtaxed the practical limits of geometrical construction by compass and straightedge.[3]

[3] As I have shown elsewhere (Yavetz 2010), a similar difficulty emerges already in Aristarchus of Samos's computation of the sizes and distances of the sun and moon. All four quantities can be found by a single construction, which is much simpler in principle than Aristarchus's complicated geometrical argument leading to the upper and lower bounds of the desired results. In practice, however, the distance from the earth to the sun must be drawn around twenty meters long on a good flat surface, in order to consistently obtain close results to the mean values between Aristarchus's boundaries.

References

Pedersen, O. (1974). *A survey of the Almagest* (pp. 273–283). Odense University Press.

Toomer, G. J. (translator) (1998). *Ptolemy's Almagest* X.7 (pp. 484–498). Princeton: Princeton University Press.

Yavetz, I. (2010). *Wandering stars and ethereal spheres: Landmarks in the history of Greek astronomy* (Hebrew) (pp. 157–160). Jerusalem: Kineret, Zmora-Bitan, Dvir/Publishing House Ltd/The Hebrew University Magnes Press.

Shaping Mathematics as a Tool: The Search for a Mathematical Model for Quasi-crystals

Henrik Kragh Sørensen

1 Introduction

The relation between mathematics and the sciences is a historically complex one that is difficult to analyze philosophically in all its nuances. Famously, in 1960 the Nobel laureate Eugene Wigner (1902–1995) considered the applicability of mathematics to the natural sciences to be miraculous since, to him, mathematics was a purely formal science which had produced results that when later used by physicists were stunningly apt to describe and understand physical phenomena and theories (Wigner 1960). During the past half-century, historians of mathematics have scrutinized Wigner's argument and, in particular, pointed out that he unduly neglected the extent to which (pure) mathematical theories were developed in interaction with their possible applications (see e.g. Gelfert 2014; Grattan-Guinness 2008; for historical perspectives on the interplay between mathematics and physics, see also Kragh 2015; Schlote and Schneider 2011). These historical analyses reveal that the application of novel mathematics in the sciences often does not take the form of *application from the shelf* but rather takes place as a dialectical and temporally extended process. Once established as a viable research strategy, the application of such mathematical theories and results may prove effective in domains outside, but typically bordering on, the domain for which they were first used. By elaborating this contrast between the development of novel mathematical theories, recognizable as such by the mathematical community, and the application of established practices in new domains, I propose to nuance the meaning of using *mathematics as a tool* in situations where the proper mathematical conceptualization is still lacking.

H.K. Sørensen (✉)
Section for History and Philosophy of Science, Department of Science Education, University of Copenhagen, Øster Voldgade 3, DK-1350, Copenhagen, Denmark
e-mail: henrik.kragh@ind.ku.dk

J. Lenhard, M. Carrier (eds.), *Mathematics as a Tool*, Boston Studies in the Philosophy and History of Science 327, DOI 10.1007/978-3-319-54469-4_5

In particular, and perhaps especially in contemporary science, the relation between mathematics and science is not always well conceptualized as "applications of (ready-made) mathematics". This paper discusses the relations between mathematics and the sciences through a case-study of the early phase in the mathematical modeling of so-called "quasi-crystals". These are materials produced in the laboratory — first by the Israeli materials scientist Dan Shechtman and his collaborators in 1982 (Shechtman et al. 1984) — which violate the age-old mathematical model of crystals, in particular the "crystallographic restriction" that rules out five-fold rotational symmetry (see below).[1] Subsequently, during the 1980s and early 1990s different approaches were pursued in order to produce a mathematical model of this new phenomenon that would satisfy the expectations of mathematicians, physicists, and crystallographers. And in so doing, the interactions between these disciplines were at many times bi-directional in negotiating the concerns arising from these different disciplinary backgrounds.

Different suggestions were brought forward from different groups of researchers. These suggestions took into account different promises such as analogy with known models, efficiency in explaining phenomena, or generalization of known methods. In the process, the very nature of the thing modeled had to be redefined — new definitions were produced for such basic notions of the theory as "crystal" and "symmetry". This in turn allowed mathematicians and scientists to pursue different avenues in modeling the new phenomenon: Some have chosen to work on the properties of these new definitions, often in a low-dimensional restriction — sometimes just one dimension. Others have sought to model how quasi-crystals grow and how their large-scale non-periodic structure comes to emerge from their regular nuclei. And yet others have tried — sometimes from a combination of these approaches — to derive interesting quantitative predictions such as melting points and the like from these models.

At present, a certain consensus seems to have been established about the basic notions, but work in the directions suggested above is still ongoing. Therefore, this paper applies a historical approach to analyze the development of mathematical models of quasi-crystals during its first decade in order to discuss roles for mathematics in the sciences that both draw upon and extend beyond the notion of "mathematics as a tool".

This paper therefore falls in three parts: In the first part, I outline a framework for analyzing ways in which mathematics enters into contemporary interdisciplinary collaborations in the sciences. In the second part, I present and discuss the case study from the mathematical modeling of quasi-crystals from which, in the third part, I draw my philosophical and historical analyses.

[1]Shechtman's discovery and the subsequent study of quasi-crystals is described and outlined in many publications, some of which are explicitly referred to in the following. The reader may also consult e.g. Brecque (1987/1988) and Steinhardt (2013).

2 Interdisciplinarity and Cultures of Mathematization

With the recent focus — institutionally, politically and philosophically — on interdisciplinary collaboration, the case of mathematization may offer a nuancing perspective. For centuries, mathematicians have applied their discipline to various sciences, and yet such mathematization is rarely described as interdisciplinary. This paper elucidates and discusses this point through a case study of a recent set of cultures of mathematization, their epistemic division of labor, and their different desiderata of mathematical models. Thus, it addresses philosophical and historical issues of what happens to scientific enquiry when mathematics is used — or rather *developed* — as a tool.

Although almost ubiquitous in modern science, the notion of mathematization is, itself, loaded with ambiguity (see, for instance, Epple et al. 2013). Mathematization shares commonalities with applied mathematics, yet it encompasses more than a body of knowledge: It is a process undertaken by disciplinary mathematicians and scientists sharing a collaborative culture. In the past, a variety of modeling and numerical perspectives were involved in mathematizing scientific problems and theories. Thus, mathematization has been a historically contingent enterprise. And at times, the processes of mathematization were opposed by existing disciplinary structures. Thus, the parties involved did not unanimously accept the drive towards mathematization. To allow for this complexity, I present in the following a preliminary framework for analyzing interdisciplinary collaborations that involve mathematics.

The history of mathematics shows that the mathematization of a problem is not a straight-forward matter and that the very desire to mathematize may encounter opposition. To recall just one example, as shown in (Abraham 2004), the efforts of Nicolas Rashevsky (1899–1972) to treat biological cell division from the perspective of mathematical physics was initially opposed by biologists who in part questioned Rashevsky's modeling for its idealizations and in part found the model to lack in the kinds of questions that it provided answers to: In particular, it was argued, cells are *not* perfect spheres, and cell division is such a complex phenomenon that is does not reduce to membrane potentials. The short-term resolution to this conflict was a hybridization of disciplines in the form of Rashevsky's efforts to institutionalize "mathematical biophysics" (see ibid.) based on the principles of mathematical physics which were eventually thwarted as a "premature birth of mathematical biology" (Keller 2002, pp. 82–89).

This paper asks a basic and rather simple question: "What happens to some of the most interesting discussions about interdisciplinarity when cases are considered where mathematics is involved?" If interdisciplinary research is simply characterized as involving multiple disciplines, the use of mathematics in other fields ought to be the primary example of interdisciplinary research — so why isn't that the first example to be come to mind?. Part of the answer may have to do with the notion of applying preexisting mathematics in the sciences which supposes a chronological ordering of theoretical mathematics and its application. In contrast to such a view, this paper seeks to understand how mathematics is employed in situations where it is not possible to use pre-existing mathematical tools and theories but rather necessary to *develop* sophisticated theoretical mathematics as the application to another field is already going on. In order to begin to analyze such a broad question, I outline what I consider a promising avenue for investigation and discuss its application to a recent case-study.

2.1 Cultures of Mathematization

In order to approach the process of mathematization as an interdisciplinary collaboration, it is fruitful to introduce the notion *cultures of mathematization* to encapsulate the *epistemic culture* involved in using mathematics as a tool for the particular problem under consideration. The philosophical study of interdisciplinary collaboration in science has been very active for the past decade, and among its achievements has been the elaboration of how such collaborations are based on epistemic divisions of labor and trust (see e.g. Andersen and Wagenknecht 2013).

The framework of cultures of mathematization in envisioned here to analyze certain kinds of interdisciplinary collaborations that involve mathematics. To enable historical and philosophical analyses, these collaborations are to be studied in their contexts, which are, however, not necessarily characterized by geographical proximity such as a laboratory. Indeed, the division of labor in these collaborations can be such that collaborators may rarely if ever meet. Nevertheless, the collaborators need to be temporally proximate for the process to involve the following two characteristics setting it apart from applications of mathematics *from the shelf:*

First, the collaboration must involve mathematics and mathematicians. These categories are used as defined by their disciplinary structure and their primary disciplinary expertise. For the collaboration to be *interdisciplinary* in nature, the separation of expertise is required. Otherwise, the collaboration will often change into a structure where the scientists *become their own mathematicians.*

Second, the involvement of mathematics must go beyond that of a *mere pre-existing tool*; in particular, the collaboration must offer roads to new mathematical insights. More specifically, for the present analyses, some of the added mathematical insights should belong to mainstream mathematics.

Although these criteria are flexible enough to be met by a great number of specific cases, some historical caution is invited by considering the various forms of applications that mathematics can have and have had to other branches of sciences.

Central to the usefulness of the framework of cultures of mathematization is the plasticity of the notion of mathematical models and, in particular, the potential of mathematical models to function as *boundary objects* bridging different cultures. When combined with the notion of *epistemic cultures* defined by Karin Knorr Cetina as "amalgam[s] of arrangements and mechanisms — bonded through affinity, necessity and historical coincidence — which in a given field, make up how we know what we know" (Knorr Cetina 1999, p. 1), the mathematical model can be seen as the nexus for a variety of cultural concerns and exchanges (see also Morgan and Morrison 1999). In his seminal book, Peter Galison explained the importance of pidgin languages and creoles in exchanges and collaborations across paradigms (Galison 1997). By constructing "contact languages", collaborating partners can "hammer out a *local* coordination, despite vast *global* differences", Galison suggests (ibid., p. 783). Although explored by Galison in relation to the construction of physical apparatus and technology, this notion of contact languages can be adapted to mathematical models, I suggest. These, too, can function as central boundary objects that enable communication between different groups with different backgrounds, interests and disciplinary matrices.

This view of mathematical models hinges on an understanding of the mathematical modeling process as an iterative and dialectic process (see Fig. 1). It consists of an underlying clockwise process during which a segment of reality is delineated, simplified, structured and abstracted into a verbal model. This verbal model is further abstracted into a mathematical description from which mathematical

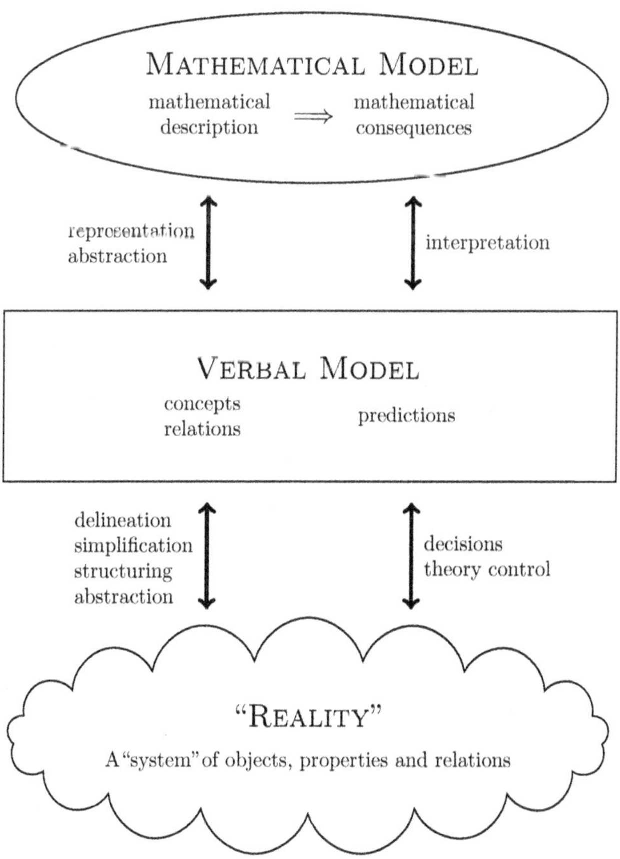

Fig. 1 Mathematical modeling as an iterative, dialectic process

consequences can be deduced, typically in the form of one or more equations linking the parameters to the modeled outcome. These mathematical consequences can, in turn, be interpreted into a verbal model, which can be used to make decisions or conduct theory control against reality. Importantly, the underlying iteration is superimposed with repeated dialectic negotiations as pragmatic and epistemic concerns and objectives are balanced. The resulting modeling process is not linear, but consists of multiple steps that go in both directions and may initiate at various points along the underlying process. On this account, the mathematical modeling process features an intrinsic verbal level, intermittent between the modeled part of reality and the formalized mathematical representation. And that verbal level, I suggest, can be adapted as the pidgin description of the mathematical model, suitable for various audiences.

2.2 Expectations and Critique of Models

In the quote given above introducing the notion of an *epistemic culture*, Knorr Cetina includes all those factors that "make up how we know what we know". In the present context, this can be specified into what may be called *cultures of mathematization* which includes those amalgams that constitute how mathematical modeling is conducted, again "bonded through affinity, necessity and historical coincidence". In particular, this allows attention to be focused on the various attitudes towards the mathematical model and towards model assessment and critique. Thus, when different cultures of mathematization critically review their boundary objects of mathematical models, they may look to different parts of the modeling process. Some will criticize the assumptions, others the idealization, yet others the mathematical apparatus and its application, and some will focus on the interpretations back into reality. Yet most or all of these aspects can coexist as cultures sharing the same boundary object.

Again, we may look to the many examples of conflicting stances towards mathematical models to illustrate that what counts as being in need of explanation and what counts as a good explanation can be highly controversial and specific to disciplines. One such example can be found in the debate over the determination of the age of the Earth as it unfolded in the nineteenth century. William Thomson's (1824–1907) (later, Lord Kelvin) mathematical model of the cooling of the Earth and his resulting estimate of the age of the Earth was steeped in religious controversy, and in the history of science it is also referred to as an example of the power of prestige in winning scientific arguments (see, in particular, Burchfield 1975). Yet, in the present context, Thomson's mathematization raises some additional issues: Thomson's model was based on an application of an already very successful technique, namely Joseph Fourier's (1768–1830) theory of heat conduction, and Thomson applied it to assumptions about the structure of the Earth that allowed him to compute the time it would have taken for the planet to cool to its present temperature known from observations. This mathematization was met with criticism from geologists as, for instance, when Thomas Mellard Reade (1832–1909) argued:

> Sir W. Thomson, however, infers, that as the Sun must have been hotter in former ages, the atmospheric agencies were then more potent; but this is all pure hypothesis, no proofs that they were being adduced. [...] [W]e know absolutely nothing of the Sun's heat, and cannot safely reason on conjectures. (Reade 1878, p. 153)

To Reade, this part of geology was concerned with explaining phenomena from empirical facts, and Thompson's mathematical model was criticized for the untestable hypotheses that it introduced.

On the other hand, Thomson's assistant John Perry (1850–1920), who as a mathematician would also criticize and eventually improve on Thomson's estimate, was also critical about the overlap of mathematics and geology (see also England et al. 2007a, b):

> I dislike very much to consider any quantitative problem set by a geologist. In nearly every case the conditions given are much too vague for the matter to be in any sense satisfactory, and a geologist does not seem to mind a few millions of years in matters relating to time. Therefore I never till about three weeks ago seriously considered the problem of the cooling of the earth except as a mere mathematical problem, as to which definite conditions were given. (Perry 1895, p. 224)

In the view of the mathematician Perry, geology was not concerned with precise quantitative descriptions such as treated by Thompson's mathematical model. Perry did not see the exercise in differential equations as having a bearing on geology until he realized that through mathematical modeling, various hypotheses for the development of the universe could be formulated and compared. This *mathematical geology* was, however, quite far from the discipline of geology that Reade had envisioned.

This brief example serves to illustrate the use of mathematics in many fields has been controversial. Part of the contention may be due to discipline formation and boundary work, but the very role of the mathematical model — its explanatory power, in particular — was also genuinely questioned as concerns and goals of the discipline were not always encoded in the model from the outset.

The two examples presented so far — the mathematization of cell division and the debate over the age of the Earth — illustrate that the notion of cultures of mathematization can capture important aspects of the disciplinary conflicts and the formation of new interdisciplinary fields. Importantly, the cases are not merely disciplinary controversies; they point to central methodological and epistemic differences in the view of mathematical modeling for the benefit of biology and geology, respectively. As such, viewing the mathematical model as a boundary object that can be appropriated and criticized enables us to appreciate the conflicts involved and points towards their resolution and the formation of new interdisciplinary fields focused on a specific culture of mathematization. For instance, Rashevsky's approach was pursued in the hybrid discipline of "mathematical biophysics" centered on his conception of the mathematical modeling process.

In both these cases, the mathematics that went into the modeling had been previously established in other fields of mathematical modeling, in particular as basic principles in physics. So they are classical examples of applications of mathematics and, as such, fail to reveal much of how mathematics is developed in the process. Therefore, the main case will address what can happen when the mathematical apparatus and even conceptual framework is not available on the shelves but has to be designed anew. This can lead to situations in which the production of the mathematical model may become a *co-production* between disciplines so as to imbue the model with greater sensitivity to the various expectations raised by different cultures.

3 Mathematical Modeling of Quasi-crystals, 1984–1995

The history of quasi-crystals could read like a standard story about a scientific revolution; and the pivotal figure, Shechtman who was awarded the Nobel Prize in Chemistry 2011, has indeed framed it in Kuhnian terms. In his Nobel lecture and in the Nobel interview, Shechtman explained his discovery in terms of a revolution against the existing, well-entrenched paradigm. Shechtman furthermore cast Linus Pauling (1901–1994) as the last protagonist of the old paradigm to die out before the discovery of quasi-crystals could be acknowledged through the Nobel Prize (see e.g. Kuhn 1962/1996, p. 151).[2]

However, as will become clear, the efforts to mathematize quasi-crystals also shed an intriguing light on the process of developing a new mathematical model to serve as the shared interface around which a new field can develop. Thus, a number of collaborative cultures have emerged with distinct, yet partially overlapping notions about the function and explanatory role of the mathematical model. This plurality of cultures of mathematization points to the centrality of a mathematical model as a boundary object while negotiating different desiderata.

3.1 *The Crystallographic Restriction*

In order to appreciate the necessity for a new mathematical model in the wake of Shechtman's discovery, it is instructive to review the standard mathematical model of crystals in terms of lattices. The modern science of crystals can be said to have begun in 1912 when scientists around Max von Laue (1879–1960) in Munich sent a narrow beam of X-rays through a crystal and onto a photographic plate (for a history of X-ray crystallography, see Authier 2013). What they discovered was a pattern of bright spots resulting from the diffraction of the beam by the constituents of the crystal. Later William Henry Bragg (1862–1942) and his son

[2]Shechtman's Nobel interview can be found online: http://www.nobelprize.org/nobel_prizes/chemistry/laureates/2011/shechtman-interview.html

William Lawrence Bragg (1890–1971) refined the method, utilizing it to determine the inner arrangements of atoms in crystals. The findings confirmed the generally held conviction dating back centuries that solid crystal structures were characterized by continuous 3-dimensional order of their constituent particles: In other terms, the atoms, ions, ionic groups or molecules inside crystals were organized in lattices that exhibit translational and rotational symmetries (see also Burke 1966, pp. 1–9). Provided with this mathematical model which identifies crystals with lattices, some mathematical theorems can be deduced; for instance, certain rotational symmetries can be ruled out by the following argument:

Assume that A and B are two points of the lattice with minimal distance between them, $|A-B|=d$, and assume that the lattice possesses a rotational symmetry with the angle α. Now, fix a plane containing A and B and the rotation that is a symmetry of the lattice, and rotate A around B through the angle α and, likewise, B around A through the same angle but in the opposite direction (see Fig. 2). The two new points A' and B' are also lattice points, and we let $d'=|A'-B'|$ denote the distance between them. It may be that $A'=B'$ (i.e. $d'=0$) which corresponds to 3-fold rotational symmetry of the lattice. If d' is non-zero, we notice that it must be an integer multiple of d, since if it is not, $0<d'-kd<d$ for some integers k, and by the translational symmetry we would be able to produce two lattice points closer to one another than A and B, which is a contradiction. Thus $d'=md$ for some integer m, and completing the rectangle in Fig. 2, we find the relation $d'=-2d\cos\alpha+d$. From this it follows that $\cos\alpha=\frac{1-m}{2}$ and thus $|1-m|\leq 2$ leading to the solutions $\alpha\in\{2\pi,\pi,\frac{2\pi}{3},\frac{\pi}{2},\frac{\pi}{3}\}$. This proves that a lattice can only possess rotational symmetries of order 2, 3, 4 or 6. In particular, rotational symmetry of order 5 is prohibited by this elementary mathematical argument which has become known as the "crystallographic restriction".

This mathematical model served the study of crystal structures successfully for more than a century after it became established, in particular through the works of René-Just Haüy (1743–1822) in the late eighteenth century. Based on it, Johann F. C. Hessel (1796–1872) in 1830 and Auguste Bravais (1811–1863) independently in 1848 could prove that only 14 Bravais-lattices and only 32 combinations of crystallographic symmetry elements are possible which provided a mathematical foundation for the classifications of crystals (Burke 1966, pp. 164–165, 171; Authier

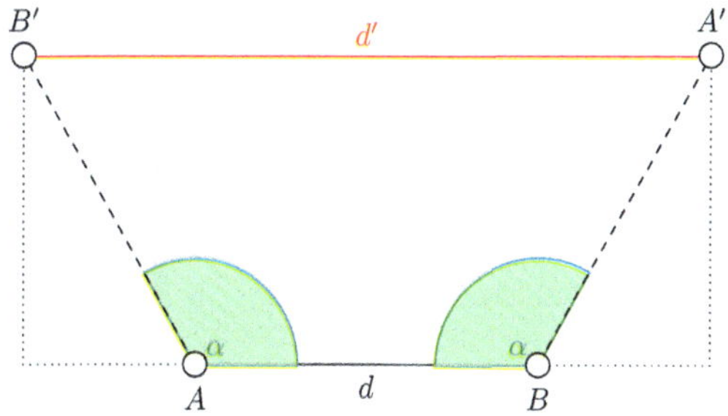

Fig. 2 The crystallographic restriction: rotational symmetries of order five are ruled out by a simple mathematical argument

2013, pp. 367–369, 375–382). Furthermore, the model supplied a foundation for explaining various physical and mechanical phenomena of crystals, for instance through considerations of density and packing factors.

3.2 *Shechtman's Discovery*

Shechtman's revolutionary experimental discovery was made in the spring of 1982 during a sabbatical which he spent at the National Bureau of Standards (NBS) in Washington, D.C. at the invitation of John W. Cahn, who is an expert on thermodynamics (see Blech et al. 2012; Hargittai and Hargittai 2012). At the NBS, research into the rapid solidification of dilute Al-Mn alloys was pursued for their potential to produce alloys free of micro-segregation. But when Shechtman chose to explore higher manganese content, the electron diffraction patterns showed the forbidden icosahedral (ten-fold) symmetry of Al_6Mn (see Fig. 3). Shechtman's observations were discussed in the group at NBS, but it would take two years before he found them ready for publication. The delay in publication was caused by a complex set of concerns, some of which were directly linked to the status of the mathematization. First, it was well known that translational symmetry would cause discrete diffraction patterns, but the converse was also widely (and falsely) believed: In particular, since his findings contradicted strongly established paradigms, alternative ways of explaining the phenomenon were considered which did not involve icosahedral symmetry. What finally convinced Shechtman to publish his results in 1984 was a mathematical model developed by Ilan Blech which could computationally produce diffraction patterns similar to those observed experimentally. The results were published in two papers — one with the entire group announcing the experimental results (Shechtman et al. 1984) and one with Blech on the mathematical model of the microstructure (Shechtman and Blech 1985).

Almost immediately after the announcement by Shechtman and his colleagues, a more comprehensive mathematical model was presented by Dov Levine and Paul Steinhardt. Among other important contributions, they defined the notion of quasi-crystals and introduced Penrose tilings into the discussion (Levine and Steinhardt 1984). Penrose tilings are non-periodic tilings of the plane or of 3-dimensional space, first discovered by the mathematician and physicist Roger Penrose in the

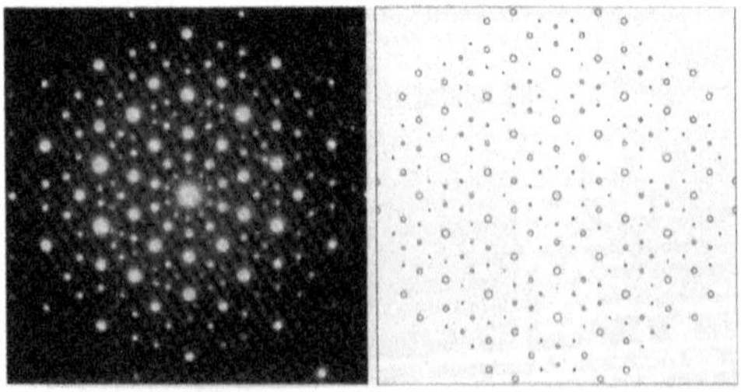

Fig. 3 Five-fold rotational symmetry as reported in 1985. Reproduced from Schwarzschild (1985, p. 17), with the permission of the American Institute of Physics

1970s (see Penrose 1974; Penrose 1979). In 1982, Alan L. Mackay proved that the diffraction patterns obtained from particles arranged in 2-dimensional Penrose tilings would exhibit ten-fold rotational symmetry (Mackay 1982; see also Mackay 1987). Although Mackay proposed to speak in terms of "quasi-lattices", the explicit combination with quasi-crystals was first suggested by Levine and Steinhardt after Shechtman's discovery was made public. Yet, although the diffraction patterns of quasi-crystals were remarkably like those obtained by Mackay, experiments soon showed that the direct generalization to three dimensions was not the proper way forward (Senechal and Taylor 1990, p. 61) Throughout the second half of the 1980s, numerous experimental studies reported producing quasi-crystals of various kinds, and sought to determine their structure (see, for instance, Bursill and Lin 1985; Coddens 1988; Lidin 1991). In order to understand the new phenomena, Penrose tilings were frequently employed, as were other mathematical tools and theories such as almost-periodic functions studied in the 1930s by Harald Bohr (1887–1951) and others.

3.3 *Redefining Crystals*

Among the flurry of activities sparked by Shechtman's announcement, certain institutional aspects merit attention. In March 1989, a ten-day conference on *Number Theory and Physics* organized at the Centre de Physique in Les Houches in the French Alpes brought together practitioners interested in quasi-crystals. The venue has since been used for further winter schools devoted to quasi-crystals, some of which have resulted in books of lectures published by Springer. The general ambition of the meetings in Les Houches is captured in the following quotation from the volume of lectures of the winter school in 1994 entitled "Beyond Quasicrystals":

> The School gathered lecturers and participants from all over the world and was prepared in the spirit of a general effort to promote theoretical and experimental interdisciplinary communication between mathematicians, theoretical and experimental physicists on the topic of the nature of geometric order in solids beyond standard periodicity and quasi periodicity. (Axel and Gratias 1995, p. v)

Among those present at the 1989 school in Les Houches was the American mathematician Marjorie Senechal, who together with Jean Taylor would report on the meeting and the state of affairs in the mathematical treatment of quasi-crystals in a short article in the *Mathematical Intelligencer* (Senechal and Taylor 1990). In that paper, they provided a subjective view of where research could be heading, for as they write:

> [L]ike the view of the Mont Blanc massif from the conference centre, the general outline and size of the problem is rather clear, but features that are prominent from our perspective may mask others, including the summit. (ibid., p. 55)

In particular, their approach was of the more abstract and general flavor as it pertained to the mathematical side of the study of quasi-crystals. They identified three different sets of questions, delineated by the disciplinary background if not by the individual interests of those who pursue them:

1. *Crystallography*: How are the atoms of real quasi-crystals arranged in three-dimensional space?
2. *Physics*: What are the physical properties of substances with long-range order but no translational symmetry?
3. *Mathematics*: What kinds of order are necessary and sufficient for a pattern of points to have a diffraction pattern with bright spots? (ibid., p. 55)

The mathematical question of characterizing the kind of order that produces discrete diffraction patterns was a general one, and in order to work on it, Senechal and Taylor suggested that mathematicians "must draw on a variety of techniques from many branches of mathematics, including tiling theory, almost periodic functions, generalized functions, Fourier analysis, algebraic number theory, ergodic theory and spectral measures, representations of *GL(n)*, and symbolic dynamics and dynamical systems" (ibid., p. 55). Thus, the study of quasi-crystals was inherently doubly interdisciplinary: On the one hand, different scientific disciplines were involved in studying the same phenomenon, and on the other hand, different branches of mathematics were expected to be crucial in formulating, characterizing, and exploring the mathematical model of quasi-crystals.

In 1991, the International Union for Crystallography (IUCr) set up a Commission on Aperiodic Crystals (CAC) and charged it with formulating a new definition of the very notion of *crystal* that would be inclusive enough to take into account the new substances discovered in laboratories all over the world following Shechtman's initial announcement (International Union of Crystallography 1992, p. 928). In a striking move to shift focus towards the instrumental detection through diffraction patterns and the highly general mathematical approach outlined above, the commission stipulated that by a crystal one should understand "any solid having an essentially discrete diffraction diagram" (ibid., p. 928). Thus, the notion of crystals was redefined so as to include quasi-crystals, and the notion of "aperiodic crystals" was introduced for those crystals in which 3-dimensional lattice periodicity is absent (ibid., p. 928). By this swift change of basic terminology — which was strongly supported by Senechal (Senechal 2006, p. 886) — the mathematical modeling question above had become the center stage for the very definition of a crystal.

3.4 Senechal's View of Mont Blanc

By 1995, the field had developed to a state such that Senechal endeavored to write her book on "Quasicrystals and geometry" (Senechal 1995). In it, she presented — again and in much more elaboration — her viewpoint as a geometer of the state of the field and the methods involved. As she wrote in the preface:

> Although the number of new and important results is increasing rapidly, there is still no agreement on how aperiodic crystals should be modeled mathematically, and most of the major problems raised by any of the proposed models are still open. (Senechal 1995, p. xii)

Nevertheless, she believed that a book was overdue as a way of explaining what is *not* known in order to attract and direct efforts into solving these problems. In order to facilitate this convergence of efforts, she presented and integrated methods and results from numerous different mathematical groups to the study of quasi-crystals. She was, however, neither willing nor able to unequivocally prefer or recommend any one approach in particular. Thus, in a sense, her previous feeling of watching the massif with the view of summit obstructed was still present, although some contours were gradually becoming discernible.

In accordance with the mathematical problem of characterizing and studying aperiodic crystals, Senechal's book starts by exploring the mathematical fundamentals of so-called *Fraunhofer diffraction patterns* derived from first principles. These diffraction patterns were, themselves, based on a mathematical model for the process of interference of waves encountering an obstacle or a slit that is comparable in size to the wave length. And Senechal showed how the diffraction pattern is the Fourier transform of the generalized density function of the object under study (ibid., pp. 86 sqq.). Thus, the problem of describing those structures that can give rise to discrete diffraction patterns becomes one of obtaining the inverse Fourier transformations for generalized functions.

The second part of Senechal's book was devoted to methods of constructing quasi-crystals — but from a very mathematical point of view. In other words, the question is how structures can be mathematically constructed which will exhibit discrete diffraction patterns. Various constructions were discussed, but among them the so-called *projection method* (also sometimes called the cut-and-project method) was prominent. Since this technique raises some issues about modeling, it merits attention here without going into any mathematical detail.

The projection technique involves projections from a higher-dimensional structure which under certain restrictions produce non-periodic structures in lower dimensions; precise criteria can be given to ensure that the resulting construction is a quasi-crystal. And examples can be described in which the projection from a 6-dimensional structure into three dimensions result in a non-periodic structure with 5-fold symmetry (see ibid., chapter 2).

Much effort has been invested in coming to understand the simplest case which is 1-dimensional quasi-crystals. Such objects can be constructed in numerous ways, both by projection and by so-called Fibonacci sequences (see ibid., chapter 4). The latter arise for instance from certain 2-letter substitution rules, and precise criteria can again be given under which quasi-crystals result.

Thus, at least two different avenues to better understand quasi-crystals were outlined by Senechal: We can try to understand real-world, 3-dimensional quasi-crystals by viewing them through the higher-dimensional formalism, or we can try to obtain a firmer grasp on simpler, 1-dimensional quasi-crystals and hope to extend such insights into the realm of physical quasi-crystals. Senechal was, herself, content to outline the mathematical theory, but she did open discussions of whether the higher-dimensional projection formalism was necessary (ibid., pp. 71–72) and whether the formation of actual crystalline aperiodic phases is in any way related to the growth of tilings (ibid., pp. 235 sqq.). These questions were, however, open and important ones at the time of her book, and to a large extent they remain so to the present (see also below).

A third part of Senechal's book — and a recurring theme throughout — was the description of what she called the "zoo" of quasi-crystals (ibid., pp. 207 sqq.). This zoo was intended to showcase the variety of aperiodic structures and tiling transforms discovered so far and function as a catalogue against which new structures could be compared for reference and classification. However, since the precise classifying principles were still not firmly established, the zoo would provisionally be ordered according to the techniques of construction and the variety of diffraction patterns obtained. In a later summary of the state of affairs in 2006, Senechal reflected how:

> Penrose tilings, the Drosophila of aperiodic order, don't tell us what the structures of real aperiodic crystals are, but they do tell us what aperiodic order can look like. (Still, we are missing something. For suitable choices of lattice, dimension, and other parameters, we get cut-and-project sets with (diffraction) rotational symmetry of any finite order. Yet the symmetries of real aperiodic crystals found so far are only pentagonal, decagonal, dodecagonal, and icosahedral. Evidently, the real crystallographic restriction is yet to be discovered.) (Senechal 2006, p. 887)

As this quotation also suggests, in the absence of a firm understanding of the underlying relations between mathematical model and physical reality, exemplars and rough classifications of the mathematical possibilities were pursued as useful and valuable for the further development of the mathematical model.

3.5 Debates Over Modeling Criteria

As indicated above, Senechal's book was intended to attract attention to the new and emerging field. And she was certainly not alone in her efforts to stimulate interest and collaboration. A NATO Advanced Study Institute was set up under the auspices of the Fields Institute devoted to the study of aperiodic long range order from perspectives of mathematics and mathematical physics. This led to a subsequent semester program in the fall and 1995, and a book was produced under the editorship of some of the other pioneers in the field:

> The goal of both events was complementary: to help establish into the mainstream of mathematics and mathematical physics this new, emerging field of study. (Patera 1998, p. xi)

This ambition to develop a mathematical framework for (mathematically) constructing and studying quasi-crystals that could be relevant both to mathematicians and to mathematical physicists draws attention to some of the features mentioned above. As later summarized by Senechal and Taylor:

> In the early days of quasicrystals, mathematicians hoped to characterize their broad structural features in one (or more) of three ways: as aperiodic tilings with matching rules; by inflation rules; or through high-dimensional interpretations. Each of these three approaches has generated much interesting new mathematics — and at least partial answers to the fundamental challenge. But only the third method has proved truly useful in describing the structure of physical quasicrystals. (Senechal and Taylor 2013, p. 4)

Yet, despite its success from a mathematical point of view, if the mathematical model was to be of value not just for the purely mathematical approach, it might be desired to address the relationship between 3-dimensional, physical quasi-crystals and the higher-dimensional spaces from which they were projected (see also Senechal and Taylor 1990, p. 63). Obviously, hidden variables were not new to physics as they had featured into many mechanistic theories of the nineteenth century. Yet, they most frequently stand out as being in need of some explanation or interpretation. Similarly, if the tiling approach was to be adopted, it would be desirable to have a better understanding of the individual tiles, since they were given no physical interpretation and thus appeared as rather artificial from the perspective of physics (ibid., p. 61).

Perhaps more importantly, the issue of explaining growth of physical quasi-crystals in terms of the mathematical model was profoundly difficult. As explained by Senechal, Penrose tilings were attractive in large part because they allowed for the adaptation of the old idea that crystals grow by accretion of building blocks under strictly local forces (Senechal 1995, pp. 235–236). Thus, this would correspond to well-entrenched notions within the physics community. Yet, this issue was clearly not settled by the mathematical treatment as it stood, and numerous other modes of growth were also being considered by the physicists working on the problem. In retrospect, Senechal explained how the interdisciplinarity of the research field had, and least by 2006, mainly been pursued through largely parallel developments of more traditional, disciplinary approaches, methods, and goals which, it was hoped, would eventually turn symbiotic:

> Meanwhile the burgeoning mathematical field of long-range aperiodic order and the experimental study of real aperiodic crystals are symbiotic and mutually stimulating. Their cross-fertilization has been more metaphorical than practical, but no less valuable for that. (Senechal 2006, p. 887)

The mathematical community subsequently addressed these challenges in a number of ways, some of which included the generalization of the model, the extension of the methods and perspectives employed, and the occasional change of perspective. For example, the idea of extending the study of aperiodic solids beyond quasi-crystals already signaled at the 1994 school in Les Houches was followed up by another school in 1998 resulting in a volume entitled "From Quasicrystals to More Complex Systems" (Axel et al. 2000).

The first quasi-crystals discovered were metastable binary alloys, but in 2007 the first complete structure solution for a icosahedral quasi-crystal was produced from a *stable* binary alloy (Takakura et al. 2007). This discovery played a central role in the retrospective article that Senechal and Taylor published in 2013, after Shechtman had also been awarded the Nobel Prize (Senechal and Taylor 2013). The structural description of the Cd-Yb quasi-crystal is, to Senechal and Taylor, particularly attractive because it answers a number of the concerns that were raised more than 25 years earlier: It not only describes where the individual atoms are placed but does so in a way that can be easily visualized based on ideas deriving from classical geometric solids (ibid., p. 3). Senechal and Taylor report on the way the structural description was obtained as follows:

> Takakura *et al.* [i.e. the authors of (Takakura et al. 2007)] worked backwards from the two observed sections (which correspond to different choices of $R_{\|}$) to estimate a single 6D periodic density giving rise to both. To get the physical structure, they used the section method, and their data, creatively. (Senechal and Taylor 2013, p. 5)

Further, Senechal and Taylor notice that the projection methods from higher-dimensional spaces ("hyperspaces") have become the standard approach:

> Materials scientists use hyperspace descriptions so frequently now that spots in the diffraction pattern are routinely given six indices instead of the once-standard three. Indeed, the section version of the cut-and-project method is so popular that we are concerned researchers may consider it to the exclusion of other types of order. (ibid., p. 5)

This illustrates that although the higher-dimensional model contains uninterpreted, hidden variables, this circumstance does not impede on its usefulness and applicability in research and modeling. Indeed, its usefulness is such that it may overshadow the provisional and plastic nature of the model, itself — or at least so it was feared.

Although mathematical frameworks for characterizing quasi-crystals were in existence by the beginning of the twenty-first century, and numerous physical exemplars had been produced in laboratories all over the world, many questions remained open. Concerning the important question of how quasi-crystals grow, one survey explained:

> Quasicrystals, like crystals, form via nucleation and growth, where a microscopic 'nucleus' of the solid phase spontaneously arises in the supercooled liquid and spreads outward, converting the system from liquid to solid. A fundamental puzzle in quasicrystal physics is to understand how the growth phase of nucleation and growth can lead to a structure with long-range aperiodicity. (Keys and Glotzer 2007, p. 1)

Ideally, the growth of quasi-crystals as well as their physical and chemical properties should be linked to their mathematical description and the mathematical model. Indeed, preferably, these questions would be answerable from a joint conceptualization of quasi-crystals encoded in a mathematical model that would possess similar potential as the discarded crystallographic restriction. However, this is no easy task, and reaching predictions and explanations of physical phenomena about quasi-crystals from the mathematical description remains a major on-going research field (see Senechal and Taylor 2013, p. 7). But combined with the advances in computer software and visualization techniques (see also ibid., p. 9), the study of quasi-crystals has not only "fundamentally altered how chemists conceive of solid matter" as the announcement stated when Shechtman was awarded the Nobel Prize. The study of quasi-crystals is also altering the way mathematicians, physicists and crystallographers look at their shared mathematical models.

4 Shaping Mathematics as a Tool

The early history of the search for a new mathematical model for quasi-crystals in the two decades following Shechtman's discovery is full of complexities. Yet, even the brief and restricted case-study outlines numerous interesting points about the epistemic relations between mathematics and the sciences.

Confronted with empirical data that refuted an old and established mathematical model, and without any viable alternative at hand, the mathematical, physical and crystallographic communities had to revise and redefine their methods and concepts. This process led to an instrumental redefinition of the very concept of crystals. It sparked new mathematical ideas and investigations into symmetry at large and aperiodic tiling patterns. And it meant the revival of interest in Penrose tilings which had hitherto mainly been considered for their aesthetic qualities. These tilings seemed promising as they exhibit high degrees of order while being aperiodic in the large, since they possess no translational symmetry. They were therefore pursued as a *fix* to the classical theory: The *new* quasi-crystals were seen and treated as exceptions that defied the characterizing crystallographic restriction. At first, the crystals, themselves, were indeed rare, but they have now been reproduced in numerous laboratories and have even been identified in very old meteors. Thus, scientists have come to recognize that quasi-crystals are not only mathematically and physically possible, they are naturally occurring and even *more normal* (in the sense of more general) than the traditional crystalline structures. In order to treat these phenomena, the theory of quasi-crystals required a comprehensive mathematical model capable of addressing a multitude of interrelated, interdisciplinary research questions.

The mathematics – the theorems and their proofs – that had gone into describing ordinary crystals was, of course, not invalidated by Shechtman's discovery. Instead, the model had proved to be insufficient in dealing with phenomena that from a physical or chemical perspective would seem to fall in its domain. Yet, perhaps because of the entrenched and paradigmatic status of the crystallographic restriction, no alternative was at hand when Shechtman made his discovery. Instead, mathematicians, physicists and crystallographers had to return to their desks and laboratories to devise new conceptualizations and mathematical models for describing and explaining the newly discovered phenomenon. In doing so, different but overlapping *cultures of mathematization* pursued the quest of modeling quasi-crystals. These different cultures originated from different concerns and disciplinary backgrounds. And they had different visions and emphases for the modeling process. In particular, views varied on the degree to which the new conceptualization should be able to answer and explain questions about physical and chemical properties of quasi-crystals. And even the fundamental physical notion that the conceptualization of quasi-crystals should be based on a model that focused on local, rather than global geometry, was suspended for some of the process as the mathematization of non-periodicity was quickly realized to be not (only) a local matter.

During the first two decades, a number of mathematical models were thus developed from different perspectives. Not only were there disciplinary differences between the approaches; there were also different mathematical theories and machineries being suggested and pursued during the early phase. Penrose tilings were but one among a number of avenues explored. And although they had an initial appeal to physical interpretation, much of the mathematical research effort was actually directed at developing construction methods for quasi-crystals that would eventually render Penrose tilings as just one (simple and beautiful) exemplar in the "zoo" of quasi-crystals.

The mathematical models were by necessity more complex than the traditional lattice model of old-fashioned crystals. And in this respect, the mathematization coincided with a general trend towards increased complexity made possible by technological developments in software and hardware and by advances in mathematics. As early as twenty years before Shechtman's discovery, the crystallographer (and historian of science) J. D. Bernal (1901–1971) had observed that the mathematical modeling of liquids was guided by pragmatic and perhaps tacit concerns over the simplicity and convenience of the model:

> There is no reason, other than convenience, for mathematical treatment of long-range order as being the only kind of order. It is clearly very much less probable than short-range order and long-range disorder. (Bernal 1964, pp. 320–321)

By the time Shechtman first saw quasi-crystal structures, the vastly increased computational power at the disposal of scientists and mathematical modelers and advances in visualization software had allowed for a new set of ambitions in modeling long-range disorder. These technical and conceptual advances allowed for the exploration of the "zoo" of quasi-crystals through new mathematical techniques and numerical simulations that could visualize their effects.

Faced with this immense and open-ended task of formulating a new theory to encompass the phenomenon of quasi-crystals, mathematicians did draw on existing knowledge applied *from the shelf*. In particular, Penrose tilings or almost-periodic functions were well-known mathematical theories that were explored in the new context. Thus, existing mathematics was of course relevant to the modeling process. Yet, impacts of the modeling approaches on mainstream mathematics are also discernible in (at least) two ways: Many of the questions that arose in the search for the new mathematical model were, themselves, of a mathematical nature. In particular, questions about relations between technical mathematical concepts such as whether all repetitive sets are Meyer sets belong to mainstream conceptual mathematics (see Senechal 2006, p. 887). Similarly, different questions presented themselves to mathematical research when concepts that were introduced for a rather specific use were abstracted and generalized. In working with an almost intractable set of questions, some of the mathematical research was pursued by restricting attention to simpler cases such as 1-dimensional quasi-crystals. This has proved to be a fruitful approach because of the relative simplicity of the model and the fact that the mathematical construction of 1-dimensional quasi-crystals was found to be related to fields of automata theory or Diophantine approximations. As such, the search for a mathematical model fed back into mathematical research in a variety of fields. But the influence also extended in a second, more social direction. Deliberate attempts were made to attract the attention of mathematicians and mathematical physicists towards the modeling of quasi-crystals. And books like Senechal's promoted the field of quasi-crystals and the related fields of geometry, generalized functions, Diophantine approximations, etc. into prominence. Thus, fields of mathematical research were given a new legitimization and new impetus through their perceived relevance to the modeling of a fascinating scientific phenomenon.

The case-study of efforts to model quasi-crystals during the first two decades after Shechtman's discovery has thus brought out issues of a richer and deeper interaction between mathematics and various scientific disciplines interested in the new phenomenon. Thus, the case illustrates that mathematics can be involved in interdisciplinary research efforts, and that "application from the shelves" does not capture all ways in which mathematization can take place. Yet, a few words of caution might be required. The chosen example of quasi-crystals is, on the one hand, a rich, complex, and on-going case where a certain division of labor has taken place from the early phase. Here, this has been illustrated mainly through analyses of the mathematical community represented by Senechal and her collaborators. Thus, at least provisionally, the main case may be rather unique. And at the same time, this case-study does not address some of the interesting features of interdisciplinary mathematical modeling present in other cases such as e.g. quantum theory or small-world networks. These other examples also show bi-directional interaction, although in different historical, institutional, and epistemic contexts. Yet, as the analysis of the case of quasi-crystals is intended to be exploratory, the following conclusions would seem not to contradict further studies along the same lines.

Firstly, through the brief presentation of two historical cases and the more elaborate example of quasi-crystals, I have presented three examples where mathematics was involved in other branches of science. Located in different social and epistemic contexts, they exhibit different set-ups for interdisciplinary collaborations, which in a sense extend the laboratory or research group. Such an extension of the notion of *collaboration* seems to be one of the necessary prerequisites for rendering interdisciplinary efforts that involve active mathematical research amenable to the vocabulary and analysis of studies of interdisciplinarity. This in turn would seem to allow for new perspectives on key questions about interdisciplinary research such as the division of labor with a formal science such mathematics, the negotiation of different desiderata for the modeling process, or the role of mathematical models as boundary objects.

Secondly, the framework of cultures of mathematization is intended as a step towards analyses focusing on the co-production of mathematical models as boundary objects in interdisciplinary collaborations that involve mathematics. Further study and discussion about the actual use of the mathematical model and its role in day-to-day research in other branches of the interdisciplinary study of quasi-crystals will be required to expand on the analysis of the model as a boundary object beyond the focus on the different desiderata (or the neglect of same) by different agents in the collaboration.

Finally, the discussion of an ongoing endeavor to shape a new mathematical model for a phenomenon that is not yet well understood has served to illustrate how the complex and open-ended process of mathematical modeling may actually take place when there is no firm candidate for a proper mathematization. In other words, the case of quasi-crystals nuances the notion of "mathematics-as-a-tool" by showing some of the dynamics involved in shaping the pieces of mathematics that are developed, placed on the shelf and, eventually, used as tools. During the extended modeling process, new conceptualizations and *mathematical* results were produced which can take up a life of their own as research objects and techniques in various branches of mathematics. As noted in the literature discussing Wigner's bewilderment at the applicability of mathematics, such phases of co-production were historically important for shaping the mathematics we know today. What this example shows is that the process involves epistemic as well as social concerns that give rise to different interests in the model. Possibly, this negotiation would not be as surprising if the domain being mathematized did not already put so much emphasis on mathematization: Various fields, at various times, have different expectations for their mathematical models. What is striking is then, perhaps, that it is was (and is) very difficult to develop a new, adequate model for quasi-crystals given that it was to replace a firmly entrenched and successful model for crystals.

This discussion and case-study has provisionally revisited the notion of interdisciplinary collaboration as it pertains to mathematics: When mathematics is involved, such collaborations are neither specifically new nor do they strictly follow the epistemic division of labor suggested by theoretical accounts of interdisciplinary expertise. If mathematics is to be seen as a *tool* in these collaborations, it is a very dynamic, plastic and evolving tool, indeed. And the best way to bring to the fore such perspectives is, I believe, the philosophically informed historical case study such as outlined here for the early history of quasi-crystals. Thus, the examples chosen and the framework employed may hopefully have added to the complication already involved in the topic of *mathematics as a tool*.

References

Abraham, T. H. (2004). Nicolas Rashevsky's mathematical biophysics. *Journal of the History of Biology, 37*, 333–385.

Andersen, H., & Wagenknecht, S. (2013). Epistemic dependence in interdisciplinary groups. *Synthese, 190*, 1881–1898.

Authier, A. (2013). *Early days of X-ray crystallography*. Oxford: Oxford University Press.

Axel, F., & Gratias, D. (Eds.). (1995). *Beyond quasicrystals. Les Houches, March 7–18, 1994*. Berlin/Heidelberg: Springer.

Axel, F., Denoyer, F., & Gazeau, J. P. (Eds.). (2000). *From quasicrystals to more complex systems. Les Houches, February 23–March 6, 1998*. Springer.

Bernal, J. D. (1964, July 28). The Bakerian lecture, 1962: The structure of liquids. *Proceedings of the Royal Society of London. A: Mathematical and Physical Sciences, 280*(1382), 299–322.

Blech, I. A., Cahn, J. W., & Gratias D. (2012, October). Reminiscences about a chemistry Nobel Prize won with metallurgy: Comments on D. Shechtman and I. A. Blech; *Metall. Trans. A*, 1985, vol. 16A, pp. 1005–1012. *Metallurgical And Materials Transactions A, 43A*, 3411–3414.

Brecque, M. L. (1987/1988). Quasicrystals: Opening the door to Forbidden symmetries. *MOSAIC, 18*(4), 2–23.

Burchfield, J. D. (1975). *Lord Kelvin and the age of the Earth*. London/Basingstoke: The Macmillan Press.

Burke, J. G. (1966). *Origins of the science of crystals*. Berkeley/Los Angeles: University of California Press.

Bursill, L. A., & Lin, P. J. (1985, July 4). Penrose tiling observed in a quasi-crystal. *Nature, 316*, 50–51.

Coddens, G. (1988). A new approach to quasicrystals. *Solid State Communications, 65*(7), 637–641.

England, P., Molnar, P., & Richter, F. (2007a, January). John Perry's neglected critique of Kelvin's age for the Earth: A missed opportunity in geodynamics. *GSA Today, 17*, 4–9.

England, P., Molnar, P., & Richter, F. (2007b). Kelvin, Perry and the age of the Earth. *American Scientist, 95*(4), 342–349.

Epple, M., Kjeldsen, T. H., & Siegmund-Schultze, R. (Eds.). (2013). *From "Mixed" to "Applied" mathematics: Tracing an important dimension of mathematics and its history* (Oberwolfach reports 12). Oberwolfach: Mathematisches Forschungsinstitut.

Galison, P. (1997). *Image and logic. A material culture of microphysics*. Chicago: The University of Chicago Press.

Gelfert, A. (2014, May). Applicability, indispensability, and underdetermination: Puzzling over Wigner's 'Unreasonable effectiveness of mathematics'. *Science & Education, 23*(5), 997–1009.

Grattan-Guinness, I. (2008). Solving Wigner's mystery: The reasonable (though perhaps limited) effectiveness of mathematics in the natural sciences. *The Mathematical Intelligencer, 30*(3), 7–17.

Hargittai, B., & Hargittai, I. (2012). Quasicrystal discovery: From NBS/NIST to Stockholm. *Structural Chemistry, 23*(2), 301–306.

International Union of Crystallography. (1992). Report of the Executive Committee for 1991. *Acta Crystallographica Section A, A48*, 922–946.

Keller, E. F. (2002). *Making sense of life. Explaining biological development with models, metaphors, and machines*. Cambridge, MA/London: Harvard University Press.

Keys, A. S., & Glotzer, S. C. (2007). How do quasicrystals grow? *Physical Review Letters, 99*(235503), 1–4.

Knorr Cetina, K. (1999). *Epistemic cultures. How the sciences make knowledge*. Cambridge, MA/London: Harvard University Press.

Kragh, H. (2015). Mathematics and physics: The idea of a pre-established harmony. *Science & Education, 24*(5), 515–527.

Kuhn, T. S. (1962/1996). *The structure of scientific revolutions* (3rd ed.). Chicago/London: The University of Chicago Press.

Levine, D., & Steinhardt, P. J. (1984, December 24). Quasicrystals: A new class of ordered structures. *Physical Review Letters, 53*(26), 2477–2480.

Lidin, S. (1991). Quasicrystals: Local structure *versus* global structure. *Materials Science and Engineering, A134*, 893–895.

Mackay, A. L. (1982). Crystallography and the Penrose pattern. *Physica, 114A*, 609–613.

Mackay, A. L. (1987). Quasi-crystals and amorphous materials. *Journal of Non-Crystalline Solids, 97–98*, 55–62.

Morgan, M. S., & Morrison, M. (Eds.). (1999). *Models as mediators. Perspectives on natural and social science*. Cambridge: Cambridge University Press.

Patera, J. (Ed.). (1998). *Quasicrystals and Discrete Geometry* (Fields Institute monographs). Providence: American Mathematical Society.

Penrose, R. (1974). The rôle of aesthetics in pure and applied mathematical research. *Bulletin of the Institute of Mathematics and Its Applications, 10*, 266–271.

Penrose, R. (1979, March). Pentaplexity: A class of non-periodic tilings of the plane. *The Mathematical Intelligencer, 2*(1), 32–37.

Perry, J. (1895). On the age of the Earth. *Nature*, 51, 224–227, 341–342, 582–585.

Reade, T. M. (1878, April). The age of the world as viewed by the geologist and the mathematician. *The Geological Magazine*. New Series, *5*(4), 145–154.

Schlote, K.-H., & Schneider, M. (Eds.). (2011). *Mathematics meets physics. A contribution to their interaction in the 19th and the first half of the 20th century*. Studien zur Entwicklung von Mathematik und Physik in ihren Wechselwirkungen. Frankfurt am Main: Verlag Harri Deutsch.

Schwarzschild, B. (1985, February). "Forbidden fivefold symmetry may indicate quasicrystal phase". *Physics Today*. News: Search & discovery, pp. 17–19.

Senechal, M. (1995). *Quasicrystals and geometry*. Cambridge: Cambridge University Press.

Senechal, M. (2006, September). What is a quasicrystal? *Notices of the AMS, 53*(8), 886–887.

Senechal, M., & Taylor, J. (1990). Quasicrystals: The view from Les Houches. *The Mathematical Intelligencer, 12*(2), 54–64.

Senechal, M., & Taylor, J. (2013). Quasicrystals: The view from Stockholm. *The Mathematical Intelligencer, 35*(2), 1–9.

Shechtman, D., & Blech, I. A. (1985, June). The microstructure of rapidly solidified Al_6Mn. *Metallurgical Transactions, 16A*, 1005–1012.

Shechtman, D., et al. (1984, November 12). Metallic phase with long-range orientational order and no translational symmetry. *Physical Review Letters, 53*(20), 1951–1953.

Steinhardt, P. J. (2013). Quasicrystals: A brief history of the impossible. *Rend. Fis. Acc Lincei, 24*, S85–S91.

Takakura, H., et al. (2007, January). Atomic structure of the binary icosahedral Yb-Cd quasicrystal. *Nature Materials, 6*, 58–63.

Wigner, E. (1960). The unreasonable effectiveness of mathematics in the natural sciences. *Communications in Pure and Applied Mathematics, 13*(1), 1–14.

Part II
Conceptual Re-evaluation

Boon and Bane: On the Role of Adjustable Parameters in Simulation Models

Hans Hasse and Johannes Lenhard

1 Introduction

Simulation brings together the important notions of model, theory, and experiment. Each of these notions has been discussed extensively in the philosophy of science. Consequently, the philosophy of simulation debates whether and how these established conceptions have changed with the rise of simulation technology.[1] We do not enter the discussion of what, in that context, an adequate conception of a simulation experiment is, nor what an appropriate notion of a theoretical model is. Instead, we focus on the interface of model and experiment. Here, adjustable parameters enter the picture. They might appear as a minor detail, a technical matter of smoothing out imperfections of a model. However, we argue that they are of central importance in simulation methodology, though they are a two-edged affair. They help to enlarge the scope of simulation far beyond what can be determined by theoretical knowledge, but at the same time undercut the epistemic value of simulation models. In short, adjustable parameters are boon and bane of simulation models.

[1] Humphreys (2004) contributed the first monograph to the field. Parker (2013) or Winsberg (2014) provide valid overview articles that include many references.

H. Hasse
Laboratory of Engineering Thermodynamics (LTD), University of Kaiserslautern, Kaiserslautern, Germany

J. Lenhard (✉)
Department of Philosophy, Bielefeld University, Bielefeld, Germany
e-mail: johannes.lenhard@uni-bielefeld.de

J. Lenhard, M. Carrier (eds.), *Mathematics as a Tool*, Boston Studies in the Philosophy and History of Science 327, DOI 10.1007/978-3-319-54469-4_6

Let us motivate this claim in more detail. Experimentation is a key element when characterizing simulation modeling,[2] exactly because it occurs in two varieties. The first variety has been called theoretical model, computer, or numerical experiments. We prefer to call them simulation experiments. They are used to investigate the behavior of models. Clearly simulation offers new possibilities for conducting experiments of this sort and hence investigating models beyond what is tractable by theoretical analysis. We are interested in how simulation experiments function in simulation modeling. Importantly, relevant properties of simulation models can be known only by simulation experiments.[3] There are two immediate and important consequences. First, simulation experiments are unavoidable in simulation modeling. Second, when researchers construct a model and want to find out how possible elaborations of the current version perform, they will have to conduct repeated experiments.

The second variety is the experiment in the classical sense. When comparing simulations to their target system, such classical experiments will usually provide the data to compare with. The situation gets interestingly complicated, since the influence of simulation on these experiments is growing. There is a beginning debate on the changing face of experimentation due to computer use (cf. Morrison 2009, 2014; Tal 2013). It is indeed striking to what extent supposedly classical experiments make use of simulation in their experimental setup; examples range from the Large Hadron Collider at Cern to scanning tunnel microscopes.

Our claim is that adjustable parameters play a crucial role in the process of building and applying simulation models. Two interconnected aspects make up our claim: First, both varieties of experiments, or if you prefer another terminology: simulation and classical experiment, cooperate. Second, this cooperation makes use of a feedback loop and works via adjusting parameters.[4]

The outline is the following. In the next section, we start with a brief introduction into mathematical simulation models, their implementation on computers and their application. In addition, equations of state in thermodynamics are introduced, as we will use examples from that area throughout the paper. We chose this field, because it is a theoretically well-founded field of both science and engineering. It provides us with excellent material to illustrate our claims as the parameters of equations of state are of very different nature. They range from the universal gas constant, which can be considered as an adjustable parameter, but one that is found to be valid in a very broad range of situations and closely linked to theory, to mere correlation parameters

[2]A variety of good motivations are given in, for instance, Axelrod (1997), Barberousse et al. (2009), Dowling (1999), Galison (1996), Humphreys (1994), Hughes (1999), Keller (2003), Morgan (2003), Morrison (2009), Rohrlich (1991), Winsberg (2003).

[3]If you want to avoid talking about experiment in this context, these properties can be known only by actually conducting simulations. Mark Bedau (2011) has highlighted properties that can be known only by actually conducting the computational process of a simulation and has aptly called them "weakly emergent."

[4]In this respect, our work elaborates the notion of "exploratory cooperation" in simulation modeling, put forward in Lenard (2007).

which are useful only in special situations and epistemologically worthless. We also argue that choosing examples from a theoretically well-founded field provides a stronger argument than choosing them from a field on which little is known and which, hence, has to rely on data-driven models. In the latter field adjustable parameters are important *a fortiori*. Thus, it is more demanding, and hopefully more fruitful, to show their role in examples taken from the former field.

The topic of parameterization of models has received surprisingly little attention from philosophers. Of course, there is an intense discussion of parameters in the context of curve-fitting and simplicity (see, for instance, Forster and Sober 1994, DeVito 1997, or Kieseppä 1997). There, parameters play the role of degrees of freedom in a quite abstract mathematical sense. In the present paper, however, we assume a physical context in which parameters might have physical meaning. Notable exceptions that discuss this (large) area are those arguing about climate science. Parameterization schemes build a main component in complex climate models (Gramelsberger 2011; Parker 2014) and contain parameters that have to be tuned. The discussion about practices of tuning is just about to start in the climate community (cf. Mauritsen et al. 2012). With thermodynamics, we add a substantially different topic to the applications discussed in this context.

We focus on the development of simulation models and on the decisive role experiments play. Here, experiments include both classical experiments, in which the real world is studied, as well as simulation experiments, in which the implementation of the simulation model on computers is investigated. The importance of the feedback loop in simulation model development is highlighted in Sect. 3, which is based on the comparison of results of computer experiments and classical experiments. This feedback loop is the means by which modeling and experimentation can cooperate closely. Many extant pictures of simulation suggest a "downward" direction (Winsberg 2014) from theoretical model to simulation, or a "bottom-up" direction from phenomena to simulation models (Keller 2003), whereas we underscore that simulation model development is a feedback loop process in which both directions interact.

Section 4 is devoted to a closer look on parameters and feedback and presents the central piece of our argument. We discuss different types of parameters and various situations in which the feedback loop is involved. Typically, simulation models are only simplified representations of their real world targets: parts of the underlying physics may be unknown or so complex that they cannot be incorporated in a tractable simulation model. Thus, workarounds are needed: these often come in the form of models in which parameters are left open – to be adjusted in the feedback loop. One could criticize this from a fundamental standpoint arguing that this is only a poor remedy for a lack of knowledge, and, hence, bane. On the other hand, one can argue that it is boon, because it allows modeling and simulation which otherwise would not be possible. This shows that a critical discussion is needed, to which we want to contribute, namely, by studying the epistemic and practical value of various classes of parameters. We also address the issue of the influence computerization has had on the use of parameters in models. It turns out that it is important and that also in this, there is boon and bane.

Finally, in Sect. 5, we sum up and draw conclusions from the fact that adjustable parameters are boon and bane of simulation. We argue further that simulation modeling adds a new experimentalist twist to mathematical modeling.

2 A Primer on Thermodynamics, Simulation, and Experimentation

Throughout the present paper, we will use examples from thermodynamics to illustrate our arguments. They are chosen from the well-known field of the so-called equations of state. We will only consider equations aiming at describing fluid states (gas as well as liquid) – but not solids. This field lends itself for this purpose as it is fundamental and well known to many scientists and engineers, and it can also be understood by others. Many other areas of science and engineering would have provided equally useful illustrations.

The best known equation of state is that of the ideal gas

$$p\,v = R\,T \tag{1}$$

where p is the pressure, $v = V/n$ is the molar volume (volume per mole of substance), and T is the temperature measured in Kelvin. All these quantities are measurable in classical experiments. R is a universal constant (8.314 J mol^{-1} K^{-1}). Equation (1) has been used before to illustrate issues of philosophy of science, for example, quite recently by Woody (2013) for discussing concepts of explanation. It is known that all substances fulfill Eq. (1) if the density $\rho = 1/v$ is low enough (or the molar volume v is large enough).

The broader concept behind Eq. (1) is that for a given amount of a given substance, p, v, and T are not independent: there is a function which describes the relation between these quantities. Hence the general form of the equation of state is:

$$\mathrm{f}\,(p, v, T) = 0 \tag{2}$$

In the low density limit the function f is given by the simple Eq. (1) which is universal in the sense that it holds for all substances. Unfortunately, the relation between p, v, and T is more complicated at higher density and different results are obtained for different substances. The reason for this is simply that at higher densities the interactions between the molecules start playing a role, and hence, the individuality of the molecules matters. We note already here, that while Eq. (1) has substance-wise the widest possible range of application and the "parameter" R has the same value for all substances, there must be ways to tune Eq. (2) to represent a given substance. That tuning is done by adjustable parameters.

Well known examples of such equations are the van der Waals Equation and the Virial Equation of state. In Eq. (2) a pair of independent variables can be chosen (e.g., p and T). The third (dependent) variable (then v) can then be calculated from Equation (2). There is a plethora of proposals for equations describing the p,v,T – behavior of substances for a wide range of conditions. Depending on the form of the function f, their evaluation may only be possible numerically. Furthermore, several solutions may be found and algorithms have to be applied to select the desired one. Different solutions may correspond to different physical states (e.g., gas and liquid) or may be unphysical.

The results from the equation of state can be compared directly to p,v,T – data obtained in laboratory experiments. A good equation of state will describe those experimental data well for a wide range of conditions. But the equation of state can do more. If it describes both gaseous and liquid states, it also describes boiling and condensation. Hence, for example, from Eq. (2) also the so-called vapor pressure curve can be found which describes the dependence of the boiling temperature on the pressure. These results can be compared to experimental data as well. The same holds for results on the heat of vaporization, which can also be obtained from Eq. (2). Calculating these properties, though well-based on general thermodynamics, usually requires algorithms, numerical schemes and a suitable implementation on computers.

It should be noted that mathematics serves as a powerful tool. Once the Eq. (2) is written down together with its parameters, which can easily be done on a piece of paper, a wealth of information on the fluid is specified, like its vapor pressure curve, or caloric quantities. The retrieval of that information can, however, be tedious. In practice, it will depend on the availability of software tools for the evaluation of the given type of equation of state, whether desired results can be obtained with reasonable effort. Although many codes in this field are well tested and considered to be reliable, there is no strict guarantee that the simulation result x^{sim} agrees with the (theoretical) model value x^{mod}. Let us move from thermodynamics to a general consideration of simulation.

Simulations are based on simulation models. We will assume here that they are given by some set of mathematical equations, which relate input to output. We acknowledge that there are other classes of simulation models, like artificial neural networks, which do not fit into that definition. Their point is exactly to connect input and output in a highly implicit way that is based on extensive parameter adjustments – "learning algorithms" – instead of explicit mathematical equations. The more standard case, where a theory in the form of mathematical equations, thermodynamics in our examples, is at hand, is discussed here. This is the harder case for our argument, because it seems to be less dependent on experimentation and parameter adjustment – but let us see.

In most cases, today, the computer is mandatory to study the model, which is too complex to yield the desired output for a given input in any other way. This is generally true already for our Eq. (2) above. Addressing problems by simulation, hence, connects three important issues: setting up the theoretical model (suitably based on the theory of thermodynamics in our examples) on the one side, and implementing and executing it on computers as well as analyzing the results on the other. The implementation includes steps like discretization, algorithms for solving the equation, coding, and compilation. In practice, there often exist different implementations of the same model on different computers. With the simulation, only the specific model implementation (on the chosen computer) can be studied. There are many situations in which the implementation must inevitably give different results compared to the mathematical model, for example, due to discretization. In other cases differences may simply result from an erroneous implementation. In many cases, the quantity x^{mod} – which results from the theoretical model for a given input – is not directly accessible and we can only retrieve numbers for the corresponding result x^{sim} of the simulation.

Scientists can vary the model input or other parameters and "observe" how x^{sim} changes. This is an experimental activity, but one that does not deal with nature or some system in the laboratory, but rather with the model implemented on a computer, that is, with the simulation model. It is important to note that what is observed in computer experimentation is x^{sim}, not x^{mod}. However, in many cases, including our case of thermodynamics, one also has a target system, that is, an object of the real world, which is described somehow by the model. Comparison with this system is not only possible, but is an essential part of the simulation activity. Only then, application becomes a topic and a potential problem. The view of simulation presented above then has to be embedded in a wider perspective which includes the real world,[5] the modeling process and the application of the simulation results (cf. Fig. 1).

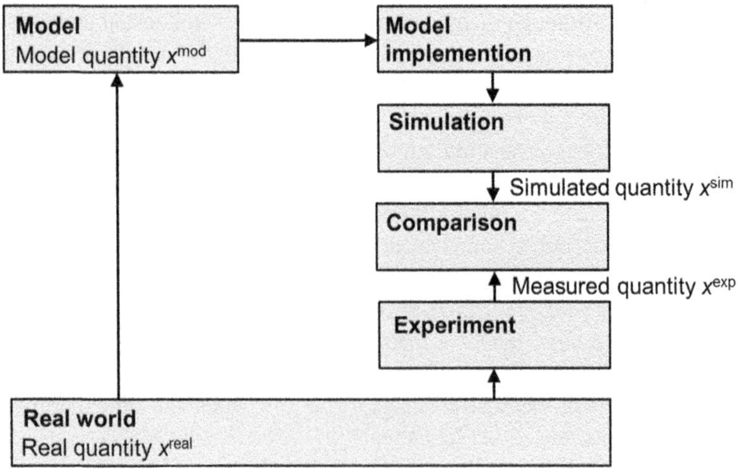

Fig. 1 Scheme showing relations between the real world, modeling, simulation, and experiments

[5]Our claim is open to many guises of how "real" is spelled out in philosophical terms. People concerned with issues of realism might want to resort to "target system," which is a less laden term (though it does not solve any of the questions).

We start with a quantity x^{real} in the real world we want to model. The corresponding entity in the (theoretical) model is x^{mod}. As the model is too complex to be evaluated directly, it is implemented on a computer and simulations are carried out as described above. These simulations yield a quantity x^{sim} which can eventually be compared to results of experimental studies of the real world x^{exp}. In general, we cannot know x^{real} nor can we know x^{mod}, we can only compare the corresponding properties of x^{exp} and x^{sim}. There are two types of experiment in play. One from "below" that provides measured values, the other from "above" that provides simulated values.[6] Up to this point, the picture coincides with prominent schemes of modeling, like R.I.G. Hughes' DDI account (1997), or Reichenbach's (1964, 102/103) appreciation how mathematical deduction (on the model level) and observation work together in science.

The discussion around Eq. (2) presented above highlights the role theory plays in this context and reminds us not to interpret Fig. 1 too literally: Eq. (2) describes a priori only p,v,T properties. But based on arguments which combine some very basic physical statements with mathematics, it can be shown that it describes also properties which are at first glance unrelated to p,v,T properties, like boiling conditions, caloric properties, and many more. It is the success of such predictions that convinces students of taking the effort of studying the theory which enables them.

Up to now, simulation was a means for revealing what the model says about the property x under investigation. This is regularly a task for which there is no alternative to using a computer. Still, the basic rationale is the standard one: The analysis and evaluation of the theoretical model via comparison to the target system.

In general, the quality of a model depends on two aspects that counteract each other. It depends both on adequacy of representation, else the model would not yield results revealing anything about the target system, and tractability, which is prerequisite for obtaining some result at all. Here is where computers have changed the picture. They can handle very long and convoluted iterative algorithms that would be intractable for human beings and, hence, make models tractable which otherwise would be useless.

Figure 1 is rich enough to account for our illustrative case. Equations of state (Eq. 2) have parameters which need to be adjusted to some data on the fluid that they are meant to describe. That data is usually taken from laboratory experiments. An alternative is results from computer experiments. In most cases, for that purpose molecular simulations based on force field are used in which the latter describes the interactions between the molecules. In the molecular simulations, the p,v,T – behavior or other macroscopic thermodynamic properties can be studied based on a model of the interactions between the molecules. The results of these simulations always fulfill Eq. (1) in the low density limit where the molecules are so far apart

[6]Addressing the intricate questions about correspondence and representation, we refer to Weisberg's recent work (2013), which offers a taxonomy for the relationships between model and target system.

that the interactions play no role. But at higher densities, when the interactions start playing a role, they deviate from Eq. (1) but open the door to formulating equations of the type (2) which depend on the interaction model. Thus, the computer experiments yield new opportunities compared to classical experiments. The type and strength of intermolecular interactions can be systematically varied and the effect of that variation on the p,v,T – behavior can be studied. This is widely used in developing new mathematical forms of equations of state (see, e.g., Wei and Sadus 2000).

3 Simulation Model Development as Feedback Loop Process

For our claim about the significance of adjustable parameters, we need a more complex picture of experimentation. In this section we highlight a particular feature of simulation model development, namely, a feedback loop of model adaptation. It is depicted in Fig. 2, which derives from Fig. 1 by adding one arrow that closes the modeling loop. It is basically a classical feedback control loop which aims at minimizing the differences between a variable (here: x^{sim}) and a set value (here: x^{exp}). The two quantities which are compared need not be scalar quantities but may have many entries or be, for example, trajectories over time. There are also many ways of carrying out the comparison.

This feedback loop easily appears as marginal, as a pragmatic handle for fine-tuning and correcting imperfections of the (theoretical and simulation) models. We argue that it is not. Adjusting parameters fulfills essential functions in simulation modeling. Repeated comparison of the two types of experiment guides the modeling

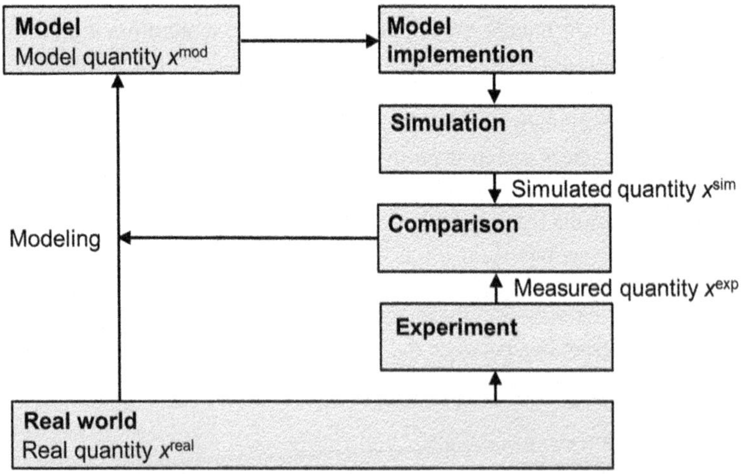

Fig. 2 Same as Fig. 1, but one *arrow* added, pointing to the *left* and closing the feedback loop of modeling

process. During this phase, the process consists in adjusting parameters. The model is explored via (simulation) experiments, motivated by comparison with (classic) experiments. We have hence a cooperation of both types of experiments that is the nucleus of model development via adjusting parameters. However, the cooperation gets even more intertwined when one takes into account that the measured quantities themselves might be partly determined with the help of simulation.

Basically two sorts of actions can be taken in modeling when the comparison of the simulation results and the experiments do not yield the desired results: a) the model structure can be adapted, that is, the set of equations is modified, for instance, to change a description or include a previously neglected effect, or b) model parameters are changed. It is this second option that we are interested in here. Parameterization schemes can be considered as a sort of auxiliary constructions that are intentionally used for dealing with missing knowledge and the inaccuracies of existing knowledge. The simulation model is designed, so that it contains parameters that can be adjusted in the course of the further development.

The remainder of the paper focuses on the role of adjustable parameters. The reasons for using adjustable parameters are discussed in more detail, and it is shown that adjustable parameters form an essential component of simulation modeling. While models with adjustable parameters have been around much longer than computers, practical hurdles had limited their use in the past. The easy availability of computers and optimization software has tremendously lowered these hurdles. It has become much easier to utilize the adaptability of models, so that it has become much more tempting to succumb to the lure of making models fit by adjusting enough parameters.

Parameter adjustment is only one way of model adjustment. Besides adjusting the model parameters, the structure of the model (the mathematical equations) can be adjusted to obtain a better fit to experimental data. The latter procedure is very closely related to parameter adjustment if the equations are changed without any other physical reasoning than obtaining a better representation of some data. The equations themselves then are seen as a sort of parameters that can be adjusted.[7] We will also address this issue in the present work.

The equations of state, which we use as examples here, contain adjustable parameters which are usually determined from experimental data. The exception is the equation of state of the ideal gas (Eq. (1)). But even in that case, it can be argued that R was once not more than an adjustable parameter. But it is no longer, R has turned out to be a universal constant. It is beyond the scope of the present work to discuss the far reaching consequences of that universality which include the definition of temperature and the atomistic nature of matter. By the way, these relations provide again an excellent example for the success of the combination of

[7]We will not discuss classes of simulation models like artificial neural networks. Arguably, they have a very generic structure and extraordinary adaptability. Essentially, they are a proposal to parameterize the entire behavior (if in an opaque, or implicit way).

mathematical deduction with observation. We will rather focus on the consequences of the adjustment of parameters in equations of state and the role the computer plays in this.

As simple examples, we use the van der Waals equation:

$$p = \frac{R\,T}{v - b} - \frac{a}{v^2} \tag{3}$$

and the Virial equation of state in the following form:

$$\frac{p\,v}{R\,T} = 1 + B\frac{1}{v} + C\frac{1}{v^2} \tag{4}$$

The researchers who have introduced these equations, J.D. van der Waals and H. Kammerlingh Onnes, received Nobel prizes in 1910 and 1913. These equations, though both with strong foundations in physics and mathematics, contain adjustable parameters, namely, a and b in Eq. (3) and B and C in Eq. (4). These parameters are needed to account for the individuality of different fluids, that is, water is different from nitrogen. The parameters are not necessarily simple numbers but can be functions of variables. For example, the theory behind Eq. (4) yields that B and C are functions of the temperature, but not of pressure. In the original version of Eq. (3) a and b were numbers. However, in later versions of Eq. (3), a was considered to be a function of temperature. Adjusting functions is obviously more flexible than adjusting numbers.

4 Adjusting Model Parameters: A Closer Look

Speaking about adjustment of parameters invokes a field of similar terms with (only) slightly differing connotations. Calibration, for instance, is used in the context of measuring instruments. Hence, using calibration of parameters makes models look a bit like precision instruments. Tuning, on the other side, has a slightly pejorative meaning, though it is used in some areas of science as the standard term. Anyway, we chose adjusting because it seems to be neutral and does not appear to be a good or bad thing from the start – though we do not claim our terminology is without alternative. Adaptation, for instance, has an equally neutral flavor.

In this section, we discuss a spectrum of situations in which parameters get adjusted. We adopt here a simple scheme of a simulation model, which is based on systems theory (cf. Fig. 3). The process model aims at describing a certain set of quantities $\boldsymbol{y}$, which we will call output variables here. The output depends on the input, which is described by another set of quantities $\boldsymbol{u}$, the input variables. Both $\boldsymbol{y}$ and $\boldsymbol{u}$ belong to the quantities which occur in the model. The latter will be called model variables $\boldsymbol{x}$ here. The set of the model variables $\boldsymbol{x}$ may contain quantities which are neither input nor output (i.e., internal variables). The question

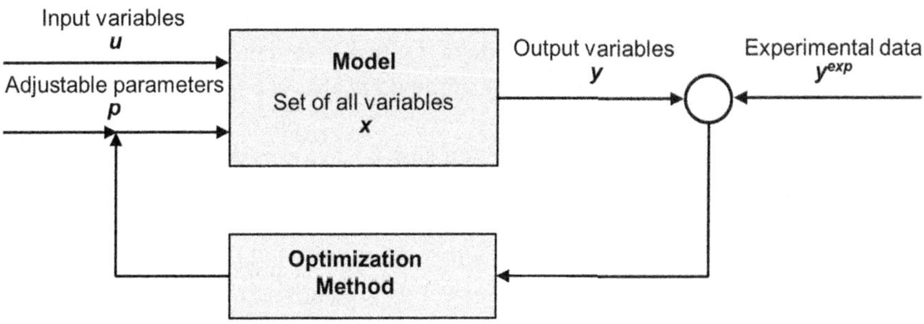

Fig. 3 Parameter adjustment in models

which subset of $\boldsymbol{x}$ is considered as input and which as output may depend on the application. In the models in which we are interested here, $\boldsymbol{y}$ and $\boldsymbol{u}$ describe properties of the target system. Ideally, $\boldsymbol{y}$ is a measurable quantity and $\boldsymbol{u}$ can be set in experiments.

4.1 Model Parameters

Besides the input variables $\boldsymbol{u}$, many models require the specification of model parameters $\boldsymbol{p}$. These do not necessarily correspond to anything in the real world. Cleverly setting the model parameters allows improving the quality of the model regarding its output $\boldsymbol{y}$. The parameter adjustment involves some kind of optimization procedure. The goal of the optimization is to improve the agreement of the model output $\boldsymbol{y}$ with some reference data, usually experimental data $\boldsymbol{y}^{\text{exp}}$ (cf. Fig. 3). We do not presuppose some elaborated formal algorithm for optimization. A simple trial-and-error method is eligible for "method," too. We should point out, however, that mathematical optimization methods reach far beyond what can be handled by simple trial-and-error. Such methods often act like black boxes for thermodynamics modelers. We leave the detailed consideration of the optimization part for another paper.

For example, in the van der Waals equation (3), the input variables may be chosen to be the temperature T and the molar volume v, and one may be interested in the result for the pressure p at those chosen conditions. The calculated result will depend on the choices made for the parameters a and b. Obviously, if some p,v,T data points are available for a given substance, the parameters a and b can be adjusted to these data. Thus, the results obtained for a and b will depend on the choice of the data set to which they are fitted, and also on the way they are fitted. For parameterizing equations of state different types of data are used (e.g., besides p,v,T data also data on vapor pressures, data on the critical point of the fluid, or caloric data). The calculation of such properties regularly involves numerical procedures and as a consequence computers are needed. This becomes especially important in the parameterization which is an optimization task that regularly involves a large number of evaluations of each property. Computers enable adjustments which were not feasible before.

On the one side, adjusting model parameters is obviously a boon, as it can make models work. In many cases it is the key to making them work. Even an otherwise poor model could be augmented by a suitable parameter fit so that it gives fair representations of $\boldsymbol{y}$. In a community which is used to judging models solely by their ability to describe certain properties $\boldsymbol{y}$, this is clearly attractive.

In the case of equations of state, there is an obvious need for an adjustment of parameters. With only a few exceptions, we are not yet capable of predicting properties of real fluids from first principles. Hence, models describing such properties must be trained by some experimental data. The way to do this is by adjusting model parameters. In the field of fluids, the predictions from first principles are presently basically limited to calculating ideal gas properties from Schrödinger's equation. But equations like (3) and (4) are far more than some mathematical form which is fitted to data. We mention only some examples:

(a) by virtue of their derivation they contain Eq. (1) as limiting case,
(b) the B parameter of Eq. (4) can directly be related to intermolecular pair interactions and was for a long time the most important source for quantitative data on them,
(c) the simple Eq. (2) predicts the existence of phenomena like critical points or metastability of fluid phases and relates them to other fluid properties in a consistent way.

These examples highlight the unifying power of the thermodynamic theory and are examples for the benefits of combining theory and experiment.

On the other side, the adjustment is a bane, since it does not remove flaws of models; it rather disguises them. Even an obviously wrong model, that is, one with internal logical contradictions, can give fair representations of $\boldsymbol{y}$ after a suitable adjustment of parameters. For a scientist, who is interested in obtaining insights from models, this is scary. He may be inclined to discard models as worthless if they only work after adjusting parameters to data which the model aims to describe. What is the use of a description of properties of something which can only be established based on the knowledge of the same properties?

This point is illustrated again using equations of state as an example. These equations can be used for describing mixtures. The key to this is finding expressions for the parameters of the equation (like a and b in Eq. (2)) which hold for the mixture. These mixture parameters are usually calculated from the corresponding pure component parameters and the composition of the mixture via so-called mixing rules. With the exception of the mixing rules for the parameters of Eq. (3), which can be rigorously determined from the principles of statistical thermodynamics, these mixing rules are empirical. They contain parameters which usually have to be adjusted to mixture data. Nevertheless, they can be submitted to some tests which can be of logical nature (i.e., if a pure component is formally split up in two identical components, the pure component result should be obtained also from the mixture model) or based on fundamental findings of thermodynamics like those from statistical thermodynamics mentioned above. It is known that mixing rules which fail both in the logical tests and those from statistical thermodynamics, can nevertheless turn out to work well in practice, if the parameters are suitably adjusted. For examples, see Mathias et al. (1991).

4.2 *Proliferation of Variants*

Moreover, adjusting parameters leads to what we call a "flood of flavors," because the results for the parameters and hence all results obtained with the model will depend on the choice of the data set to which the parameters are fitted, and on the way they are fitted. The flood may turn into a deluge if also variants of the mathematical form are included, which are only introduced to improve some fits and have no other basis.

In fact, computers have opened the gates to that flooding. The possibility to easily create and check variants of some model on empirical grounds is at first sight positive. Upon closer inspection, the picture changes. Firstly, the plethora of variants of a given model will rarely have epistemic value. But even from an entirely instrumental standpoint, it may be detrimental. A plethora of versions of a model will create an obstacle for anybody who wants to use the model. Which one to choose? By facilitating the creation of sprawling mutations of models, computers have contributed to the fragmentation of research.

Let us only consider the van der Waals equation, Eq. (3), as an example. It was developed in 1873. Meanwhile there are more than 400 equations of state (so-called cubic equations of state) which can be considered to be variants of that single equation (Valderama 2003). While this gives, of course, enormous credit to the ground-breaking work of van der Waals, it is also distressing. The variants can hardly be classified on theoretical grounds. Rather, historical (when were they developed?), sociological (how well are they received?) or pragmatic arguments (what practical benefits are offered?) and classifications are used. There are some very successful variants which are widely used, and there are certainly elder versions which have technical drawbacks, but there is a plethora of variants that are very similar. Many of these have been used only by the group which has proposed the equation. This danger has nicely been captured by D. Frenkel in his paper on the "dark side" of simulations: "In the past, we had to think about the role of simulations because they were expensive, now we have to think because they are (mostly) cheap" (2013).

Note that the above discussion only addresses the number of mathematical forms of the equation. For each of these there exists a plethora of specific variants. For example, for describing mixtures, one can combine a given equation of state with many different mixing rules. Due to the combinatorial explosion, most of the options have never been explored and never will. And there is very likely no loss in not doing so. Furthermore, we have not addressed that, even in the simpler case of a single pure component, there is a practically unlimited number of options for choosing the data set used for the parameterization – each of which will yield a different set of model parameters.

This sprawling of variants cannot be solely attributed to the use of computers but it is certain that computers have strongly accelerated that development. They have also favored the increase of the number of parameters in a model of a given object. While the van der Waals equation (Eq. 3) only has the two parameters a and b, modern equations of state may have 30 or more adjustable parameters.[8]

4.3 Necessity of Adjusting Model Parameters

What are the reasons that make this parameterization problem so endemic and in a sense unavoidable? In general, any mathematical model presents an idealized version of the real world target system. There is always more abundance in the target system than in some mathematical model equations. Mathematics can be seen as a science which works with objects that are so simple that they can be fully described – which is never possible for an object of the real world.[9] Hence, there may be unknown properties of the target system that should be included in the model, but are not. Leaving open some model parameters and adjusting them to experimental data can be considered as a pragmatic remedy for this.

Even if all properties of the target system which have an influence are known, it can still be prohibitive to explicitly account for their influence in the model. There may simply be a lack of theories, or existing theories might be so complex that they would make the model intractable. Adjustable parameters are of prime importance in this context. They enable using simplified but tractable models. Such models may only be loosely related to the real object and may be obvious over-simplifications. But leaving open some parameters in such models and adjusting them in a clever way can make them work. This is at the core of engineering. Engineers look for simple models which will "do the job," that is, produce good results in a certain application. Their main interest is in the application, not in the physical interpretation. Carefully parameterized simple models can give astonishingly reliable and useful results. As the parameterization involves some comparison with experimental data, it even guarantees a certain quality of the model (at least for the representation of the data that was used for the parameterization). All this is relevant not only in engineering but in many parts of science. Furthermore,

---

[8] The coincidence of computer modeling, exploratory setting of parameters, and proliferation of models has been discussed by Lenhard (2016) in the context of computational chemistry.

[9] Actually, even the objects of mathematics kept ready surprises. The development of the discipline has been accompanied by an extraordinary – and often unexpected – malleability of objects.

accuracy matters and for the reasons mentioned above even good models will never be perfect. Parameterization can be used for alleviating this too.

For example, in the van der Waals equation (1), the parameters a and b have a physical meaning. They are associated with attractive (a) and repulsive (b) interactions between the particles. It is well known that there are many different types of attractive forces, which are all lumped into the a parameter. It can, hence, be considered as an "effective" parameter. Such parameters are meant to describe the influence of a certain class of physical phenomena within a given model. In addition, the parameter b can be considered as such an effective parameter describing repulsion. Despite the crude simplifications in the assumptions on the intermolecular interactions, the van der Waals equation and its mutants have been extremely successful in describing real fluids. There are two main reasons for this. The first is that the structure of the equation (which comes from theory) is able to reproduce qualitatively the most important features of the behavior of fluids like the coexistence of vapor and liquid at certain conditions, the ideal gas limiting behavior etc. The second reason is that the equation contains the parameters, which can be suitably adjusted. Both reasons act together.

Above, we have discussed how simulation takes advantage of the possibility to iterate the feedback loop. It is the very point of the feedback modeling loop that the model is adapted to yield some global behavior. Consequently, the parameters which are used to achieve this do not necessarily follow their physical interpretation – and they do not even need to have such an interpretation at all.

4.4 Parameters with and Without Independent Physical Meaning

In principle, any variable in a model can be used as adjustable parameter. Two cases should be distinguished, depending on the question whether the parameter has an independent physical meaning or not. Independent physical meaning is used here in the sense that there is a physical interpretation outside of the context of the parameter fitting. For reasons of illustration, consider a model for describing the dependence of a physical property y (output) on some other physical property u (input). In the model, it is simply assumed, without any physical argument, that the relation is linear, hence:

$$y = a\,u + b \tag{5}$$

where a and b are adjustable parameters. The parameter a describes the sensitivity of y on changes of u, which is a physical property in the example. Whether the parameter b has a physical interpretation depends on the range that the input values for u have. If $u = 0$ is a physically reasonable scenario, then b is simply the result for y for that case. All this is trivial and not the case we want to make, because here the physical interpretation is no more than some curve discussion in high school mathematics. The case we are interested in is when the linear relation of Eq. (5) is resulting from some physical theory; and there could be a possibility for calculating a from properties that are not y and u. Still, a could be used as adjustable parameter in the fit using data on y and u.

We use the van der Waals equation for a further illustration of the above: assume its parameters are fitted to experimental p,v,T - data of some liquid. On closer inspection of Eq. (2), one finds that the liquid density at high pressures is determined by the b parameter. Hence, one can physically interpret the b parameter as describing the liquid density at high pressures. This is considered here as an interpretation in the context of the parameter fitting, and hence not an independent physical interpretation. However, as stated above, by virtue of the derivation of the van der Waals equation, the b parameter has a deeper meaning. It describes repulsive intermolecular interactions. These obviously become very important in liquids at high pressures, where the distances between the particles in the fluid become very low. Repulsive interactions can in principle also be determined independently, namely, from quantum chemistry. Unfortunately, the derivation of the van der Waals equation is based on such crude simplifications that there is no way to relate or predict the b parameter from independent sources of information, like quantum chemistry.

The above shows different things: while it is fair to say that b is related to repulsive interactions, there is no way to establish such a correlation quantitatively. An important consequence of this is that the numbers for b obtained from fitting should not be over-interpreted as carrying useful quantitative information on the repulsive interactions. That this is not possible becomes also evident when considering that the numbers obtained for the b parameter of a given real fluid will depend strongly on the choice of the data set used for the fit. Nevertheless, it is obviously a merit of the van der Waals equation that it gives structural insight into the importance of certain interactions at certain conditions, in our example the repulsive interactions in liquids at high pressures. We now return from the example to our main line of argumentation.

First, consider the case where the variable, which is used as parameter, has an independent physical meaning. By using it as adjustable parameter that physical meaning is given up in the first place. A number is assigned to that variable in the feedback loop based on pragmatic considerations of the overall model quality, and disregarding the physical interpretation that the resulting number may have. However, one may try to recover the physical interpretation after the parameterization by comparing the result with some independent information on the property, if such information is available. The result of the comparison may well be disastrous without compromising the usefulness of the overall model. But such an outcome will shed a bad light on the explanatory power of the model. On the other hand, it might turn out that the fit has produced a number which is "physically reasonable," that is, which meets some expectations based from considerations that were not included in the fit. This would be a clear sign of the epistemic value of the model, even in a strong sense where it not only predicts physical phenomena qualitatively but also quantitatively.

If independent information on a variable (parameter) is available, one may ask why that independent information was not used right away in the model. A good answer to that question would be the lack of accuracy of the information. If the output of a model strongly depends on a variable which cannot be measured accurately, the variable cannot be used as input variable straightforwardly. In such a situation, the procedure which we have called parameter fit here could be a part of a scheme for data estimation. Pushing this point further, the use of physical variables as adjustable parameters can be considered as a part of a measuring scheme for the associated properties which involves both classical experiments, modeling and simulation.[10]

Let us now turn to the second case where the variable which is used as parameter has no independent physical meaning. At first glance, that case may seem to be trivial. One simply obtains some numbers from the fit and there is no need nor possibility to interpret the results for these numbers. All there is to do is to check the overall model quality. Maybe some kind of curve discussion of the fit can be added (cf. the example around Eq. (5)).

Things become more interesting if we consider why adjustable parameters without physical meaning are used. Basically, this results from operative requirements in the modeling process in which highly complex target objects have to be described with models that are still feasible. This may make it attractive to represent a complex real situation by a model which was initially developed for describing a much simpler physical situation. These models have parameters which describe physical quantities in the simple context for which they were initially developed. They therefore also carry names of physical properties. But in the complex context in which they are used as fitting parameters, the original physical meaning is lost, they degenerate to empirical parameters. However, this is disguised by the fact that they carry names of physical quantities. This has caused much confusion. One should refrain from physical interpretations of results of fits of such parameters, despite the fact that they often carry names of physical variables. An example might clarify the point.

For instance, for modeling liquid mixtures, there is an entire class of models, the so-called lattice models, in which the liquid is represented by particles at fixed positions on a lattice. In the simplest version, each particle occupies one lattice site. This picture is taken from an ideal crystal. For describing liquids, in which molecules are constantly moving around, changing their neighbors, this is a bold simplification. The key parameters in such models are those describing the interaction energies of neighbors on the lattice. For example, in a mixture of two components A and B, there are three such energies, those of the AA, the BB, and the AB interaction. If such simple models were applied to describe some crystal solid, one could hope to interpret the results for the interaction energies determined from a fit to some data in a physical way and compare them to independent data.

[10] Here, our paper ties in with recent accounts of how simulation influences the standard notion of experiment and measurement (cf. Morrison 2009, 2014, Tal 2013).

In the context of modeling liquids, there is no hope in such an endeavor. The numbers obtained for the interaction energies from fits to liquid mixture data have no meaning, even though they are still called interaction energies.

On the other hand, parameters of models which were chosen for entirely mathematical reasons (e.g., coefficients of Taylor series expansions) may turn out to have a strong independent physical meaning. For example, Eq. (3) can be considered as a Taylor series expansion around the state of the ideal gas and B and C are just the first two coefficients of that expansion. The theory of statistical mechanics shows that these coefficients are directly related to the energy of pair interactions in the gas.

In developing computer models, also hybrid schemes are applied in which model parts with a strong physical background and parameters that have an independent physical meaning are combined with empirical parts. Here, the parameters are merely there to improve some model results, which without the use of these parameters would be inacceptable.

Again, the van der Waals equation provides an example. Originally, the parameter a was a number with a certain value for each fluid. But it was soon realized that for accurate descriptions of the thermodynamic behavior over a large temperature range, namely, of the vapor pressure curve, substantial improvements could be achieved by allowing for a temperature dependence of a. The mathematical forms for describing $a(T)$ are empirical and so are the adjustable parameters in these forms.

4.5 Parameters in the Implementation

So far, we have only discussed model parameters. We have neglected the fact that the (theoretical) models often cannot be studied directly. They first have to be implemented on computers. Different implementations of the same model will usually not yield exactly the same results. As a consequence, the implementation, which is a part of the feedback loop of modeling, will influence the model parameterization. Aside from implementation errors, the differences between different implementations of one model are luckily often small enough to be neglected. Model parameters determined in one study are regularly and successfully used in other studies, even though the model implementations differ. However, there is no guarantee that this is the case.

There is more concern about parameters which occur in the model implementation. Prominent examples of such parameters are those used in the discretization of models or those used to control numerical solvers. Ideally, these parameters are chosen from ranges in which the influence of the parameter on the simulation result is negligible (e.g., the grids used for the discretization must be "fine enough"). But it may be very difficult to guarantee this.

When there is an influence of such parameters on the simulation results, they can actively be adjusted in the modeling feedback loop to improve the simulation results. This is much more problematic than adjusting model parameters, as it is implementation-dependent. It forecloses the discussion of the model without referring to the specific implementation. It also may be misused to feint a success of the model, which cannot be attributed to the model but just to a deliberately tuned implementation.

We think that adjusting parameters of the implementation should always be done based on the consideration of minimizing the influence of those parameters at acceptable simulation effort. It must never be used for tuning simulation results in the modeling feedback loop.[11]

4.6 Models and Correlations

Both the term "model" and the term "correlation" are used for referring to descriptions of objects of the real world. Model has a better reputation than its counterpart correlation. Correlations are often considered as "some kind of (empirical) model," but one where physical theory is not invoked. Rather, statistical considerations play the leading role – largely independent from the physical properties of the particular target system under investigation.

In the framework that we have presented here, a correlation is just an extreme version of a model. In Fig. 3, the term model could be replaced by correlation and nothing else would have to be changed. The feedback loop is even more essential for the correlation than it is for the model. This is due to the fact that the correlation relies on adjustable parameters, either fully or at least in essential parts. It does not even have to have any physical background. A correlation can, for example, just be a mathematical form, which is taught to describe a physical situation by adjusting its parameters. Artificial neural networks provide a telling example, since they work with a generic structure, while the parameter adjustments determine the behavior nearly completely. It is noted here that most physical phenomena can be described by suitably parameterized mathematical expansions around a given point (Taylor series, Fourier series etc.).

From this we conclude that there is a continuous transition between physical models and correlations, depending on the degree in which they rely on adjustable parameters and whether the parameters of correlations are open to a physical interpretation as in the van der Waals equation (see 4.4. above). The number of adjustable parameters in a physical model should never exceed the number of adjustable parameters in a purely mathematical (statistical) correlation of the studied phenomenon, which yields a similar quality of the description. Else the physical theory does supposedly not work.

[11]Cf. Oberkampf and Roy (2010, section 13.5.1) for a systematic proposal of how parameters influence the validation of simulations from an engineering perspective.

5 Conclusion: Boon and Bane

The cooperation of experiments, that is, of simulation and classical experiments, plays a crucial role in simulation modeling. This cooperation thrives on the feedback loop in modeling which provides the basis for adjusting parameters. Notably, we monitored the significance of parameter adjustments even in the area of thermodynamics and equations of state where theory is highly developed and well-grounded. Adjustable parameters will arguably not be less significant in fields with less support from theory.

Adjusting parameters is often the clue that makes a model work, it is a *boon*. Using the term applicability in an engineering sense, which can broadly be identified with usefulness for solving practical problems, it is fair to say that adjusting parameters is often the prerequisite for the applicability of a model. At the same time, the adjustment of parameters limits the applicability of models. The model will often only be useful for describing scenarios which are not "too far away" from the scenarios that were used for the fit. We cannot enter into the interesting discussion of this in detail here and just mention that the question how far a model carries beyond the range where it was parameterized is closely related to the quality of the theory behind it, that it is by no means trivial to establish metrics to measure what "far" means, and that the answers to the latter question will be strongly case-dependent. In any case, it must be clear that the adjustment of parameters can simply not replace a sound theory.

Let us consider the equation of state of the ideal gas (Eq. 1) as an example. We start by simply considering it as a model to describe p,v,T data of gases at low densities, and state that the region of the applicability of the model is extremely large as it holds for all substances. The model allows far-reaching predictions as the "parameter" R does not depend on the substance. Equation (1), which is often also called "Ideal Gas Law" would obviously lend itself to a closer discussion of the relation of the terms model and theory, but we must refrain from this here. We rather move to other equations of state, for which the picture changes. Let us use the van der Waals equation (Eq. (3)) as an example and assume first that we are merely interested in using it for describing properties of a certain pure fluid. For this, we need to have numbers for the parameters a and b of that fluid. They must be obtained from an adjustment to some experimental data. Once this is done, we can make all sorts of predictions using Eq. (3) but the quality of these predictions will strongly depend on the relation between the data that were used for the fit and the data which are to be predicted. Interestingly, there are common notions in the thermodynamic community as to which data are to be used for parameterizing equations of state for fluids. For example, as a rule, experimental data on the so-called critical point are used, if they are available. The reason is that parameterizations based on such data are found to be more broadly applicable than competing parameterizations. To summarize, the following issues are inextricably entwined: the model with its parameters, the way the parameters are determined, and the applicability of the model in certain situations. Taking this into account, a scheme for comparing the quality of different models would be to use the same data for the parameterization, to apply the models for studying the same quantities, and to compare the quality of the results.

The limitation of the applicability of a model by its parameterization is not a bane in itself. But it becomes a bane when it is overlooked. We must get used to never think of models without considering the way they were parameterized.

In a sense, adjusting parameters is strongly guided by predictive quality over a certain – and maybe very limited – range. While enabling application, this procedure diminishes the explanatory capacity of simulations, because the iterated adjustments tend to obscure or convolute what can be attributed to general theory, parameterization strategy, and particular parameter adjustment. In brief, adjustable parameters are boon and bane of simulation. Empirical, theoretical, and computational aspects complement each other.

Though our results belong to the philosophy of simulation, they point toward a more general philosophical perspective. Let us take simulation in terms of mathematical modeling. Simulation then does not merely extend mathematical modeling, but adds a new twist to it. Now classical and simulation experiments cooperate, building on the feedback loop and on adjustable parameters. Our investigation thus adds a new twist to the so-called "new experimentalism" of Hacking, Franklin and others. They highlighted the importance of experimental (instead of solely theoretical) knowledge in the sciences. Philosophy of science would then examine how experimental knowledge actually arrived at and how this knowledge functions. The rationality of science then is not distilled from some logic, but from actual scientific practices. Our paper contributes to this line of reasoning and extends to practices of simulation. The interesting twist then introduces an empiricism that is different from earlier accounts of modeling in an important way. It neither explores the theoretical model, nor inserts measured values (as in semi-empirical methods) for parameters hard to calculate. The cooperation of experiments and the exploratory/ empiricist nature of adjusting parameters extend well into the home territory of theory.

Acknowledgments H.H. gratefully acknowledges support of this work by the Reinhart Koselleck program of Deutsche Forschungsgemeinschaft.

J.L. gratefully acknowledges support of this work by DFG SPP 1689.

References

Axelrod, R. M. (1997). Advancing the art of simulation. In R. Conte, R. Hegselmann, & P. Terno (Eds.), *Simulating social phenomena* (pp. 21–40). Berlin: Springer.

Barberousse, A., Franceschelli, S., & Imbert, C. (2009). Computer simulations as experiments. *Synthese, 169*, 557–574.

Bedau, M. A. (2011). Weak emergence and computer simulation. In P. Humphreys & C. Imbert (Eds.), *Models, simulations, and representations* (pp. 91–114). New York: Routledge.

DeVito, S. (1997). A gruesome problem for the curve-fitting solution. *British Journal for the Philosophy of Science, 48*(3), 391–396.

Dowling, D. (1999). Experimenting on theories. *Science in Context, 12*(2), 261–273.

Forster, M., & Sober, E. (1994). How to tell when simpler, more unified, or less ad hoc theories will provide more accurate predictions. *British Journal for the Philosophy of Science, 45*(1), 1–35.

Frenkel, D. (2013). Simulations: The dark side. *The European Physical Journal Plus, 128*(10). doi:10.1140/epjp/i2013-13010-8.

Galison, P. (1996). Computer simulations and the trading zone. In P. Galison & D. J. Stump (Eds.), *The Disunity of science: Boundaries, contexts, and power* (pp. 118–157). Stanford: Stanford University Press.

Gramelsberger, G. (2011). What do numerical (climate) models really represent? *Studies in History and Philosophy of Science, 42*, 296–302.

Hughes, R. I. G. (1997). Models and representation. *Philosophy of Science*, *64*(Proceedings), S325–S336.

Hughes, R. I. G. (1999). The Ising model, computer simulation, and universal physics. In M. Morgan & M. Morrison (Eds.), *Models as mediators* (pp. 97–145). Cambridge: Cambridge University Press.

Humphreys, P. W. (1994). Numerical experimentation. In P. Humphreys (Ed.), *Patrick Suppes: Scientific philosopher* (Vol. 2, pp. 103–121). Dordrecht: Kluwer.

Humphreys, P. W. (2004). *Extending ourselves: Computational science, empiricism, and scientific method*. New York: Oxford University Press.

Keller, E. F. (2003). Models, simulation, and 'computer experiments'. In H. Radder (Ed.), *The philosophy of scientific experimentation* (pp. 198–215). Pittsburgh: University of Pittsburgh Press.

Kieseppä, I. A. (1997). Akaike information criterion, curve-fitting, and the philosophical problem of simplicity. *British Journal for the Philosophy of Science, 48*(1), 21–48.

Lenhard, J. (2007). Computer simulation: The cooperation between experimenting and modeling. *Philosophy of Science, 74*, 176–194.

Lenhard, J. (2016). Computer simulation. In P. Humphreys (Ed.), *Oxford handbook in the philosophy of science* (pp. 717–737). New York: Oxford University Press.

Mathias, P. M., Klotz, H. C., & Prausnitz, J. M. (1991). Equation-of-state mixing rules for multicomponent mixtures: The problem of invariance. *Fluid Phase Equilibria, 67*, 31–44.

Mauritsen, T., et al. (2012). Tuning the climate of a global model. *Journal of Advances in Modeling Earth Systems, 4*, M00A01. doi:10.1029/2012MS000154.

Morgan, M. S. (2003). Experiments without material intervention. Model experiments, virtual experiments, and virtually experiments. In H. Radder (Ed.), *The Philosophy of Scientific Experimentation* (pp. 216–235). Pittsburgh: University of Pittsburgh Press.

Morrison, M. (2009). Models, measurement, and computer simulation. The changing face of experimentation. *Philosophical Studies, 143*, 33–57.

Morrison, M. (2014). *Reconstructing reality. Models, mathematics, and simulations*. New York: Oxford University Press.

Oberkampf, W. L., & Roy, C. J. (2010). *Verification and validation in scientific computing*. Cambridge: Cambridge University Press.

Parker, W. S. (2013). Computer simulation. In S. Psillos, & M. Curd (Eds.), *The Routledge companion to philosophy of science* (2nd ed.). London: Routledge.

Parker, W. S. (2014). Values and uncertainties in climate prediction, Revisited. *Studies in History and Philosophy of Science, 46*, 24–30.

Reichenbach, H. (1964). The rise of scientific philosophy (11th print). Berkeley/Los Angeles: University of California Press.

Rohrlich, F. (1991). Computer simulation in the physical sciences. In A. Fine, F. Forbes, & L. Wessels (Eds.), *PSA 1990* (Vol. 2, pp. 507–518). East Lansing: Philosophy of Science Association.

Tal, E. (2013). Old and new problems in philosophy of measurement. *Philosophy Compass, 8*(12), 1159–1173.

Valderrama, J. O. (2003). The state of the Cubic Equation of State. *Industrial Engineering Chemistry Research, 42*(7), 1603–1618.

Wei, Y. S., & Sadus, R. J. (2000). Equations of state for the calculation of fluid-phase equilibria. *AIChE Journal, 46*(1), 169–196.

Weisberg, M. (2013). *Simulation and similarity. Using models to understand the world.* Oxford: Oxford University Press.

Winsberg, E. (2003). Simulated experiments: Methodology for a virtual world. *Philosophy of Science, 70*, 105–125.

Winsberg, E. (2014). Computer simulations in science. *The Stanford Encyclopedia of Philosophy* (Fall 2014 Edition), E. N. Zalta (Ed.), http://plato.stanford.edu/archives/fall2014/entries/simulations-science/

Woody, A. I. (2013). How is the ideal gas law explanatory? *Science and Education, 22*, 1563–1580.

Systems Biology in the Light of Uncertainty: The Limits of Computation

Miles MacLeod

1 Introduction

In many fields there is a growing expectation that the power of modern computation will supplant the limitations of traditional modeling methods when dealing with highly complex phenomena like biological and climate systems. Computational methods, and the mathematical methods on which they are based, are conceived of as a powerful resource for stepping around human cognitive limitations, and also practical limits on experimentation, through combinations of brute numerical force and subtle mathematical analysis. The methods being developed are increasingly more than just extensions of traditional modeling methods, often employing novel types of argumentative and inferential strategies which help derive information from complex uncertain phenomena. These strategies include for instance ensemble strategies. Ensemble strategies were impossible before accessible high-powered computation became available. As in the case of climate change (Parker 2010a, b; Lenhard and Winsberg 2010) many of these have yet to be properly studied and their limits understood. Yet it is imperative that computational fields properly adapt their aims and goals to fit realistic expectations about what these new tools can do.

One field for which this imperative is particularly relevant is computational systems biology. In systems biology the rhetoric of what computational and mathematical methods can achieve has arguably got ahead of what is actually being accomplished. The promise that such methods could provide large-scale predictively

M. MacLeod (✉)
University of Twente, Schubertlaan 23, Enschede, 7522JN, The Netherlands
e-mail: m.a.j.macleod@utwente.nl

J. Lenhard, M. Carrier (eds.), *Mathematics as a Tool*, Boston Studies in the Philosophy and History of Science 327, DOI 10.1007/978-3-319-54469-4_7

robust models of complex metabolic and gene-regulatory systems, suitable for medical and other purposes, still seems far off despite at least 15 years of work. In this chapter we will investigate one of main limitations on computational and mathematical methods in systems biology, which makes it particularly difficult for modelers to produce the high-validity models required for predictive purposes. This is the problem of measuring and globally fitting parameters of large-scale models of biological systems. The uncertainty over parameter values that results from imprecise and incomplete data, and bio-variability – the natural variability of biological systems – creates substantial uncertainty over the ranges for which models are valid. Researchers are developing ways of working around this problem using computational and mathematical techniques such as sensitivity analysis and ensemble reasoning. However many questions can be raised about whether these can and do always improve model validity.

These problems suggest there are limits to how well current models using computational and mathematical methods really can produce predictively robust models of highly complex biological systems. The field itself has not yet engaged in significant open discussion of these issues. I will show that even though computational methods like ensemble methods do not necessarily grant accurate enough representations for robust prediction, they can nonetheless be used to contribute to the investigation and understanding of biological systems. These uses have less to do with the brute force construction of widely valid representations of biological systems and more to do with the powerful roles methods like ensemble methods can have facilitating search and discovery. These uses suggest that in order for philosophy of science to develop more realistic images of what computational and mathematical methods can do in systems biology, and elsewhere, we should concentrate on their roles as *tools of investigation*. The concept of tool provides important metaphorical resources for developing our understanding of what computational resources contribute to scientific practice. Importantly it helps identify the roles that research contexts and users play in methodological design and decision-making.

Some of the insights raised in this paper are drawn from the research and educational literature in the field, and also from a 4-year ethnographic study of model-building practices in two systems biology labs led by Professor Nancy Nersessian. Our group performed laboratory observations and interviews of lab participants, some over the course of their graduate research projects. Some of the case examples drawn on below are taken from this study. The names of these researchers have been encoded to protect their identities.[1]

[1]This research was funded by the US National Science Foundation which requires following human subjects' regulations. Identities are concealed for this reason.

2 Systems Biology: Aims and Goals

There is substantial enthusiasm in the life sciences for computational approaches, and their potential to implement more complex statistical and mathematical models, and more sophisticated methods for analyzing such models. Systems biology is just one of a set of recent fields that includes genomics, bioinformatics and synthetic biology, which treat computation as essential to tackling biological complexity. There has been no shortage of articles heralding the novelty, power and necessity of a computational systems approach to modeling the dynamic properties of large complex human gene regulatory, metabolic and cell -signaling systems (see for instance Hood et al. 2004; Kitano 2002; Ideker et al. 2001). Modern systems biology, often called computational systems biology to distinguish it from older non-computationally intensive attempts to model biological systems, aims to build dynamic mathematical models of such systems. These systems are identified as the units of organization that control and implement particular biological functions. They are generally composed of large networks of interacting genes and metabolites. They are for the most part highly nonlinear. Elements within networks often play multiple functions, and there are frequent feedback and feed-forward interactions. A key argument for a computational and mathematical approach is that only quantitative modeling and analysis is sensitive and powerful enough to represent these complex dynamical relationships amongst system elements.

Modelers however are not typically biologists but come from engineering and other quantitative fields. Systems biology borrows concepts and techniques in turn from engineering, some of which we will talk about below like sensitivity analysis. The field positions itself against molecular biology, which is characterized as reductionistic and overly qualitative in its approach to investigating biological networks and pathways through, primarily, controlled experimentation (Westerhoff and Kell 2007).

In the language of the field systems biology aims at a "systems-level understanding". The concept is a multifarious and somewhat vague concept (see MacLeod and Nersessian 2015). However one of its most important dimensions is encapsulated by the idea that computational modeling will enable the accurate manipulation or control of biological networks (Kitano 2002). Control requires the generation of predictively robust models, or models that have a wide-range of validity and remain accurate when perturbed along dimensions which push a system outside its natural operation. Mathematical analysis of such models can then yield predictions of where and how to intervene on these networks to engineer desirable results in order for instance to prevent or cure disease or to produce a desirable chemical (Hood et al. 2004). The belief that models of the order of validity required to achieve these aims are now possible given the power and availability of modern computation, is the basis of some of systems biology's most profound and significant claims. These claims have undoubtedly played a part in its ability to establish itself in the academic landscape.

3 Building a Model and the Problems of Parameters

However these predictive goals face a substantial obstacle in the form of parametric uncertainty. To understand why parameter uncertainty is such an issue in systems biology it is wise first to have some idea of how models are constructed in the field. There are different approaches amongst systems biologists and we concentrate only on one, the bottom-up approach. Bottom-up systems biologists start from available experimental data and knowledge of particular biochemical interactions to assemble biological pathways and construct models. Top-down systems biologists, although they do rely on experimental data and information, ostensibly reverse engineer network structures themselves using high-powered data collection techniques and analysis (see Westerhoff and Kell 2007). For the most part modelers in systems biology (of at least the bottom-up version) aim to construct ordinary differential equation (ODE) models of their systems. The components of these models represent the regulation states or chemical concentrations of particular elements in a cell like genes and metabolites over time. A system is first visually represented as a pathway diagram that captures the specific interactions between these elements and the order in which they happen. One choice a modeler has to make once he or she has assembled a pathway is how to mathematically represent these interactions as terms within an ODE system of equations. Choices range from more mechanistic to less mechanistic representations. For instance in the case of metabolism, Michaelis-Menten or Hill representations of enzymatic kinematics are considered more mechanistic options, being derived from models of the processes by which an enzyme converts a substrate molecule to another. Other more complex options also exist that can incorporate the dependence of enzymatic reactions on multiple substrates (rather than just one) like ordered bi-bi or Ping-Pong representations. On the other hand more abstract generic or canonical representations like power law representations are available that can capture a wide range of possible interaction behavior within the degrees of freedom of the representation's parameters (Savageau 1969; Voit 2000). These representations ignore the mechanistic details behind for instance enzymatic interactions, but are more mathematically tractable. They afford better possibilities for mathematical analysis and are easy to optimize thus reducing the computational work required in parameter fitting processes. For many systems biologists, the only way to achieve high-validity models is to build detailed mechanistic representations of biological systems that capture well the underlying processes. At the same time computational and mathematical tractability considerations, particularly with respect to parameter fitting, force them to walk a tightrope between building the detailed representations required and relying on mathematical strategies of abstraction and simplification that remove detail that may, in the end, be relevant to the overall predictive robustness of a model, as we will see below.

Once representations are chosen it is an easy task to translate a pathway diagram into a set of ordinary differential equations. The main remaining task is to calculate the parameters governing each interaction in the model. When modeling metabolic systems, interactions are governed by kinetic parameters (e.g. rate constants, kinetic orders, Michaelis-Menten constants and other parameters), depending on the representations chosen. Parameter estimation and fitting are steps around which much error and uncertainty over the validity of these models is introduced.

Many of the problems of deriving parameters have been identified in our group's ethnographic studies (see MacLeod and Nersessian 2013a, b, 2014). They are common throughout the field and are often commented on (for instance see Wahl et al. 2006; Turkheimer et al. 2003; Brown et al. 2004). Here is a list. In the first place the experimental record constructed by molecular biologists often does not contain specific parameter values, but just measurements from which modelers must estimate individual parameters using regression or graphical estimation techniques. The data is often noisy, part of which is caused by bio-variability (see below) and part by experimental error. Secondly the measurements that are available have often been performed *in vitro*. Effects that occur at very low metabolic concentrations may not be replicable in a test-tube. But *in vitro* measurements are problematic for systems biologists. Most advocates of systems biology believe that parameter values are partly determined by the systemic contexts in which the interactions they govern occur. In vitro testing cuts these interactions out thus producing skewed measurements that do not necessarily reflect reality. Further measurements that have been made are often on related but not necessarily the same cell-types or on cells from related but different species. These measurements may have been made using different experimental protocols.

All these situations mean that parameters that have been estimated have both measurable (in form of error bars) and unmeasurable or unknown degrees of uncertainty associated with them. Finally the experimental record is often very incomplete, leaving many parameters impossible to calculate from the data. Sometimes this is because the interactions involved have not been of interest to molecular biologists. Sometimes it is because accurate measurements of the reaction kinetic profile of single parts of a biochemical network are technically impossible with current experimental techniques (Wahl et al. 2006). Modelers gather whatever they can find from whatever sources they can get them. These problems are hard to resolve given present technology and are accentuated by the fact that systems biologists generally do not have the access to the experimentation they require to reduce parameter uncertainty. As one lab PI we interviewed put it...."and I still maintain, I've said it for 20 years, you need 10 experimentalists for every modeler". This is certainly not going to happen anytime soon.

The modeling strategies modelers use are almost completely dependent on the data they have available and modelers look for schemes that are tractable given the available data. If only steady-state and flux-data are available then modelers will build steady state models and use linear techniques like flux balance analysis and metabolic control analysis to estimate unknown parameters. The former works for instance by imposing plausible biological optimization constraints on networks and finding an optimal solution for parameters that satisfies these constraints. Such results can only be helpful however for finding certain parameters. Parameters that control dynamic behavior cannot be recovered this way. In general much hard work goes into understanding a problem and then transforming it into a form that can actually be solved given the data.

Nonetheless the lack of measurements which modelers can use for even approximate estimation of parameters leaves them with many undetermined parameters, sometimes as many as 50. A fraction of labs, such as one our group studied, are fully equipped wet-labs. Modelers in these labs may develop their own experimental skills, and thus can more directly resolve such parameter problems. For the most part modelers in the field however have to rely considerably on computationally intensive global fitting algorithms to determine remaining parameters. A global fitting process will sample combinations of unknown parameters across a prescribed range (in this case a range which is considered biologically plausible) and try to identify the combination that gives the best fit between the model and the available data on how the system actually behaves. Often some of this data is left out of fitting process, in order to test the model later. Various basic algorithmic techniques for exploring parameter spaces are adapted for this problem, including comprehensive grid searching techniques, genetic or evolutionary algorithms, and simulated annealing.[2] These techniques try to balance the computational costliness of exploring a parameter space against the risk of ending up in a local minimum which is not in the neighborhood of the best fit. A significant proportion of researchers in fact work specifically on improving algorithmic techniques for systems biological models.

In theory if the model is accurate in its structural details the best fit should serve as a mechanistic representation of the underlying system. However global fitting, and the moves required to get it to work computationally, introduce considerable uncertainty. Firstly modelers usually try to estimate search regions of their complex parameter spaces in order to limit their search, often relying on biological information about what might be plausible. Such judgments can be wrong given the complexity of these spaces. Secondly these spaces may contain many local minima that produce models similarly consistent with the data and there is no guarantee that a parameter combination will be found that represents the best fit. Part of the reason for this is parameter compensation, which is a common aspect of nonlinear ODE models and nonlinear systems. Within a certain range pairs or other combinations of parameters can be varied such that the model will produce the same fit with the data. Unfortunately the problem of such multiple fits is that there is no guarantee that the models will still give the same predictions outside the range of the available data. In philosophical terms the correct or most accurate model is underdetermined by the available data, even if the structural account of the network is right (see further the problem of sloppiness below).

These various levels of parametric uncertainty compound with the result that model validity becomes difficult to assess and reliable models difficult to find.

[2] I use "parameter space" here to refer to the space that plots parameters values against performance or fitness (measured by a set of performance conditions like fit to the data).

For instance uncertain estimates of parameters affect the global fits that are found for the remaining parameters. Uncertainty intervals create space for parameters to compensate each other in more alternative ways. Models might be well fitted to the data, and may even give some correct predictive results. Yet given the very real possibility of alternative fits within the range of parameter uncertainty that can do the same job (and make the same predictions), there is no guarantee that the model found is the best or most reliable option over a wider range of perturbations. It is thus very difficult to produce models that be relied upon as predictively robust.

Additional to all these sources of uncertainty is the basic problem of biovariability. Parameters formulated with respect to one cell-line or cells within one individual organism may not capture the right parameter sets for others. Whereas parameter sets derived using parameters estimated from data from different organisms or cell-lines (using average or median values for instance) may not represent a population well (Marder and Taylor 2011). Unless the range of variability is well understood, can be difficult to generalize a successful model as representative of the population at large.

These problems make the task for modelers of finding predictively robust models that work for individuals or populations highly challenging and frustratingly difficult even with the aid of computation (see Hasse and Lenhard, chapter Boon and Bane: On the Role of Adjustable Parameters in Simulation Models, this volume). In the next section we will focus on two kinds of novel computational and mathematical methods we have observed that modelers use to try and overcome these uncertainty problems and generate robust solutions.

4 Negotiating Parameter Uncertainty – Sensitivity and Ensemble Techniques

The two techniques I will describe here operate at two different stages of the model building process. The first is often used to reduce the complexity of the parameter fitting task such that more robust best-fit solutions can be found without getting stuck with a computationally intractable problem or a bumpy parameter space. The second is a technique that is more novel in systems biology. It is used to try and generate sound inferences from multiple models that fit the data. One of its purposes is improving prediction. We consider whether such methods really do help achieve the kinds model validity needed for robust prediction.

4.1 Sensitivity Analysis

Sensitivity analysis broadly construed is not a new technique. It has a long history in engineering and other fields, and has multiple purposes including for instance the analysis of a model to identify sensitive parts of a system or to measure uncertainty in predictions from a model. In systems biology it is also increasingly used as a technique for simplifying model-building tasks by trying to filter out just those parameters that have most effect or control over the particular dynamic relationships researchers might be interested in. Such techniques enable modelers to confine the parameter-fitting problem to a more limited set of parameters.

Sensitivity analysis applied this way is very common and almost essential to the model building process for many modelers. As mentioned, with such complex systems fitting all the parameters might be computationally intractable with the resources (usually PCs) modelers have available. Given the complexity of parameter landscapes resulting from these systems one can be much less confident that a parameter-fitting algorithm will find a good fit and not get stuck in inadequate local minima. This kind of analysis should help construct models that can capture the dominant dynamic behavior of a system at least.

One modeler we interviewed for instance, G10, was modeling lignin synthesis in two plant species. Lignin, a structural material in cells, interferes with attempts to get plant metabolism to produce biofuel chemicals from biomass. G10's goal was to try to understand how to effectively control lignin production to make biofuel production more efficient. His experimental collaborators provided him with some data, but he had to assemble the synthesis pathways himself. His original pathways represented something like a consensus of biological opinion on lignin synthesis. During the model-building process he inferred more possible interactions in the network that seemed plausible in order for the model to be capable of capturing the system. For one modeling task, performed on the poplar tree species, his network contained about 20 interacting metabolites. He aimed specifically to discover robust dynamic relationships between particular input variables and particular output variables, which should guide effective manipulation and control of the lignin system. The ratio of two lignin monomers S and G were his target output variables. He had a total of 27 unknown parameters. To reduce this problem he calculated how variations in individual parameters (associated with particular network elements or variables) across his network affected the output from the model (in the form of the S/G ratio), running his model with thousands of different parameter sets (sampled using Monte Carlo techniques). He calculated the Pearson correlation coefficient (a measure of sensitivity) for the effect of variations of each parameter on variations in the S/G ratio. This informed him of which parameters were most crucial to fit well with respect to getting a good model fit. With this information he was able to simplify his parameter-fitting problem significantly to fitting just 8 parameters. The other insensitive parameters were set to biologically reasonable values. As a result he was able to find equations connecting target input variables in the system to the S/G ratio that fit the available data very well.

Such approaches represent creative and powerful uses of computation for finding routes to simplify complex problems. However while approaches like these certainly do generate easier parameter fitting problems, optimizing a model in this way is likely to have consequences for how valid a model is when trying to predict a broader range of responses of a system to perturbation. Sensitivity analysis so used is designed to simplify the problem of getting a model to replicate a limited set of system functions or relationships among only particular variables in the system. The model is abstracted and simplified to meet this goal. As we saw in G10's case the statistical analysis he uses contrives to select-out just specific parameters from the network model that reproduce the experimental behavior of just one variable, the S/G ratio, rather than other variables in the system. Only the uncertain parameters that are sensitive with respect to those relationships are fitted. The rest are just very roughly approximated. This builds quite a significant constraint on the parameters selected and fitted, and those that are not. Such a technique can work very well at getting a model to fit a set of data that map how this variable changes over a certain domain. But there may well be other sensitivities amongst other variables in the system, which are lost due to this narrow selective procedure. These methods raise very reasonable questions about the extent to which they might be geared to find fits that do not represent a system well overall when larger ranges of manipulation and perturbation are required.

4.2 Ensemble Reasoning

Borrowing from cases in other sciences, like climate modeling, systems biologists are developing ensemble methods in order to draw inferences given both structural and parametric uncertainty (see Wahl et al. 2006; Tran et al. 2008; Turkheimer et al. 2003). G10 used his own ensemble strategy, which he hoped would improve the predictive accuracy of his models despite the parameter uncertainty he was confronted with. Having performed the sensitivity analysis mentioned above to reduce the number of unknown parameters that needed to be calculated, G10 ran a simulated annealing process to fit these significant parameters. In his words he "collected 20 GMA [generalized mass action] models with similar training errors by running the optimization 20 times as there is no guarantee that a global optimum could be found" (dissertation presentation). These models had different parameter calculations but these were nonetheless thought to cluster on the same fitness peak in the parameter space, varying according to the width of that peak. G10 calculated the error bars for the mean values of these parameters of about 33% of the parameter value for most parameters to within a 95% certainty. These models all gave nonetheless highly consistent predictions of the S/G ratio in response to particular experimental observations of its response to changes in the concentrations of five enzymes in the system. Further all ensemble members predicted the results of two experiments that were not used to fix the model. Using this ensemble G10 calculated a strategy for minimizing the S/G ratio of the monolignol biosynthetic pathway in order to obtain a higher yield of the desired biofuel component xylose. "Now, since we had a valid model, we could use it to find out which enzymes should be targeted to, for example, decrease the S/G ratio. And here's what we found. So, if you knock down just one enzyme, you could get a 25% decrease in the S/G ratio. But if you're allowed to change the expression of as many as three enzymes, you could get a 38% decrease in the S/G ratio, which is huge compared to the natural variation that is about 5 to 10%." (dissertation proposal). The ensemble technique thus allowed G10 to illustrate that his predictions were robust despite uncertainty over the correct parameter values. Modelers also have the option of averaging over ensemble results or using other statistical techniques to try to produce predictions that represent at least approximately what would be expected of the system in the wild (Turkheimer et al. 2003).

In the case of ensemble modeling, justification of how and why ensemble modeling can be relied upon to overcome parameter uncertainty and enable robust prediction is still limited. Most new ensemble methods are tested on published models from one or a few case studies rather than justified using mathematical arguments. "Experimental" strategies like these are common throughout systems biology as a way to demonstrate the efficacy of a new parameter or structure fitting mechanism, but are not necessarily good bases for drawing conclusions about the general applicability of ensemble methods. The nature of a problem can indeed play a large part in how well a method might perform. For example an ensemble technique might be designed to give sets of models that, when averaged over, provide predictively robust results for a given system. When tested against an uncertain parameter space – a space with many unfitted parameters and few constraints on their potential values – its performance may look very poor if the parameter landscape is flat or very bumpy, and very good if the landscape has one well-defined peak. Testing on one or a few cases might not be strongly informative about how good a method generally is.

However the central problem is that while it is intuitive to expect that a convergent ensemble might help unearth classes of models that fit the data equally well within relatively confined parameter ranges this does not necessarily imply that all models in the ensemble will maintain consistent behavior under perturbation. Variations between parameter sets may still have consequences when a model is pushed outside the domains used to test model performance. Indeed ensemble ranges even within convergent ensembles may still vary significantly. As mentioned, G10 estimated the mean values of his significant parameters to vary by on average 33% within a 95% confidence interval. He does not however provide in his dissertation or interview comments any analysis that suggests that we should not expect, given these parameter ranges, that model behavior might vary substantially outside the data used for fitting and testing.

As it turns out finding a set of models that have relatively consistent behavior through collective fitting within some domain is not necessarily a good indication that such models will behave similarly outside that domain (Brown et al. 2004; Gutenkunst et al. 2007; Apgar et al. 2010). Recent work by these authors analyzes the so-called sloppiness – as opposed to stiffness – of many systems biology models. The sloppiness of a model, which is a model robustness concept, measures the degree to which a model is insensitive to certain parameter changes or combinations of changes (defined by vector directions in parameter space) by measuring the degree to which a model maintains the same overall behavior despite these changes. Stiff directions in a parameter space represent directions of parameter change for which model behavior is highly variable. Sloppiness is calculated by the measuring the divergences in model behavior from a "true model" over parameter space. The variations in model behaviors with respect to this model are summed for each different parameter set in the space. A hessian matrix then records how responsive model behavior is to variations in pairs of parameters over the space. For systems biology models this matrix elicits sets of ellipsoidal eigenvectors over the parameter space that represent spaces in which two parameters can vary with respect to one another while model-behavior remains almost invariant (given some upper bound on variation). The long and short axes of an ellipsoid represent stiffness and sloppiness respectively. Along sloppy directions parameters can vary over substantial orders of magnitude and still compensate each other to achieve similar model behavior. Gutenkunst et al. find few eigenvectors in their test case that run along the directions of single parameters, illustrating the importance of compensation relationships as the basis of sloppiness. "The ellipses are skewed from single parameter directions." (p. 1873)

Gutenkunst et al. analyzed 17 published models from systems biology published in the top journals of the field. They treated each as the "true model" for the purposes of this analysis, and used their analysis to explore variation in model behavior for parameter sets around the "true set". They discovered that all these models contained sloppy parameter relationships that would produce consistent behavior with the collectively fit true model. Hence given the same set of data, numerous models with different parameter sets could be found that accurately gave approximately the same output. These parameter sets are the sets which ensemble methods like G10's are primed to pick out. Critically however Gutenkunst et al. found that if certain measured parameters are known with less certainty then this uncertainty creates space for more parameter combinations which fit the model to the data, but do not produce consistent behavior. Parameter set combinations found through collective fitting will no longer align along a sloppy direction, but rather spread out along the uncertainty intervals of the most uncertain parameters. Hence predictions from such models are no longer reliable. This raises large questions about the robustness of many models in systems biology, which invariably do involve parameter measurements of different certainty. If a model can yield many parameter-fits with given data, which yield different model behavior otherwise, then it becomes entirely problematic to assume that these models will be capable of giving accurate robust predictions of the kind systems biology needs.

This problem is not an easy one to resolve. One might expect, as Gutenkunst et al. state, that stiff directions in parameter space should represent the most important for model-building, since changes in them affect the error of the model substantially. Rather than using ordinary sensitivity analysis model parameters could be reduced effectively by aggregating parameters along these directions. However discovering these directions requires having fitted models in the first place. This is the problem that initially inspired doing this kind of sensitivity analysis in the first place. Large-scale models in most cases cannot be fit without simplification.

As an upshot it seems unlikely that reasoning with ensembles of models fitted with different parameters really improves the chances of finding high-validity models. If some parameters are known with less certainty than others, then it is likely that consistent collective fits will not align along a sloppy direction, and predictions from different parameter combinations within the ensemble will vary despite fitting the data equally well. The result, inevitably, is large uncertainty over the validity of the model. Averaging over an ensemble given this variation will be no necessary indication of what the best option is or even that an averaged parameter set will be found within the set of best fits.

Researchers like G10 however often motivate ensemble techniques alternatively on more explanatory or representational (rather than strictly predictive) grounds. Such ensembles capture or represent well individual variability in populations.

"This notion of finding not just a single best model, but an entire class of competent fits, is inspired by the argument that inter-individual differences among organisms are reflected in slightly or even moderately different parameter profiles . . . .. The search for classes of solutions has also been supported in other scientific domains as diverse as simulations of climate change, . . . and models of gene regulatory networks . . . . . . and cell signaling pathways . . . ..." (G10: dissertation)

In this way the sloppiness properties of models explain the physical capability of biological systems to produce similar behavior despite variations between individuals in many cases. Ensemble results help illustrate this robustness (Brown et al. 2004). However these arguments, which are mainly motivated by pointing to the sloppy structure of models, do nothing to help ensure that the models being produced generate parameter ensembles that co-occur because they capture the underlying robustness of systems rather than because parameter-uncertainty facilitates many compensating solutions on any collective fitting task.

Issues like these raise definite concerns over the ability of systems biology to use computational and mathematical methods to overcome uncertainty and complexity, and achieve robustly predictive models. There are limitations to the degree to which computational and mathematical methods really can overcome parametric uncertainty (see also Lenhard and Winsberg 2010). If we take these kinds of concerns with uncertainty seriously, then there is a question mark over the purpose and value of computational modeling in any field. However it is possible to identify practices in systems biology that work productively within these limitations by applying computational modeling more pragmatically as a tool of investigation.

5 Systems Biology in an Uncertain World

While systems biology may have built its status and importance at least partially around prediction and control goals, systems biologists generally fashion more limited goals in the contexts of their own projects. In fact researchers frequently demonstrate in their unpublished work how to use novel computational techniques like ensemble methods, to extract useful information about a system out of a model despite parameter uncertainty. These kinds of insights provide a good basis for building a more refined perspective on what computational and mathematical methods actually can bring to the study of highly complex biological systems.

Indeed many modelers in the context of research papers promote the idea that the value of their work is not in the short term at least to provide high-validity models which can be used for predictive purposes. Instead their more explicit aim is to use computation to augment investigation of system structure. For example Kuepfer et al. (2007), in a study using ensemble methods to model the structure cell signaling systems, "stress that this study does not present a single 'correct' model", and should not be interpreted that way producing, "instead a family of improved models as a tool for further detailed elucidation of TOR signaling." (p. 1005). Wahl et al. (2006), when proposing their own method for testing model ensembles, describe their goals as a "systematic model-building process for data-driven exploratory modeling", that aims to discover "essential features of the biological system." (p. 283) Testing large numbers of models can help reveal "the most probable candidates for designing further experiments." (p. 283).

Pronouncements from these authors are invariably accompanied by statements that true models do not exist and that best-fit selection processes are unreliable or "intrinsically unstable". (Turkheimer et al. 2003, p. 490). Viewpoints like these see the role of models as a valuable part of an "explorative cooperation" between models and experimentation (see Lenhard 2007; MacLeod and Nersessian 2013b). Models help identify and weigh up the plausibility of different hypotheses, which can guide experimenters towards efficient routes of investigation. Such philosophical views have become embedded quite deeply in the way some labs are organized and operate in systems biology. For instance one of the labs our group studied is a fully equipped wet-lab. Modelers still come from engineering but most of them learn experimental skills over the course of their graduate studies and do their own experimentation (MacLeod and Nersessian 2014). Our analysis of their practices and objectives put them closer to this goal of explorative cooperation. Models are used to guide and augment experimental decision-making (MacLeod and Nersessian 2013b). For these purposes models do not need to achieve the standards of predictive robustness to be informative. What they need to do rather is help filter out plausible from implausible possibilities for making cost-effective decisions about likely areas in which current pictures of pathways break down and more experimental work is required. Systems biology has a definite contribution to make here because models of these complex systems can support inferences about system structure that would be impossible otherwise given system complexity. This way of operating shifts the focus from treating computation as a brute force way to construct high-validity representation, to treating it as a sophisticated way to identify, explore and assess hypotheses.

For example Kuepfer et al. (2007) in a paper titled "Ensemble modeling for analysis of cell signaling dynamics", motivate the need for ensemble modeling due to structural uncertainty. Using an example they illustrate how reasoning with ensembles can help resolve hypotheses about the structure of pathways. Their case was that of the TOR pathway in *S. cerevisiae* the regulatory mechanism of which is not completely known, although the elements involved in the pathway mostly are. They posed three ensemble groups (with 19, 13 and 13 models respectively). Each group represented an important possible difference in the overall mechanism of regulation. Within each group other variational possibilities were represented. Each of these were weighted according to how well they could be fit to the data and how well the fits performed in predictive tests. The different hypotheses about the regulatory mechanism which each ensemble represented were then be tested by comparing how these ensembles performed overall. As they carefully emphasize, their results should only be taken to single out hypotheses for further experimental testing. They claim to provide no single valid representation of the pathway.

Another example of using ensemble reasoning in such a manner comes from the work of G12, another researcher we tracked (see MacLeod 2016). One of G12's problems were data sets on specific variables which were inconsistent with one another, suggesting there was either a problem with her underlying pathway representation or a problem with one or more of her data sets. These variables represented elements of a signaling cascade, occurring in sequence upstream and downstream of one another. The signaling cascade was integrated with a much larger model of a regulatory metabolic system (called the Nox1 system). To debug this problem she built three different models, two of them truncated by starting the cascade at the elements downstream. For each model she combined the available data set on each element with data on the entire metabolic system. Given the amount of unknowns, G12 was in no position to find best fits of each model and then hope to test each model's performance. Instead G12 decided to examine each model's performance over a large parameter space, to see how many model candidates representing each alternative could replicate behavior well, and how robustly they could do so. This space of possible parameters was widely explored using Monte Carlo methods with 10,000 samplings of the parameter space for each model. Each alternative model was then tested against four (biologically plausible) conditions that captured how Nox1 regulation was expected to operate, as well as against two conditions ensuring accuracy with the available data on system operation. Over this parameter space only one of the model alternatives was able to produce fits which satisfied all these conditions. G12 concluded that only this alternative correctly captured at least that part of the signaling cascade it represented. She hypothesized in turn that to build a complete picture of the signaling would require adding unknown elements into the account of that cascade to capture its mechanism and bring the data into alignment. She posited one herself which gave good simulation results.

It is very important to note that G12 was left without a single model or best fit of the alternative that had good solutions. She had derived twelve parameter set candidates from the Monte Carlo sampling some differing with respect to some significant parameters. Her intention however was not to use these to somehow derive predictions of manipulations to the Nox1 system. In this respect it was not so important that she had potentially divergent solutions. What mattered was that her computational exploration of parameter space told her that only one of her representations of the signaling cascade was capable of producing a model that could fit the data and the performance conditions she imposed. The problem that her parameters were uncertain did not prevent her reaching her goals in this case. G12 contrived to draw her inferences based on massive computational work sampling large ensembles of alternatives, with very little substantial data and considerable uncertainty.

Uses of computational modeling like these concentrate on sorting through hypotheses about system structure and using computation to decide what is possible and what might not be. Other researchers also shy away from representing the value of computational and mathematical methods strictly in terms of the production of high-validity representations. They rationalize the current value of modeling rather as the generation of abstract schematic representations that can build understanding and focus a modeler's attention on the incremental steps needed to improve model performance (MacLeod 2016). Models produced this way have been called "mesoscopic" (Voit et al. 2012; Voit 2014). These models represent all individual biochemical interactions in a network using one tractable mathematical formulation, thus abstracting away much lower-level detail about each specific interaction. At the same time they only aim to approximate certain aspects of system function. Not all interactions within a system are represented, but only those thought most dominant for a particular set of functions. Hence a mesoscopic model is a very simplified, but very tractable representation. Such models Voit and his collaborators argue represent in fact the vast majority of models in current systems biology. Their value as they see it lies not in their high fidelity but in their ability, when constructed with the right mathematical frameworks, to facilitate a middle-out modeling process. As long as these models are partially accurate, modelers can treat them as heuristic platforms that allow them to understand how to develop these models to provide more mechanistically accurate and complete accounts of their systems. They can do this for instance by expanding the model internally, by taking apart interactions black-boxed in the network or by adding to the network more external components. Such reasoning processes are facilitated by the experimental platform a working model provides, allowing a modeler to manipulate variables and simulate different additions or changes to the network representation over large ensembles if required and thus explore effective ways to improve the model. This can be done, as in the G12 case, without needing to work with just one parameter set, but with a variety that can represent genuine uncertainty about parameter values.

In these cases the value of computational methods (and the underlying mathematical methods on which they depend) lies not with the ability of computation to produce high-validity representations. Rather it lies with its ability to help modelers draw out information from their models or classes of models about various hypotheses they have that shed light on the problems they are interested in addressing. I have called these elsewhere "heuristic uses" (MacLeod 2016). Computational models can help do this without needing to satisfy the rigorous constraints required for robust prediction in order to be successful, but simply by providing the power to examine large numbers of alternatives.

6 Analysis – The Tool Perspective

It should not be doubted that computational and mathematical methods are powerful tools for biological research the potential of which is just starting to be fully realized, but the ways in which they are effective tools is not so visible nor publicly understood both in the field and outside it. This paper has been an attempt to show that the common expectation that raw computation will allow us to crash through complexity and build high-validity predictively robust models is problematic and hampered by some substantial constraints, especially in the form of parameter uncertainty. While powerful methods like ensemble and sensitivity methods are continually being developed, the problem of limited data and parameter uncertainty cannot be easily resolved through computational power alone. There are deep questions to be answered about just how possible it might be to build models reliable for predictive purposes for complex nonlinear systems given that data uncertainty creates such large underdetermination problems. When scaled up to apply to complex systems through computation, mathematical modeling gets more slippery than we might have expected.

However when practices in the field are examined closely we find a variety of uses of computational simulation models for which the aim is not so much to fashion highly accurate representations, but rather to bring computational and mathematical power to bear on investigation by allowing modelers to derive and search through hypotheses about system structure in ways impossible without this power. These uses suggest that in order for philosophy of science to develop realistic images of what computational and mathematical methods can do in systems biology, and indeed elsewhere, we should concentrate on their roles as *tools of investigation* rather than as representational machines that can capture or emulate through their sophistication and power any level of complexity and detail. The faith in prediction and control as reasonable goals for systems biology, and throughout science in general, no doubt has gained some of its credence from beliefs in the power of computational and mathematical methods, and the belief that modern computational power alone can wash away many long-standing obstacles to building accurate models of complex systems. This attitude has put systems biology in particular in a somewhat awkward position, since despite almost 20 years of activity, models are not being produced to anything like the fidelity required for instance to capture disease processes (Voit 2014).

Recasting computational and mathematical methods in systems biology as primarily "investigative tools" is an important philosophical step in this respect, and wherever computation is being applied to generate leverage over complex systems with limited and incomplete data. Philosophers can help replace naïve and optimistic presuppositions about the reach and validity of computational simulation models, a problem policy makers are often confronted with (see for instance Petersen 2006), with more reasonable accounts of what computational models can do. Further we can provide computational researchers like systems biologist philosophical means to present and understand more accurately what the principal benefits of these methods are to themselves and to outsiders like experimental biologists.

However treating simulation models in this context as investigative tools has further importance. Part of the aim of contemporary philosophy of science is not just to deconstruct and analyze the epistemological or ontological elements underlying a theory or method, but to understand why that method or theory is chosen and designed the way it is given the context of practice in which it occurs. If we wish to explain or rationalize a chosen method or theory we should not neglect the fact that some of this rationalization depends on how well the method or theory accommodates local scientific practices composed of teams of human agents with particular expertise and technological resources, each bound by various ontological and epistemological commitments.

Support for this broader explanatory agenda in philosophy of science comes from Chang (2014). Scientific actions according to Chang occur within systems of practice, sets of interconnected epistemic activities (physical and mental operations) which are coherent with respect to given sets of purposes and constraints. Purposes are defined by the system, and the particular role a methodology is supposed to play within it. However a methodology should also be adapted to (designed for) both its purpose but also the capacities of its users. Indeed when a methodological strategy is described as an investigative tool it evokes investigative purposes as a central part of its rationale or design, but also the capacities of its users. Both factors should play a role when it comes to rationalizing methodological choices. Yet while philosophy of science is used to talking about whether particular theories or methods meet their purposes on the basis of epistemological principles, they are much less used to seeing where the user fits in. Chang (2014) for instance has argued that in our accounts of scientific practice the user or the human or epistemic agent is often neglected or characterized as just a vessel of beliefs, rather than a creature with practical and cognitive capabilities and limitations (p. 70). In the context of computation, it is somewhat natural to construe computation as an independent automatic process free of human limitations. Therein lies its essential power – the ability for instance to do numerical calculations at enormous speeds impossible for humans. But humans need to be able to understand what a computer is doing to some degree in order to for instance to debug problems and verify the operations are working correctly, but also arguably to interpret what the computer is finding. Machine learning techniques are often criticized as removing human control and insight from the problem-solving process. For example one of the researchers studied by our group when asked about machine learning labelled it a "blind process" of optimization insufficient for producing the understanding required to model systems effectively.

"So you gotta have insight, then there's a lot of–as [lab director] puts it, 'elbow greasing'...... Cause it's easy to fit everything in and say, this is how it works. But then if you really want to get the results afterwards, like have the model – let the model have the predictive power you want it to have, you gotta be sure about what you are doing."

"Being sure about what you are doing", requires the development of computational methods that meet the constraints of users enough to give them control and insight over how a computational process explores a problem and the information it reveals about the operation or organization of the system it represents. As Humphreys puts in most simulation model-building situations, "one cannot completely abstract from human cognitive abilities when dealing with representational and computational issues." (Humphreys 2011, p. 314).

Although we can only touch on these issues briefly, it is worth reflecting on how the concept of investigative tool invites deeper insights in the design of computational strategies in systems biology. Indeed when it comes to unpacking and understanding the uses of computational simulation models in systems biology, both the context and user are essential factors in their construction and operation. Firstly as tools *of investigation* they are integrated into systems of practice which are designed to expose or shed light on a problem. Carusi (2014) and Carusi et al. (2012) have argued that in systems biology the roles and uses of simulation and mathematical models are often constructed within fluid research contexts in which the constraints and structure of a problem are not well-understood. Biological variability and mismatches between the capabilities of experimentations and the demands of models, as well as different scales of biological organization, require strategies of computational simulation which are tailored to serve roles as parts of complex problem-solving strategies which shift back and forth between computational simulation, experimentation and other epistemic resources. These roles aim to bring forth information about a system to help simplify aspects of a problem and calibrate the relations between different epistemic resources (such as existing data sets, experimental design, existing models, statistical techniques). They may be used as we saw to test the consistency of data sets, to test parameter-fitting methodologies, optimize experimental procedures, or explore the consequences of an abstract model. Simulation models are constructed and applied narrowly and technically in these contexts, without the end of producing high fidelity predictions. In one case Carusi documents the use of model ensembles to establish "comparability between the variability in the population of models, and that in the experimental data set", through repeated process of simulation, parameter fitting and experimentation (Carusi 2014, p. 33).

The uses of computational strategies we documented above all have the feature of being designed for just particular investigative functions which contribute to larger problem solving strategies. Indeed researchers like Voit conceive of mesoscopic modeling in just this way as a step in an ongoing process, not as a final outcome. For philosophy of science then it is important when understanding and evaluating computational models to be aware that their uses and value may be defined by research contexts and problem-solving processes, and the various constraints, affordances and uncertainties of these. These uses may be technical and narrow, designed to serve particular investigative functions. These functions may fall well short of representing phenomena with high fidelity, and cannot properly be rationalized with respect to such a goal.

Secondly, computational strategies in systems biology are designed with the cognitive constraints and capacities of the user in mind. The uses cited above by G12 and those advocating mesoscopic modeling are designed to facilitate cognitively manageable processes of investigation and discovery. Voit and his collaborators advocate mesoscopic modeling precisely because it facilities what they call "hierarchical learning." We learn best by beginning from simplified representations and building in complexity in a step-wise fashion. Mesoscopic models are simplified enough that modelers can learn the dynamic relations within the model and in turn recognize or predict pathways to improvement. In this respect they are tools optimized for their users for developing better representations over long courses of investigation by negotiating the tightrope between uninformative abstraction and uninterpretable complexity. The goal is to work towards high fidelity representations of systems, but only in the long-run.

Both these aspects of tool-use and tool-choice or design can be studied further in the context of systems biology and elsewhere. These are only the briefest of hints of what this kind of analysis might tell us. The importance of the tool conception is that it helps identify the roles that research contexts and users play in methodological design and decision-making. While I do not want to suggest that all uses of computational models in science should be analyzed as investigative tools, where it is appropriate it promises a rich and more informative framework for analyzing computational methods and understanding their function and design.

Acknowledgments The research for this paper was supported by an US National Science Foundation (DRL097394084), as well as by a Postdoctoral Fellowship at this Academy of Finland Centre of Excellence in the Philosophy of the Social Sciences and a position at the University of Twente. I would like to thank the editors of the volume in particular for their helpful advice in the development of this paper.

References

Apgar, J. F., Witmer, D. K., White, F. M., & Tidor, B. (2010). Sloppy models, parameter uncertainty, and the role of experimental design. *Molecular BioSystems, 6*(10), 1890–1900.

Brown, K. S., Hill, C. C., Calero, G. A., Myers, C. R., Lee, K. H., Sethna, J. P., & Cerione, R. A. (2004). The statistical mechanics of complex signaling networks: nerve growth factor signaling. *Physical Biology, 1*(3), 184.

Carusi, A. (2014). Validation and variability: Dual challenges on the path from systems biology to systems medicine. *Studies in History and Philosophy of Science Part C: Studies in History and Philosophy of Biological and Biomedical Sciences, 48*, 28–37.

Carusi, A., Burrage, K., & Rodríguez, B. (2012). Bridging experiments, models and simulations: An integrative approach to validation in computational cardiac electrophysiology. *American Journal of Physiology-Heart and Circulatory Physiology, 303*(2), H144–H155.

Chang, H. (2014). Epistemic activities and systems of practice: Units of analysis in philosophy of science after the practice turn. In L. Soler, S. Zwart, M. Lynch, & V. Israel-Jost (Eds.), *Science after the practice turn in the philosophy, history, and social studies of science* (p. 67). New York: Routledge.

Gutenkunst, R. N., Waterfall, J. J., Casey, F. P., Brown, K. S., Myers, C. R., & Sethna, J. P. (2007). Universally sloppy parameter sensitivities in systems biology models. *PLoS Computational Biology, 3*(10), e189.

Hood, L., Heath, J. R., Phelps, M. E., & Lin, B. (2004). Systems biology and new technologies enable predictive and preventative medicine. *Science, 306*(5696), 640–643.

Humphreys, P. (2011). Computational science and its effects. In M. Carrier, & A. Nordmann (Eds.), *Science in the context of application: Methodological change, conceptual transformation, cultural reorientation* (pp. 131–142). Springer.

Ideker, T., Galitski, T., & Hood, L. (2001). A new approach to decoding life: Systems biology. *Annual Review of Genomics and Human Genetics, 2*(1), 343–372.

Kitano, H. (2002). Looking beyond the details: A rise in system-oriented approaches in genetics and molecular biology. *Current genetics, 41*(1), 1–10.

Kuepfer, L., Peter, M., Sauer, U., & Stelling, J. (2007). Ensemble modeling for analysis of cell signaling dynamics. *Nature Biotechnology, 25*(9), 1001–1006.

Lenhard, J. (2007). Computer simulation: The cooperation between experimenting and modeling. *Philosophy of Science, 74*(2), 176–194.
Lenhard, J., & Winsberg, E. (2010). Holism, entrenchment, and the future of climate model pluralism. *Studies in History and Philosophy of Science Part B: Studies in History and Philosophy of Modern Physics, 41*(3), 253–262.
MacLeod, M. (2016). Heuristic approaches to models and modeling in systems biology. *Biology and Philosophy, 31*(3), 353–372.
MacLeod, M., & Nersessian, N. J. (2013a). Building simulations from the ground up: Modeling and theory in systems biology. *Philosophy of Science, 80*(4), 533–556.
MacLeod, M., & Nersessian, N. J. (2013b). Coupling simulation and experiment: The bimodal strategy in integrative systems biology. *Studies in History and Philosophy of Science Part C: Studies in History and Philosophy of Biological and Biomedical Sciences, 44*(4), 572–584.
MacLeod, M., & Nersessian, N. J. (2014). Strategies for coordinating experimentation and modeling in integrative systems biology. *Journal of Experimental Zoology Part B: Molecular and Developmental Evolution, 322*(4), 230–239.
MacLeod, M., & Nersessian, N. J. (2015). Modeling systems-level dynamics: Understanding without mechanistic explanation in integrative systems biology. *Studies in History and Philosophy of Science Part C: Studies in History and Philosophy of Biological and Biomedical Sciences, 49*, 1–11.
Marder, E., & Taylor, A. L. (2011). Multiple models to capture the variability in biological neurons and networks. *Nature Neuroscience, 14*(2), 133–138.
Parker, W. S. (2010a). Whose probabilities? Predicting climate change with ensembles of models. *Philosophy of Science, 77*(5), 985–997.
Parker, W. S. (2010b). Predicting weather and climate: Uncertainty, ensembles and probability. *Studies in History and Philosophy of Science Part B: Studies in History and Philosophy of Modern Physics, 41*(3), 263–272.
Petersen, A. C. (2006). Simulation uncertainty and the challenge of postnormal science. In J. Lenhard, G. Küppers, & T. Shinn (Eds.), *Simulation: Pragmatic constructions of reality – Sociology of the sciences* (pp. 173–185). Springer: Dordrecht.
Savageau, M. A. (1969). Biochemical systems analysis: I. Some mathematical properties of the rate law for the component enzymatic reactions. *Journal of Theoretical Biology, 25*(3), 365–369.
Tran, L. M., Rizk, M. L., & Liao, J. C. (2008). Ensemble modeling of metabolic networks. *Biophysical Journal, 95*(12), 5606–5617.
Turkheimer, F. E., Hinz, R., & Cunningham, V. J. (2003). On the undecidability among kinetic models: from model selection to model averaging. *Journal of Cerebral Blood Flow & Metabolism, 23*(4), 490–498.
Voit, E. O. (2000). *Computational analysis of biochemical systems: A practical guide for biochemists and molecular biologists*. Cambridge: Cambridge University Press.
Voit, E. O. (2014). Mesoscopic modeling as a starting point for computational analyses of cystic fibrosis as a systemic disease. *Biochimica et Biophysica Acta (BBA)-Proteins and Proteomics, 1844*(1), 258–270.
Voit, E. O., Qi, Z., & Kikuchi, S. (2012). Mesoscopic models of neurotransmission as intermediates between disease simulators and tools for discovering design principles. *Pharmacopsychiatry, 45*(S 01), S22–S30.
Wahl, S. A., Haunschild, M. D., Oldiges, M., & Wiechert, W. (2006). Unravelling the regulatory structure of biochemical networks using stimulus response experiments and large-scale model selection. *IEE Proceedings-Systems Biology, 153*(4), 275–285.
Westerhoff, H. V., & Kell, D. B. (2007). The methodologies of systems biology. In F. Boogerd, F. J. Bruggeman, J.-H. S. Hofmeyer, & H. V. Westerhoff (Eds.), *Systems biology: Philosophical foundations* (pp. 23–70). Amsterdam: Elsevier.

The Vindication of Computer Simulations

Nicolas Fillion

1 Introduction

The relatively recent increase in prominence of computer simulations in scientific inquiry gives us more reasons than ever before for asserting that mathematics is a wonderful tool to acquire knowledge of the world. In fact, a practical knowledge (a 'know-how') of scientific computing has become essential for scientists working in all disciplines involving mathematics. Indeed, a scientist can only reap the benefits of the use of simulation with a practical understanding of the basic numerical algorithms used to implement computational models in computer codes, such as the basic methods to evaluate polynomials and truncated series, to find zeros of functions, to solve systems of linear equations, to solve eigenvalue problems, to interpolate data, to differentiate and integrate functions on small domains, and to solve ordinary differential equations. Understanding even a small, select group of algorithms to accomplish such tasks numerically can lead to astounding results, that have seemed "unreasonably effective" to even accomplished numerical analysts:

> My first real experience in the use of mathematics to predict things in the real world was in connection with the design of atomic bombs during the Second World War. How was it that the numbers we so patiently computed on the primitive relay computers agreed so well with what happened on the first test shot at Almagordo? There were, and could be, no small-scale experiments to check the computations directly. [...] this was not an isolated phenomenon—constantly what we predict from the manipulation of mathematical symbols is realized in the real world. (Hamming 1980, p. 82)

The successes of mathematics are numerous and impressive but, especially when it comes to computer simulations, we should not succumb to an "optimization of

N. Fillion (✉)
Department of Philosophy, Simon Fraser University, 4604 Diamond Building, 8888 University Drive, V5A 1S6, Burnaby, BC, Canada
e-mail: nfillion@sfu.ca

J. Lenhard, M. Carrier (eds.), *Mathematics as a Tool*, Boston Studies in the Philosophy and History of Science 327, DOI 10.1007/978-3-319-54469-4_8

enthusiasm and misjudgment" (Kadanoff 2004). Indeed, any good discussion of the astounding success of mathematics in science is not unmitigated, and an examination of its less successful or failed employments is a fruitful avenue for dissipating the "unreasonability" of the successes.[1] Indeed, despite their incontestable success, it must be emphasized that the numerical methods underlying simulations provide at best approximate solutions and that they can also return very misleading results. Moreover, an overconfidence in simulation and computational methods can have dramatic effects. The delicate task of designing simulation models must thus go hand-in-hand with an ever more delicate task of analysis of the simulation results.

Accordingly, epistemological sobriety demands that we clarify the circumstances under which the simulations can be relied upon. Even if many scientists and philosophers of science are now familiar with some of the basic computational methods I have referred to above, it remains an incontestable fact that the circumstances underlying the success and failure of simulations is much less well understood. After clarifying the nature of the problem in the next section, I will examine many ways in which this question can be approached, including the appeal to data and physical intuition, the use of benchmark problems, and a variety of a priori and a posteriori methods of error analysis. This will lead to a realization that a sound perspective on the justification of results obtained by computer simulation rests on a sound balance between qualitative asymptotic analysis and quantitative a posteriori measurement of error.

2 Justifying Computer Simulations

Since the United States' Department of Energy and the Department of Defence adopted the terminology in the 1990s, it has become standard to distinguish two aspects of the justification of computer simulations known as *verification* and *validation*.[2] There are minor variations in the definition of those two justificatory processes in various publications, but the core idea remains by and large the same. Oberkampf et al. (2004) define verification as the "assessment of the accuracy of the solution to a computational model" and validation as the "assessment of the accuracy of a computational simulation by comparison with experimental data." As they explain, in verification "the relationship of the simulation to the real world is not an issue," although in validation "the relationship between computation and the real world, i.e., experimental data, *is* the issue" (p. 346). They acknowledge, however, that "the state of the art has not developed to the place where one can clearly point out all of the methods, procedures, and process steps that *must* be

[1]For more details on the relation between the unreasonable effectiveness of mathematics and computational issues, see Fillion (2012).

[2]Note however that despite the broad use of this terminology in computational mechanics, it remains comparatively rare in the field of numerical analysis.

undertaken for V&V." In what follows, I will discuss a number of options that are at one's disposal in thinking about V&V of simulations. In particular, I will focus on the stage of verification since it has received little attention in the philosophical literature. I will adopt a numerical analysis point of view to examine verification in a way that can enlighten aspects of the philosophy of computer simulations.

Some philosophers have made the apt suggestion that the justification of simulations can be thought of exactly as we think about the justification of inferential patterns more generally (e.g., Wilson 2006). At least, it seems apt to the extent that one focuses on the process of verification as defined above, whereas validation is more plausibly understood as a judgement about the truth or accuracy of a statement. Let us explore the basic implications of treating simulations as inferences.

The fundamental epistemological function of good arguments is that they give us the knowledge that something follows from a set of premises, or at least that it is made more likely by the premises. In deductive logic, we have grown accustomed to showing that an argument is good by indicating that it is an instance of an argument form that can be shown to be valid by means of a deduction that relies on primitive rules of inference or by a soundness proof. This procedure must of course be altered significantly to analyze non-deductive arguments. Nonetheless, in both cases, we are typically given a set of premises and a conclusion, and the task consists in justifying the inferential step. However, in the context of simulations, we are not given the conclusion because the mathematical models we seek to analyze typically fail to be inferentially transparent. To illustrate this point, suppose that we are given an initial-value problem

$$\mathbf{x}'(t) = \mathbf{f}(\mathbf{x}, t) \qquad \mathbf{x}(t_0) = \mathbf{x}_0$$

for which no elementary or convenient closed-form solution is known on a given time interval $[t_i, t_f]$.[3] In such a case, we hope that our use of a numerical method will give us two things:

1. a computed solution $\hat{x}(t)$ (the conclusion we seek);
2. some guarantee that $\hat{x}(t)$ approximates $x(t)$ to some degree of accuracy (provided $x(t)$ exists and is unique; if it doesn't, we would appreciate that the method point it out to us with an error flag).

Thus, in this case, the *discovery* of the conclusion is often obtained with a method that also provides the *justification* of the inferential step. In a good-case scenario, it is plausible that the context of discovery and the context of justification be handled all at once. As we will see, the situation is rarely that simple, and yet it must be acknowledged that the discovery of a conclusion and the justification of the inferential step are closely intertwined when it comes to simulations.

[3]For an excellent discussion of such complications, see Borwein and Crandall (2010). For a more philosophical discussion, see Fillion and Bangu (2015).

This difference being understood, Reichenbach's (1949) famous distinction[4] between two types of justification will help us to clarify the respects in which it is appropriate to discuss various approaches to justifying the results of computer simulations (and, in particular, its verification step). He introduced the term "vindication" to describe the process of justification of a rule, as opposed to the justification of a factual proposition by showing that it is true or probable. Interestingly, Feigl (1950) introduced the term 'validation' for the second type of justification, which echoes the V&V terminology. Since vindication consists in justifying not a proposition but instead an inferential practice, Reichenbach considered this type of justification to be fundamentally a pragmatic affair. For instance, a rule of deductive inference would be justified pragmatically by means of an argument showing that it leads us to true conclusions, provided that the premises are themselves true. Similarly, his distinction was introduced as part of an argument meant to show that using a certain rule of induction would lead to certain benefits characterized in terms of predictive accuracy. If it is correct to treat particular simulations as inference and numerical methods as inferential patterns, then it appears that discussion of their justification would also take the form of a pragmatic vindication.

Even if the point will be substantiated later, I also want to point out that the pragmatic nature of inferences arising from simulations is more thorough than what relates to Reichenbach's problem. Indeed, in both cases the justification is pragmatic in that the justification is that of a practice and not of a proposition. However, in the case of simulations, whether a particular inference is to be deemed successful is also a pragmatic affair; different instances of a pattern might be assessed differently depending on various contextual elements, and this sets simulations apart from deductive inferences. In fact, the very criteria used to assess the quality of computed solutions have a context-sensitivity to the modelling situation that adds an unavoidable pragmatic dimension to the vindication of inferences resulting from simulations.

3 Vindication by Data and Physical Intuition

The status of the justification being clarified, let us now turn to various approaches employed to vindicate simulations. A very natural proposal is to simply appeal to 'physical (or biological, chemical, financial, etc.) intuition' or to compare the simulated results with data; these are two senses in which simulations are assessed on the basis of empirical information. The latter approach suggests that it is beneficial to think about simulations in analogy to experiments, since they are justified in similar ways. I will argue that to the extent that verification is concerned, this is incorrect. Moreover, to the extent that validation is concerned, the experimental justification for the model equations has very little to do with the

[4]See also, e.g., Feigl (1950) and Salmon (1991).

epistemology of computer simulations per se. However, in addition to V&V, I will identify a third stage of model development and analysis for which the comparison of simulations to experiments is beneficial. The other approach will instead assess the correctness of an inferential pattern based on its agreement with physical intuition. Whereas the first approach sheds little light on the problem of vindication I have outlined above, the second does shed some light, although in a limited way.

3.1 Simulation as Experiments

To begin, a common idea in the philosophical literature is to suppose that the results of computer simulations could be justified on the basis of various methods that have traditionally been central to confirmation theory. Thus, for the same reasons that the scientific method prescribes that we test whether hypotheses, predictions, theories, etc., are true based on comparison with experimentally acquired data, we should assess the justification of the results of simulations based on a comparison with data. However, even if this is something that scientists certainly do, it cannot be the whole story. Some have emphasized that computer simulations are particularly important when data are sparse (e.g., Humphreys 2004; Winsberg 2010). But in such cases simulations can obviously not be vindicated by a comparison with data. A particularly compelling example is proposed by Winsberg: If we simulate convective flow within a star, surely we will not be able to physically probe the convective process, so the result of the simulation must be sanctioned independently. Another more earthly example was noted above in the quote from Hamming.

This suggests that we should examine the justifiability of simulation results in a different way. However, even if the particular mathematical descriptions that result from simulations cannot always be assessed by direct comparison with data, there is another sense in which simulations (as inferences) can be justified in a way that is analogical to experiments. This is in fact what Winsberg seems to have recently proposed. He suggests that computational models are *self-vindicating*, in a way analogous to the claim by Hacking (1992) that experiments are self-vindicating:

> Whenever these [simulation] techniques and assumptions are employed successfully [...]; whenever they make successful predictions or produce engineering accomplishments—their credibility as reliable techniques or reasonable assumptions grows. (Winsberg 2010, p. 122)

It is certainly the case that the justification invoked for the results of simulation is sometimes analogous to experiments as explained. However, I maintain that when a simulation is successful in one kind of applications, it says very little about whether they will be successful in other applications as well.

If we focus on the vindication of computer simulations in the sense explained in the previous section, it must be acknowledged that past successes in themselves are insufficiently projectble to provide the sort of inferential justification we seek. The results of a (collection of) simulation(s) may be sufficiently accurate to lead to strikingly accurate predictions and to excellent engineering accomplishments, *even when* the computational method used to solve a particular set of model equations is very poor. Indeed, if the equations are extremely stable (as is the case for the so-called "stiff" problems, for instance) under all kinds of perturbations—whether they are physical or due to numerical error—huge amounts of computational error will be simply dampened and the computed results will still be accurate. However, using the same method for a moderately (un)stable problem would lead to unacceptable results.

The problem comes from the fact that an argument based on success in applications does not allow one to infer that the very same computational method would pass the verification test in another application. It is possible to say that the computational inference accomplished by the simulation is vindicated for the set of applications for which it is known to be successful, since it certainly satisfies our pragmatic objectives. Nevertheless, to the extent that Winsberg's approach is about verification—and I must admit that it is not entirely clear to me to which extent this is the case even if he uses this terminology—it aims to justify not only particular instances of the more general inferential pattern, but the inferential pattern itself. If it is to guarantee success across applications, the vindication of an inferential pattern must be established independently, by examining the *mathematical* circumstances in which the computational methods used can be expected to be equally successful. This, however, relies on rigorous mathematical error analysis, and not on past successes. I will have more to say about this sort of analysis in Sect. 4.

But before we move to this topic, there is another aspect of the justification of the results of computer simulations that is by-and-large independent of validation and verification, namely, *calibration*. The step of validation often focusses on the justification of the use of certain equations as schemata for the construction of our model. For a collection of intended applications, we would for example show that it is reasonable to treat a certain fluid flow as incompressible, as having no drag, as being inviscid, etc. This is typically done with perturbation methods that characterize the local asymptotic behaviour of a fluid satisfying a certain set of idealizing conditions (see, e.g., Batterman 2002a,b, 2009). To justify his use of idealizations in physics along such lines, Galileo (1687, p. 67) claimed that he was following the example of Archimedes who made the same false assumptions "perhaps to show that he was so far ahead of others that he could draw true conclusions even from false assumptions." In fact, perturbation theory could be very aptly described as a methodology to draw true (or accurate) conclusions from (sometimes unabashedly) false assumptions. As a method of rational justification, this sets it apart from logic and probability.

However, this step concerns predominantly the justification for representational schemata, and determining the actual value of many of the parameters still has to be done. Determining such parameters is what I am referring to as 'calibration.' Calibrating a model typically requires many rounds of simulation with updating based on previous successes (Oberkampf et al. 2004), and there are many similarities between experiments and calibration (see, e.g., Tal 2011). Nonetheless, it is typically the case that a certain success in validation and verification must be achieved before a model can be successfully calibrated.

So, to summarize, of the three aspects of the justification of a computer simulation, i.e., validation, verification, and calibration, the latter is the most similar to experiments, and verification is the least similar. However, to the extent that 'vindication' is used to justify an inference pattern (in a pragmatic way), the only aspect of the justification that the term 'vindication' applies to is verification. As a result, there are important limitations to the claim that computer simulations are self-vindicating in a way that is analogous to experiments. This point will be further exemplified with the case of numerical integration in Sect. 4.2.

3.2 *Disagreement with Physical Intuition*

Experience is sometimes used in a different way to assess the quality of numerical solutions. Instead of comparing the dynamical behaviour predicted by the simulation of the calibrated model to actual data, scientists often reject solutions on account of a disagreement with "physical intuition." The notion of physical intuition is often readily admitted to not be prima facie a very rigorous one. In such cases, intuition is often associated with an understanding, based on experience, of the more general qualitative dynamical features of the system's behaviour. Typically, this approach will thus be used as a way of discarding results rather than as a way of vindicating them. Let us examine an example of this mode of vindication, namely, the solution of Duffing's equation.

Duffing's equation characterizes a weakly nonlinear unforced oscillator. The situation is similar to that of a simple harmonic oscillator in which Newton's second law is combined with a Hookean linear restorative force $F = -kx$, except that there is a cubic component εx^3 added to the restorative force (Bender and Orszag 1978). So, Duffing's equation is

$$x'' + x + \varepsilon x^3 = 0\,, \tag{1}$$

with initial conditions $x(0) = 1$ and $x'(0) = 0$ and it is assumed that $0 < \varepsilon \ll 1$.

A natural way to solve this equation would be to use series methods and to use the first few terms of the asymptotic expansion for the sake of numerical computation. The classical perturbation analysis supposes that the solution to this equation can be written as the power series

$$x(t) = x_0(t) + x_1(t)\varepsilon + x_2(t)\varepsilon^2 + x_3(t)\varepsilon^3 + \cdots\,. \tag{2}$$

Substituting this series in Equation (1) and solving the equations obtained by equating the coefficients of matching powers, we find $x_0(t)$ and $x_1(t)$ and we thus have the first-order solution

$$\hat{x}(t) = \cos(t) + \varepsilon\left(\frac{1}{32}\cos(3t) - \frac{1}{32}\cos(t) - \frac{3}{8}t\sin(t)\right)\,. \tag{3}$$

The problem with this solution is that it contains a term (namely, $t \sin t$) that approaches ∞ as $t \to \infty$; such a term is known as a *secular* term. The fact that $\hat{x}(t)$ contains a secular term conflicts with physical intuition. More precisely, the presence of this secular term shows that our simple perturbative method has failed since the energy conservation prohibits unbounded solutions.

A slightly different argument used to discard this solution is that, mathematically, the secular term $t \sin t$ implies that our method has failed since the periodicity of the solution contradicts the existence of secular terms. Either ways, we assume some knowledge of the qualitative character of the system, i.e., the global boundedness or the periodicity of the solution. Many working scientists base these assessments directly on physical intuition, although they may also be based on mathematical theorems. So, this characterization is correct, but it requires foreknowledge of what is physically meaningful or of whether the solutions are bounded. This is not a categorical objection to this approach, since it is an incentive to better understand qualitative features of the systems studied. However, in practice, it is often better to rely on a direct assessment of the error resulting from a given computational method.

4 Vindication by Error Analysis

The discussion above suggests a number of reasons for which numerical analysts do not rely on the self-vindicating virtues of simulation techniques on the basis of empirical information, but rather on rigorous error analyses of computational methods founded on one of the most important branches of applied mathematics, namely, perturbation theory. Accordingly, it is customary to accompany the design of a numerical algorithm and the decision to adopt it for the simulation of the behaviour of a system with theorems bearing on the convergence of the method for the relevant parameters, on the numerical stability of the implementation of this algorithm in computer arithmetic, and on the sensitivity of the model equations to perturbations of arbitrary nature.

There are to my knowledge only three ways of measuring computational error.[5] Mathematical problems can be thought of as maps from a set of input data to a set of output data, i.e., as $\varphi : \mathscr{I} \to \mathscr{O}$. The mapping itself will typically be given as

$$x \xrightarrow{\varphi} \{y \mid \phi(x, y) = 0\}\,, \tag{4}$$

[5]For a similar claim, see Grcar (2011). For a more extensive explication of the notions introduced below, see Corless and Fillion (2013). Higham (2002) and Deuflhard and Hohmann (2003) are also excellent alternative presentations.

where x is the input data, y is the solution sought, ϕ is the defining function, and $\phi(x, y) = 0$ is the defining equation (or condition). Here are two examples of how this notation works:

1. You want to solve a system of linear equations $\mathbf{Ax} = \mathbf{b}$ for the unknown vector $\mathbf{x}$. The input data is the matrix $\mathbf{A}$ and the vector $\mathbf{b}$, and the output is the vector $\mathbf{x}$:

$$\langle \mathbf{A}, \mathbf{b} \rangle \xrightarrow{\varphi} \{\mathbf{x} \mid \mathbf{Ax} = \mathbf{b}\}$$

2. You want to solve a system of differential equations $\mathbf{x}'(t) = \mathbf{f}(\mathbf{x}, t)$ with initial value $\mathbf{x}(t_0) = \mathbf{x}_0$ for the unknown $\mathbf{x}(t)$. The input data is the differential vector field $\mathbf{f}$ and the initial condition $\mathbf{x}_0$, and the output is the solution vector $\mathbf{x}(t)$: $\mathbf{x}(t)$:

$$\langle \mathbf{f}, \mathbf{x}_0 \rangle \xrightarrow{\varphi} \{\mathbf{x}(t) \mid \mathbf{x}'(t) = \mathbf{f}(\mathbf{x}, t)\}$$

Even if it may seem a little pedantic, this notation makes clear the three possible ways of measuring error, as we will see shortly.

Now, when we use a computer for computations, we *engineer* the problem so that it can be implemented and efficiently executed on a digital computer. Thus, in our first example, instead of actually computing $\mathbf{x} = \mathbf{A}^{-1}\mathbf{b}$, we would solve a modified problem $\hat{\varphi}$ that we could define as follows:

$$\langle \mathbf{A}, \mathbf{b} \rangle \xrightarrow{\hat{\varphi}} \left\{ \hat{\mathbf{x}} \,\middle|\, \begin{array}{c} \hat{\mathbf{x}} \text{ is the result of Gaussian elimination without} \\ \text{pivoting in double-precision floating point arithmetic} \\ \text{for the floating-point representation of } \mathbf{A} \text{ and } \mathbf{b} \end{array} \right\}$$

Essentially, this engineering process transforms an exact algebraic problem involving real or complex numbers into a computationally simpler series of finitary operations of limited precision but greater tractability. Notice that $\mathbf{x}$ and $\hat{\mathbf{x}}$ will only exceptionally be identical. In general, for a problem such as the one in Equation (4), the difference between the exact solution y and the approximate solution $\hat{y}$ will be denoted Δy and will be called the *forward error*. The concept of forward error is typically the one that is referred to when people talk about error; when they refer to a computed solution as being "accurate" or "approximately true," one means that the forward error is small. This can be represented in a diagram as follows:

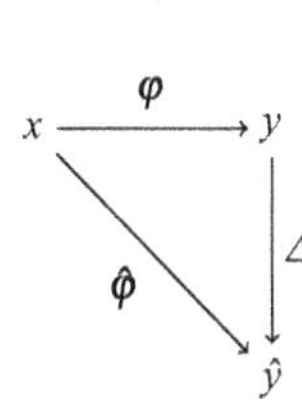

Even if the forward error is the most common measure of error, in many contexts this is by no means the most insightful, and so there are two other ways of measuring error. When we compute $\hat{y} = \hat{\varphi}(x)$, we may ask: when I computed $\hat{y}$, for what data $x + \Delta x$ have I actually solved the problem φ? Again, a convenient graphic representation is as follows:

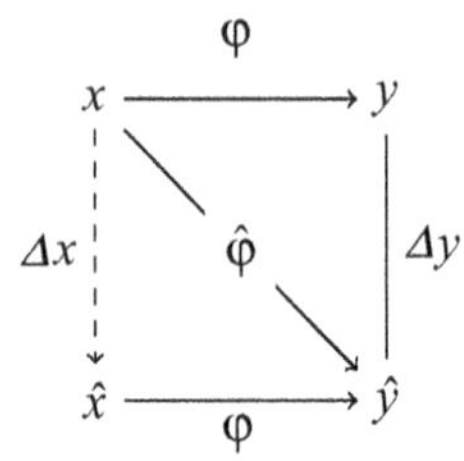

Δx is known as the *backward error*. In effect, we are asking: when I modified the problem φ to get $\hat{\varphi}$ and a typically unknown error Δy resulted, to what perturbation of the input data was this equivalent? If the backward error is smaller than the modelling error and the experimental error that we are aware of, then for all we know, $\hat{y}$ could be exactly representing the target system. This approach allows us to determine not whether the solution is "approximately true" but rather whether it is *true to our assumptions*, i.e., compatible with what we know to be the case. This measure of error is advantageous, since it is directly interpretable in the modelling context.[6]

Here is an example. Consider again the solution of a system of linear equation $\mathbf{Ax} = \mathbf{b}$ using Gaussian elimination without partial pivoting, and suppose that it leads to a computed vector $\hat{\mathbf{x}}$. In a landmark result that illustrated the power of backward error analysis, Wilkinson (1963) showed that there exists a matrix $\mathbf{E}$—an "error" matrix—with "relatively small" entries such that

$$(\mathbf{A} + \mathbf{E})\hat{\mathbf{x}} = \mathbf{b} \, .$$

That is, the method exactly solved a slightly different problem. As a result, instead of saying that we have *approximately* solved $\mathbf{Ax} = \mathbf{b}$, we say that we have *exactly* solved the *modified problem* $(\mathbf{A} + \mathbf{E})\mathbf{x} = \mathbf{b}$. Here, you could interpret the entries of $\mathbf{E}$ either as a computational error or as a physical error, so the computational error is tied back to modelling error. If $|e_{ij}| < 10^{-8}$, say, and the a_{ij} were measured with 4 significant digits, then for all you know, *you might have found the exact solution for the actual system*! Clearly, in this case, there is a context-dependence to the modelling context that adds an additional pragmatic dimension to the vindication of simulations by error quantification. That is, in addition to recognizing that the vindication of such a computation is pragmatic in the sense that it is goal-directed, it is essential to emphasize that the modelling context plays an essential role in

[6]For more details on this, see Fillion and Corless (2014).

determining whether the goal is achieved. As such, the vindication of computer simulations importantly differs from the vindication of other kinds of inference (e.g., deductively valid inferences).

A third measure of error is known as the *residual*. I will have more to say about it toward the end. For now, let it suffice to say that when we have a mathematical of the form $\varphi : x \rightarrow \{y \mid \phi(x, y)\}$, but such that y cannot be obtained in any straightforward way, we can instead use an approximate method that will compute some "solution" $\hat{y}$. Using the residual, we do not need to antecedently justify the use of our method; rather, after the fact, we will now compute

$$r = \phi(x, \hat{y}) ,$$

i.e., the quantity that results from replacing the exact solution y by another value $\hat{y}$ in the defining function. The quantity r is what we call the residual. The smaller the residual, the closer $\hat{y}$ is to actually satisfying the defining equation of the problem. In the case of our system of linear equations, the defining equation is just $\mathbf{Ax} - \mathbf{b} = 0$, and so the residual is the vector

$$\mathbf{r} = \mathbf{A}\hat{\mathbf{x}} - \mathbf{b} .$$

In addition to the three measures of error introduce above, it is important to draw a distinction between two *modes* of error analysis: a priori and a posteriori error analysis. An a priori error analysis consists in finding bounds on the maximum error that can result from using a certain method by examining the details of the algorithm and of its implementation. Thus, we can estimate the maximum error *prior* to the actual computation or simulation. Forward error analysis and backward error analysis are usually understood to be of this kind, but as we will see in our first subsection, there are more or less satisfactory ways to do a posteriori forward error analysis. In a posteriori error analysis, we do not attempt to estimate the error beforehand. Instead, we use a readily available method to compute something, and then after the fact we use the something in question and attempt to determine how close to the exact solution it is. The most natural measure of error for a posteriori analysis is the residual.

4.1 Direct Assessments of the Forward Error

A useful method to assess the reliability of computational methods in terms of their forward error is to simply compare the results it provides to carefully selected problems for which the exact solution is known. Such problems are known as *benchmark* problems. This approach is very widespread. In fact, Oberkampf et al. (2004) claim that verification is done "primarily by comparison with known solutions." Nonetheless, this approach to verification has significant weaknesses. Decades ago, Clifford Truesdell had already identified the problem:

> [...] benchmark problems are simple. The fact that a code is accurate for a known and simple instance then breeds confidence, merely emotional, that it will be accurate also for hitherto unsolved problems. The risk of such inference is plain and great. The true solution of the unsolved problem need not be even roughly like the solution of the simple, classical instance. The unsolved problem may well be unsolved just because its solution is in essence different, far more complicated and far more delicate. (Truesdell 1980, p. 598)

Comparison with benchmark solutions may sometimes provide a gain in confidence that is more than emotional, but only to the extent that it is juxtaposed with a more systematic type of analysis.

What such a more systematic analysis must look like, however, is often misunderstood by working scientists. Let me use an example in which a naive approach was used in an attempt to vindicate a method to interpolate data and subsequently solve a quadrature problem in medical sciences. The data in question is a collection of measured glucose levels in patients afflicted with diabetes; the area under the curve that interpolate those data points is used to compute metabolic indices that help clinicians in various tasks, e.g., to determine treatment dosage.

Tai (1994) introduce a formula humbly called "Tai's mathematical model" to find such areas under curves. For a collection of points (x_i, y_i), $i = 1, 2, \ldots, n$, the proposed formula is as follows:

$$\text{Area} = \frac{1}{2}\sum_{i=1}^{n}(x_i - x_{i+1})(y_{i+1} + y_i)$$

Tai claims that "other formulas tend to under- or overestimate the total area under a metabolic curve by a large margin." Moreover, it is claimed that "Tai's model was developed to correct the deficiency of under- or overestimation of the total area under a metabolic curve."

Strikingly, neither Tai nor the reviewers seem to have noticed that this formula is just the venerable trapezoidal rule for quadrature, already known in Newton's time. I could only verify one of the alternative formulas discussed by Tai, the one proposed by Wolever et al. (1991), and it is just the Midpoint Rule, which was known even earlier. As Corless remarks, it "might seem humorous that someone could think reinventing a method that is at least hundreds of years old was publishable, but it seems less funny when you realize that she was correct, and count the number of citations" (Corless and Fillion 2013, p. 421).

This being said, the question remains: can the formula in question do what is claimed? As I will explain below, there is a rigorous asymptotic type of a priori error analysis that can be used to settle the issue. But instead, the author used a different, a posteriori calculation of the forward error:

> The validity of each model was verified through comparison of the total area obtained from the above formulas to a standard (true value), which is obtained by plotting the curve on a graph paper and counting the number of small units under the curve. The sum of these units represent the actual total area under the curve. (Tai 1994)

The author makes her case by looking at a data set and compare the result with the "true value" by the method above, and conclude that Tai's model is doing better. This would be amusing if the method had not been used in medical research, where one would hope that higher standards are in place. Shortly after the publication, Allison et al. (1995) correctly identified the rule as the trapezoidal rule, and criticized Tai's "error analysis":

> Tai offers a "proof" entailing a comparison with areas computed by addend up boxes on graph paper. However, this is unnecessary. The trapezium method is closer to a "gold standard" than the box-counting approach, which she uses as a validation.

Thus, a danger with such naive a posteriori methods is that, in fact, they rely on methods for the estimation of error that are even less reliable than the method that is being tested. Therefore, it cannot seriously be considered as a sound way of vindicating the result of numerical computations.

4.2 Asymptotic Error Analysis

To introduce the methodology of error assessment that is based on asymptotic error analysis, I will follow up on the problem that Tai attempted to tackle. The paper demonstrates the technique for a set of data that appears to be sampled from a downward parabola. Can such a set of data show that the trapezoidal rule reduces the error, and in particular the overestimation error?

One of the alternatives that Tai compares the trapezoidal rule to is the midpoint rule, according to which the area under the curve is estimated by the formula

$$\int_a^b f(x)dx \approx (b-a)f(\frac{a+b}{2}) \, . \tag{5}$$

In contrast, the trapezoidal rule says that

$$\int_a^b f(x)dx \approx (b-a)\frac{f(a)+f(b)}{2} \, . \tag{6}$$

Assuming that f is convex upward and sufficiently differentiable, one easily can show that

$$(b-a)f(\frac{a+b}{2}) \leq \int_a^b f(x)dx \leq (b-a)\frac{f(a)+f(b)}{2}, \tag{7}$$

by simply noticing that on $[a, b]$, $t(x) \leq f(x) \leq s(x)$, where $t(x)$ is the tangent at the midpoint and $s(x)$ is the secant connecting $(a,f(a))$ to $(b,f(b))$. This is known as the Hermite-Hadamard integral inequality (see Fig. 1).

Fig. 1 The Hermite-Hadamard inequality

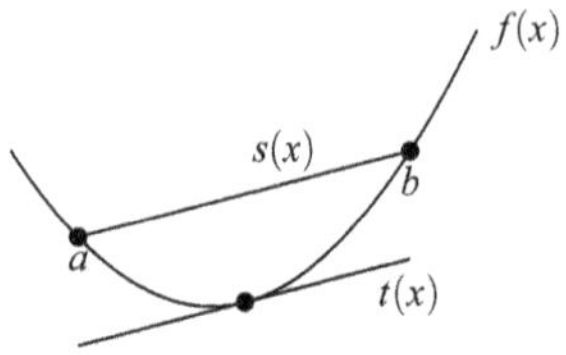

In other words, for a convex upward curve $f(x)$ (on the interval $[a,b]$), the midpoint rule *underestimates* the area under the curve while the trapezoidal rule *overestimates* the area under the curve. However, Tai's data is sampled from a convex downward curve; but the same result holds, except that "$\leq$" must be replaced by "$\geq$" in Equation (7), in which case the midpoint rule overestimates the area under a curve and the trapezoidal rule underestimates it. So, for the data she used, Tai was right that the midpoint rule generally overestimates the area under the curve, and that the trapezoidal rule does not. However, this is certainly not the case for a general curve f, since such a curve will alternate between convex upward and convex downward intervals, making it impossible in general to say whether or not the midpoint or the trapezoidal rule will overestimate the are under the curve. Thus, we can expect Tai's results not to be replicable in general, and we have a principled criterion to do so. Such considerations show clearly that mathematical results can be much more powerful than past successes at showing when we can expect a numerical method to perform well.

But what can we say about the error in itself, whether it is an overestimation or an underestimation? Perhaps it is still true that the trapezoidal rule does better? To demonstrate such a fact, however, no empirical information or a posteriori error analysis is required. The vindication of a method as compared to another one can be done without any appeal to past successes or similar modes of justification that have been discussed before. In fact, it is common in computational mathematics to perform an *asymptotic analysis* of the error generated by a numerical method. It is worth examining the details of the methodology used to prove such results as it does not seem to have penetrated the epistemology of computer simulation literature to a sufficient degree, but we relegate it to "Appendix: An Example of a Comparative Asymptotic Analysis of the Error of Computational Methods". In the case of the midpoint and trapezoidal rules, such an analysis reveals that both rules are accurate to order $O(h^3)$—i.e., the error equals $k \cdot h^3$ where k is some coefficient depending on the derivatives of f and h is the abscissa spacing between data points—but the coefficient of the dominant term for the error of the midpoint rule is smaller than that of the trapezoidal rule by a factor of 2 in favour of the midpoint rule. So, on this account, Tai's claim is also shown to be incorrect.

Note, however, that this does *not* mean that the error will be sufficiently small for some intended application, since the derivatives of f at 0 can be very large. Thus, it is important to not only know the order of accuracy of a method, but also the estimate of the error. Such methods of error analyses tell us which methods give comparatively better results, but as is the case for all methods of forward error analysis, *they do not tell us when the error is small enough.* This is why such analysis should be supplemented by a backward error analysis that will find the size of physical perturbations of the input data that would have the very same effect. Having such a measure of error thus allows us to interpret the error in the modelling context and determine, based on our modelling objectives, whether the error is small enough. This, again, is the second pragmatic dimension of the vindication of computations in simulation that I have alluded to earlier.

For such methods of error analysis, whether or not the series converge is mostly irrelevant. What we need to have is a residual or a forward error that goes to zero as the interval becomes increasingly small, approaching zero. Whether there is convergence far away from the point of expansion is irrelevant. This is why the limit of what happens to the truncated series as $N \to \infty$ has little relevance for this sort of methods. Instead, we use the idea of homotopy (or analytic continuation) and reapply our procedure based on an asymptotic expansion about a new point, and slowly march along with the dependent variable to find an approximate computed solution.

However there is another big family of numerical methods known as iterative methods for which convergence matters. The problem will arise if the function we are attempting to compute using approximation schemes is not sufficiently smooth, or when the convergence is limited to very small radii of convergence. This is why proving that a method converges for a certain collection of circumstances is a useful way of vindicating computations; it provides an a priori sufficient condition of success, i.e., a condition that must be satisfied for the reliability of the error estimates generated by series methods.

But what happens when convergence fails to obtain? Is there a way to assess the error in the hope of vindicating computations or of showing them to be untrustworthy? In his otherwise excellent discussion of the importance of appreciating the role of numerical methods for a sound philosophical appraisal of science, Wilson (2006, p. 217) suggests that mathematicians may be "unable to certify [their] conclusions" when the aforementioned convergence conditions fail, and that "they have no other means of guaranteeing that its calculations will be accurate." This remark ignores the fact that convergence conditions are (sometimes, but not always) only sufficient conditions of success; they are not necessary conditions. One can always proceed to a direct quantitative, a posteriori assessment of the magnitude of the error using the residual measure of error. In some cases, even when there is a failure of convergence, the residual will turn out to be small enough for the demands of the modelling context.

5 Conclusion

We have seen that it is important for the epistemology of computer simulation to characterize the circumstances under which the results of simulations can be considered trustworthy. Using terminology commonly used in the field of computational mechanics, we have emphasized three aspects of the justification of simulations, namely, validation, verification, and calibration. I have stressed that to the extent that verification is concerned, it is beneficial to realize that the computational steps involved in simulations are inferences. Thus, the problem of justifying computation in simulations is analogous to the justification of inferences and inferential patterns in general. Using Reichenbach's distinction between two types of justification—validation and vindication—I have argued that what we seek is a pragmatic vindication of computations in simulation.

We have seen that the vindication of computations in simulation can be approached in many ways. In particular, there are three different measures of error that we can attempt to gauge in order to establish that the results of a simulation are reliable, namely, the forward error, the backward error, and the residual. Given that the pragmatic dimension of the vindication of computer simulation demands that we assess the results within a modelling context, we have seen the advantage of backward error analysis in that it provided physically interpretable measures of error. On the other hand, direct assessment of the forward error, though often useful and readily available, are often advantageously substituted for more principled forms of error analysis. In particular, asymptotic perturbation methods typically provide the deepest insights in the general reliability (or lack thereof) of a computational method.

This multifaceted depiction of computational error analysis shows that, from the point of view of the epistemology of simulations, the computational methods simulations rely on can be vindicated or disproved in a number of ways, and that the rich circumstances that can be encountered make it unlikely that we will ever have "one vindication to rule them all." Nonetheless, it appears that the best approach is to recognize that a sound balance between asymptotic error estimate supplemented by an a posteriori estimate of the error (using, for example, a calculation of the residual) is in most cases the most promising approach. The analogy between simulations and experiments, though useful in some situations, should accordingly not be seen as a pivotal element in the vindication of computer simulations.

Appendix: An Example of a Comparative Asymptotic Analysis of the Error of Computational Methods

The following appendix exemplifies how numerical analysts quantify the error resulting from computational methods by means of asymptotic analyses. Let us return to the two methods of numerical integration discussed by Tai. To find the error in the midpoint approximation M, we take the Taylor expansion about the midpoint of the interval $[-h/2, h/2]$, which is 0 (note that for convenience we make the standard change of variable $a = -h/2$ and $b = h/2$):

$$I = \int_{-h/2}^{h/2} f(x)dx = \int_{-h/2}^{h/2} \sum_{n=0}^{\infty} \frac{f^{(n)}(0)}{n!} x^n dx = \sum_{n=0}^{\infty} \frac{f^{(n)}(0)}{n!} \int_{-h/2}^{h/2} x^n dx$$

$$= hf(0) + \sum_{n=1}^{\infty} \frac{f^{(n)}(0)}{n!} \int_{-h/2}^{h/2} x^n dx$$

Note that, when n is odd, $\int_{-h/2}^{h/2} x^n dx$ vanishes to 0 (by symmetry about the point 0). Thus,

$$I = hf(0) + \frac{f^{(2)}(0)h^3}{24} + \frac{f^{(4)}(0)h^5}{1920} + O(h^7)$$

Now, for this interval, the midpoint approximant M is $hf(0)$. Thus, the error $E_h^{(M)}$ resulting from the midpoint rule on an interval of width h is

$$E_h^{(M)} = I - M = \sum_{\substack{n=2 \\ \exists k \in \mathbb{Z}.2k=n}}^{\infty} \frac{f^{(n)}(0)h^{n+1}}{(n+1)!2^n}$$

Taking the first two terms of $E_h^{(M)}$, we have

$$E_h^{(M)} = \frac{f^{(2)}(0)h^3}{24} + \frac{f^{(4)}(0)h^5}{1920} + O(h^7)\,.$$

Now, we will find the trapezoidal error, $E_h^{(T)} = I - T$, by expressing T as a combination of Taylor series. Let us first find the Taylor series about $x = 0$ (the midpoint) for $f(-h/2)$ and $f(h/2)$:

$$f(-\frac{h}{2}) = \sum_{n=0}^{\infty} \frac{f^{(n)}(0)}{n!}(-\frac{h}{2})^n = \sum_{n=0}^{\infty} \frac{f^{(n)}(0)(-1)^n h^n}{n!2^n}$$

$$= f(0) - \frac{1}{2}f^{(1)}(0)h + \frac{1}{8}f^{(2)}(0)h^2 - \frac{1}{48}f^{(3)}(0)h^3 + \cdots$$

$$f(\frac{h}{2}) = \sum_{n=0}^{\infty} \frac{f^{(n)}(0)}{n!}(\frac{h}{2})^n = \sum_{n=0}^{\infty} \frac{f^{(n)}(0)h^n}{n!2^n}$$

$$= f(0) + \frac{1}{2}f^{(1)}(0)h + \frac{1}{8}f^{(2)}(0)h^2 + \frac{1}{48}f^{(3)}(0)h^3 + \cdots$$

Because of the alternating sign in $f(-\frac{h}{2})$, we find that the odd powers of n vanish when we calculate T:

$$T = \frac{h}{2}\Big(f(-\frac{h}{2}) + f(\frac{h}{2})\Big) = h\Bigg(\sum_{\substack{n=0 \\ \exists k\in\mathbb{Z}.2k=n}}^{\infty} \frac{f^{(n)}(0)h^n}{n!2^n}\Bigg)$$

$$= hf^{(0)}(0) + \frac{1}{8}f^{(2)}(0)h^3 + \frac{1}{384}f^{(4)}(0)h^5 + \cdots$$

As a result,

$$E_h^{(T)} = I - T = -\frac{f^{(2)}(0)h^3}{12} - \frac{f^{(4)}(0)h^5}{480} - O(h^7)\,.$$

So, both rules are accurate to order $O(h^3)$, but the coefficient of the dominant term for the error of the midpoint rule is smaller than that of the trapezoidal rule by a factor of 2.

References

Allison, D. B., Paultre, F., Maggio, C., Mezzitis, N., & Pi-Sunyer, F. X. (1995). The use of areas under curves in diabetes research. *Diabetes Care, 18*(2), 245–250.

Batterman, R. W. (2002a). Asymptotics and the role of minimal models. *British Journal for the Philosophy of Science, 53*, 21–38.

Batterman, R. W. (2002b). *The devil in the details: Asymptotic reasoning in explanation, reduction, and emergence*. Oxford: Oxford University Press.

Batterman, R. W. (2009). Idealization and modeling. *Synthese, 169*(3), 427–446.

Bender, C., & Orszag, S. (1978). *Advanced mathematical methods for scientists and engineers: Asymptotic methods and perturbation theory* (Vol. 1). New York: Springer.

Borwein, J., & Crandall, R. (2010). Closed forms: what they are and why we care. *Notices of the American Mathematical Society, 60*, 50–65.

Corless, R. M., & Fillion, N. (2013). *A graduate introduction to numerical methods, from the viewpoint of backward error analysis* (868pp.). New York: Springer.

Deuflhard, P., & Hohmann, A. (2003). *Numerical analysis in modern scientific computing: An introduction* (Vol. 43). New York: Springer.

Feigl, H. (1950). De principiis non disputandum…? In *Inquiries and provocations* (pp. 237–268). Dordrecht: Springer. 1981.

Fillion, N. (2012). *The reasonable effectiveness of mathematics in the natural sciences*. PhD thesis, London: The University of Western Ontario.

Fillion, N., & Bangu, S. (2015). Numerical methods, complexity, and epistemic hierarchies. *Philosophy of Science, 82*, 941–955.

Fillion, N., & Corless, R. M. (2014). On the epistemological analysis of modeling and computational error in the mathematical sciences. *Synthese, 191*, 1451–1467.

Galileo (1687). De motu. In I. Drabkin (Ed.), *On motion and on mechanics*. Madison: University of Wisconsin Press. 1960.

Grcar, J. (2011). John von Neumann's analysis of Gaussian elimination and the origins of modern numerical analysis. *SIAM Review, 53*(4), 607–682.

Hacking, I. (1992). The self-vindication of the laboratory sciences. In A. Pickering (Ed.), *Science as practice and culture*. Chicago: University of Chicago Press.
Hamming, R. (1980). The unreasonable effectiveness of mathematics. *The American Mathematical Monthly, 87*(2), 81–90.
Higham, N. J. (2002). *Accuracy and stability of numerical algorithms* (2nd ed.). Philadelphia: SIAM.
Humphreys, P. (2004). *Extending ourselves: Computational science, empiricism, and scientific method*. New York: Oxford University Press.
Kadanoff, L. P. (2004). Excellence in computer simulation. *Computing in Science & Engineering, 6*(2), 57–67.
Oberkampf, W., Trucano, T., & Hirsch, C. (2004). Verification, validation, and predictive capability in computational engineering and physics. *Applied Mechanics Review, 57*(5), 345–384.
Reichenbach (1949). *The theory of probability: An inquiry into the logical and mathematical foundations of the calculus of probability*. Berkeley: University of California Press Berkeley.
Salmon, W. C. (1991). Hans Reichenbach's vindication of induction. *Erkenntnis, 35*(1–3), 99–122.
Tai, M. M. (1994). A mathematical model for the determination of total area under glucose tolerance and other metabolic curves. *Diabetes Care, 17*(2), 152–154.
Tal, E. (2011). How accurate is the standard second? *Philosophy of Science, 78*(5), 1082–1096.
Truesdell, C. (1980). Statistical mechanics and continuum mechanics. In *An idiot's fugitive essays on science* (pp. 72–79). New York: Springer.
Wilkinson, J. H. (1963). *Rounding errors in algebraic processes* (Prentice-Hall series in automatic computation). Englewood Cliffs: Prentice-Hall.
Wilson, M. (2006). *Wandering significance: An essay on conceptual behaviour*. Oxford: Oxford University Press.
Winsberg, E. (2010). *Science in the age of computer simulation*. Chicago: University of Chicago Press.
Wolever, T. M., Jenkins, D. J., Jenkins, A. L., & Josse, R. G. (1991). The glycemic index: Methodology and clinical implications. *The American Journal of Clinical Nutrition, 54*(5), 846–854.

Empirical Bayes as a Tool

Anouk Barberousse

1 Introduction

Within the mathematical realm, statistics looks as if it were a tool *par excellence*. It does not seem to have a proper object, as number theory or complex function theory do, their objects being respectively natural numbers and complex functions. It is *used* in a bunch of applications without being defined by its own domain: it is often associated with probability on the one hand and with theories of inference on the other, thereby fluctuating, as it seems, between mathematics per se and logic, or even applied logic. As a result, it looks as if it could adapt to a variety of tasks, as a good tool that would not be too specialized.

The purpose of this paper is to investigate the hypothesis that statistics is a tool by focusing on a recent trend called "Empirical Bayes". Empirical Bayes is a set of statistical methods that rely on machine computation. Its recent development and success has partly transformed the whole field of statistics as several scientific domains, like phylogenetic, image analysis, and climate science make a heavy use of these new methods. Even though part of its inspiration is Bayesian, its connection with historical Bayesianism, as a philosophical doctrine, or even with Bayesian statistics as it has developed before the computer age, is rather loose.

In order to assess whether Empirical Bayes is a tool, I will first give a non-technical presentation of these methods. Then I will illustrate them by presenting how they are used in climate statistics. In the third part of the paper, I will discuss the global hypothesis of mathematics as a tool and argue that in the case of Empirical Bayes, what can be said a tool is a more complex set both including a mathematical part and a stochastic model.

A. Barberousse (✉)
Université Paris-Sorbonne, 1 rue de la Sorbonne, 75005, Paris, France
e-mail: anouk.barberousse@paris-sorbonne.fr

J. Lenhard, M. Carrier (eds.), *Mathematics as a Tool*, Boston Studies in the Philosophy and History of Science 327, DOI 10.1007/978-3-319-54469-4_9

2 Empirical Bayes

Empirical Bayes (EB), also called "empirical Bayesianism" or "pragmatic Bayesianism" is a new trend in statistics characterized by a heavy reliance on the use of computers. It might also be called "computerized Bayesianism", except that, as we shall see, its relationship with older forms of Bayesianism is not straightforward at all. Moreover, EB can appear under different forms according to the scientific domains to which it is applied. The expression "EB" is more an umbrella covering a multiplicity of particular, rapidly evolving methods than a unified approach. As a result, it is not easy to present it in any simple way.

In this section, I first present what is common to EB and other forms of Bayesianism, namely, Bayes updating. Then, I give a brief and simplified account of EB techniques. This allows me to enter into some details in the question of the elusive foundations of these techniques. Switching from foundational issues to practical ones, I then emphasize the role stochastic computer models play within EB, and at last I present its main advantages.

2.1 Empirical Bayes and Other Varieties of Bayesianism

In order to introduce EB, it may be useful to recall that as its name indicates, it is inspired by older forms of Bayesianism, even though it is different. What is common to EB and older forms of Bayesianism is Bayesian conditioning, according to which the probability attributed to a hypothesis has to evolve when new relevant information occurs. The crudest form Bayes conditioning rule can take is:

$$P(H|E) = P(E|H).P(H)/P(E) \tag{1}$$

where P is the probability function, $P(A|B)$ is the conditional probability of A given B and is defined as $P(A \wedge B) / P(B)$, H is the hypothesis under consideration, and E is the newly available information or evidence that is viewed as relevant to H. Bayes conditioning has a long history in classical science where it has been used, especially by Laplace, to find out causes when only their effects were known. In the trends of Bayesianism that have developed at the beginning of the twentieth century, Bayes conditioning is only one part of Bayesianism, a whole philosophical doctrine that may also include views about beliefs and rationality. These views, however popular they may have been, are absent from most versions of EB. As the name "pragmatic Bayesianism" indicates, its proponents, like Jim Berger, José M. Bernardo, Andrew Gelman, Robert E. Kass, Christian Robert, George Casella, and others, are not keen on philosophical debates but rather look for efficient statistical methods and operational techniques. This is the reason why they are eager to implement Bayes conditioning (under more sophisticated forms) on computers.

"Bayesian" is also the name of a currently popular theory of confirmation. Does EB have anything common with the Bayesian theory of confirmation? The answer to this question is twofold. First, as a set of statistical methods, it is not unified enough to pretend to the title of a "theory". But second, it has do to with confirmation of hypotheses, and its main tool is one form or other of Bayes conditioning. More precisely, within EB, hypotheses are attributed probabilities and these can evolve as new evidence occurs. Thus, within EB, confirmation has to do with probability of hypothesis going up or down in face of new evidence, not with statistical testing-even though, as we shall see, some EB versions are so eclectic as to include elements of classical statistics.

2.2 An Overview of EB Techniques

Now to the point. The aim of EB is to perform statistical inferences, namely, to assess the plausibility of hypotheses when confronted with data. Here is a general description of the problem EB methods are designed to solve. Based on a sample of n observations whose sampling distribution is described by the known function $f(y_1, \ldots, y_n \mid \theta)$, one wants to estimate the value of the unknown parameter θ. As is well-known, the field of statistics divides up according to the way θ is conceived of. In classical statistics, θ is assumed to have one true value that the statistician tries to estimate as closely as possible given available data. Within the Bayesian camp, θ is conceived as a random variable whose variability can be described by the distribution π (θ).

Usually, π (θ) depends on other parameters λ, so that π (θ) is expressed as π $(\theta \mid \lambda)$, from which a posterior distribution π $(\theta, \lambda \mid y_1, \ldots, y_n) \propto \pi$ $(y_1, \ldots, y_n \mid \theta)(\theta \mid \lambda)(\lambda)$ can be inferred. This procedure, which can be iterated, ends up with *hierarchical Bayes models*. The process reaches an end when the prior distribution does not depend on any previously unmentioned parameter. As clear from this description, the notion of a hierarchical Bayes model has no formal definition. Any such Bayesian network with more than three levels of random variables is usually called a hierarchical model.

For readers who are familiar with the usual Bayesian framework, the most important point to add to the above description is that the last introduced parameter is estimated *classically*. EB can thus be analyzed as as an approximation to a fully Bayesian treatment of a hierarchical model as the highest-level parameters are given their most likely values, instead of being integrated out.

Let us now go into the details of one of the procedures, called "maximum marginal likelihood", that is used to estimate an unknown parameter θ. We start from a variable x whose likelihood function is $p(x \mid\theta)$. Let us suppose that θ's prior distribution is $g(\theta \mid \lambda)$ where λ is a higher level parameter. $E(\theta \mid x)$ is the quantity to minimize; it is called "Bayes estimator". First, Bayes formula (1) is used to determine the posterior distribution:

$$p(\theta|\mathrm{x},\lambda) = p(\mathrm{x}\,|\theta)\,g(\theta \mid \lambda)\,/m(\mathrm{x} \mid \lambda) \tag{2}$$

where $m(\mathrm{x} \mid \lambda) = \int p(\mathrm{x} \mid \theta)\, g(\theta| \lambda) d\theta$ is the marginal distribution of x.

Then $E(\theta \mid x) = \int \theta p(x \mid \theta)g(\theta \mid \lambda)d\theta / \int p(x \mid \theta)g(\theta \mid \lambda)d\theta$.

When θ is unknown, a new distribution $h(\lambda)$ is introduced, as mentioned above, so that the posterior distribution on θ is obtained by marginalizing on λ:

$$p(\theta|\mathrm{x}) = \int p(\mathrm{x}\,|\theta)\, g(\theta \mid \lambda)\, h(\lambda)\, d\lambda / \int\int p(\mathrm{x}\,|\theta)\, g(\theta|\lambda)\, h(\lambda)\, d\lambda d\theta \quad (3)$$

It is usually difficult to integrate this quantity. For this reason, the marginal distribution of x is often used in order to estimate λ with the help of the marginal maximum likelihood estimation (MMLE) of $\hat{}\lambda$, which is a classical way to estimate $\hat{}\lambda$. Thus (3) allows one to draw inferences on $p(\theta\ |\mathrm{x}, \hat{}\lambda)$. To put it in a nutshell, a marginal posterior distribution is set on θ by taking the weighted average of estimated parameters on the whole range of possible values for all noise parameters, the weight being given by the posterior distribution (a "noise" parameter in this context is a parameter one is not directly interested in estimating but whose value is unknown or uncertain). The main advantage of the MMLE method is that the estimate is robust with respect to the uncertainty affecting the noise parameters of the problem.

To sum up, EB approaches constrain parameter values by combining prior distributions that account for uncertainty in the knowledge of parameter values with information about the parameters estimated from data (Kennedy and O'Hagan 2001). Whereas the first part is common with older forms of Bayesianism, the last part is distinctive of EB, whose proponents are willing to draw on all available means to fix priors, as we shall see in the next section.

2.3 Unclear Foundations

The MMLE method is just one example of a large variety of techniques for parameter estimation that are currently used within EB. Other examples involve techniques that would be conceived as anti-Bayesian in more strict Bayesian frameworks, for eclectism is the key word in EB. For instance, whereas the "old Bayesians" rejected confidence intervals and statistical significance, EB proponents include them in their toolkit because they commonly use them in practice when no other means is available to perform parameter estimation. Thus, from a practical point of view, "everything goes" for EB proponents, who are willing to describe their new way / approach as based on the common ground between classical and Bayesian statistics. Once this common ground is identified, methods can be built up that benefit from both approaches. As one of the first defenders of EB has it, "EB methods attempt to borrow the strengths of each [classical and Bayesian] approaches"; EB is "a compromise between classical and Bayesian approaches" (Casella 1992). Others also claim that the old debate between orthodox Bayesianism and classical (frequentist) statistics is outdated. In the field, the practices are not as divided as the terms of the old debate suggest. Maximum (of) likelihood and Bayesian estimators are often used together: "confidence, statistical significance, and posterior probability are all valuable inferential tools." (Kass, 2011). From the perspective of the debate between classical and Bayesian statistics, this patchwork of practices seems inconsistent. This is the reason why Deborah Mayo (2013), for one, points out the "conceptual confusion" with respect to the foundations of EB, which do not match the quickly evolving practices of the field.

A striking example of the surprising way EB relates to older debates is the attitude of its proponents toward the interpretation of probability. A major feature of older forms of Bayesianism was that they were strongly associated with either a subjective or an epistemic interpretation of probability. By contrast, proponents of EB now consider that the question of how to interpret the probability concept can be neglected. For sure, most of them conceive of probability as a natural measure of uncertainty or plausibility, although/but they do not automatically adopt a subjective interpretation of such a measure. On the contrary, even though they oppose the view that probability is given its meaning by frequencies in an ideal, infinite population from which our observed data are a sample, they may adhere to the view that the meaning of probability is given by frequencies emerging when the statistical procedure is applied many times on the long run. This attitude is grounded in the idea that the use of epistemic probability is not in principle incompatible with the wish to reach good frequentist properties. Among the frequentist properties that are considered important within EB, calibration plays a major role (Robert 2016). When applying EB procedures, one hopes that the frequency (in a series of applications of the statistical procedure) at which the value will fall within the confidence interval will be at least 95% and that the average error within a series of applications of estimator A will never be systematically higher than the average error with any other estimator B. As we shall see in the next section, computer simulations of stochastic models embody the very essence of the frequentist calibration of epistemic probability.

As should now be clear, EB is by no means a unified field, whose main tenants could be easily identified. However, when one does not try to characterize EB from a statistical point of view but discover how it is implemented in practice, a major feature can emerge, namely, *explicit modelling* (cf. Kass 2011, p. 6). For sure, EB aims at performing statistical inference, but not only by designing new statistical techniques that would have a unique justification and efficiency. On the contrary, EB is mostly efficient because of the *modelling practice* it is based upon. Proponents of EB consider it an efficient statistical method not so much because it uses a variety of statistical techniques in a new way, but above all because this method forces scientists to be as careful and explicit as possible in their modelling decisions. The scientific burden is thus switched from statistical inference to explicit modelling (Kass 2011, p. 7).

2.4 Modelling as a Main Component of the Method

As explained above, parameter estimation is obtained from available data. The heart of the method is to draw as much information as possible from the data and this is obtained by building up stochastic models allowing for further averagings among new "data", i.e., model-generated data. As it is necessary to explicitly formulate what the components of the stochastic model are, all the assumptions of the analysis are themselves made explicit. This is a well-known advantage of Bayesian methods in general, because once assumptions are made explicit, they can be discussed, criticized, and improved.

The stochastic models involved in EB methods represent the exhaustive collection of all possible scenarios that are a priori (i.e., before conducting statistical inference) conceivable about the investigated phenomenon. Each scenario corresponds to a set of definite values for the model parameters. These stochastic models are computer models. EB is indeed a computer intensive method whose development is mainly due to the recent availability of computational power. Computational power allows for large sample sizes, repeated Bayesian updating, and removing the restriction to conjugate priors that was holding when computer power was lacking. Whereas updating is often difficult to process by hand, its being implemented on computer makes it an easy and rapidly performed task. This in turns allows for the computation of hierarchical models with several layers.

EB scientists like Christian Robert, Andrew Gelman and George Casella insist that the computational difficulty of Bayesian updating was a major obstacle to the development of Bayesian methods in the past. Now that this obstacle has been overcome, this has opened an opportunity for the invention of new algorithms. At the heart of most algorithms are Monte Carlo Markov Chains (MCMC) modules that allow for a posteriori sampling of the distribution of interest. Sampling is done by rejecting every simulation (i.e., implementation of the stochastic model) whose outcomes are incompatible with observations. The effect of stochastic modelling via MCMC modules is thus to expand the set of "data" that are compatible with observations. Here, the "data" are the outcomes of the simulations that are consistent with observations. They can be statistically processed in the same way as data provided by observations. The availability of these new "data" allow statisticians to focus on the frequency of occurrence of certain parameter configurations within the remaining set of possibilities. The realization that Markov chains can be used to model a large variety of situations has be a turning point in the development of EB.

2.5 Advantages of EB Methods

Proponents of EB describe an impressive list of advantages that are specific to these methods and are mostly dependent upon the fact that EB allows for different modelling strategies. Thus the *flexibility* of this approach is often put forward as an opportunity to try out different models and test them for efficiency. Flexibility mostly depends on the modular character of the approach, which allows for the combination of hierarchical models, thus permitting the articulation of different levels. This possibility not only occurs when the investigated phenomenon lends itself to decomposition into hierarchical layers, but also when it does not at first sight. The opportunity offered by the approach invites scientists to find out how to model complicated situations by means of a hierarchy of simple layers. This modelling strategy, which calls for the specification of simpler sub-models, can result in estimates with improved precision.

Associated with modularity is the possibility of representing various sources of uncertainty, a major topic in the study of complex systems. This in turn allows for more robust estimations when the models include a large number of parameters. The reason for this comparative advantage is as follows. In maximum of likelihood methods, the treatment of noise parameters is done by co-maximisation: likelihood is first maximized with respect to all parameters at the same time, and then one gets rid of noise parameters. This amounts to betting on a *unique value* for noise parameters, that is, to ignoring that they may be uncertain. This may be unsafe in case there are many noise parameters. The end result is that the potential richness of the models is limited. By contrast, EB techniques allow for flexibility in the treatment of uncertainties.

The representation of uncertainty is a central topic in older Bayesianism as well, where it was sometimes criticized as being a door open to subjective elements creeping in scientific activity. For non-Bayesian statisticians, the main problem of older Bayesianism is the interpretation of prior distributions, supposed to be unscientific, or at least scientifically suspect if they are understood as related to degrees of belief or of ignorance. Some EB proponents, accordingly, have tried hard to construe prior distributions as some sort of mathematical quantities used to obtain posterior distributions. This trend has led to the development of "conventional priors", namely, priors that have the least influence on resulting inferences, letting the data dominate. The debate over how prior distributions are chosen and interpreted is still open, even within the context of EB.

To sum up, MCMC modules provide EB modellers with a generic, powerful computational framework, which allows for the relatively easy implementation of a wide range of models. Last but not least, computational power allows for much more data to be statistically processed. All these features explain why EB methods are rapidly developing in the following scientific domains: phylogenetic reconstruction, climate study, interpretation of neuroscience imaging, etc.

3 Empirical Bayes in Climate Science

In order to illustrate how EB methods are used in scientific practice, I will now focus on the example of climate statistics. This domain is especially exciting for anyone interested in EB because climate science, as it is nowadays carried out, contains a huge statistical part. For sure, climate scientists rely on many physical laws in order to build up their models, in particular hydrodynamical laws, but also on vast quantities of observational data. The latter need to be processed statistically. Statistics are not only necessary for input data, though, for the outputs of the models also have to be subjected to statistical analysis or they would tell nothing understandable.

As it happens, even though most statistics performed in climate science are classical, EB methods are rapidly developing, as testified by the increase in references to EB papers in the last two IPCC assessment reports. EB methods are used in three main domains within climate modelling: detection and attribution of climate change, estimation of climate sensitivity, and regional projections. I will briefly present each domain and describe how EB methods are actually implemented to solve the statistical problems at issue.

3.1 Detection and Attribution of Climate Change

The problem of the detection of climate change is to find out a very weak signal in the global evolution of climate indicating that its recent evolution has departed from its previous course. For sure, in order for this task to be feasible, it is necessary to have some knowledge of ancient climate evolution. Now, observation data like temperature and humidity records are only available since a very short time, compared to the relevant temporal scales. Knowledge of past climate is however rapidly growing so as to constitute a sufficient basis to find out differences between present-day and past evolution. This knowledge both comes from indirect observations (e.g. observation of ice cores) and from simulations of past climate, whose outputs are statistically processed.

The problem of attribution is to identify the causes of climate change, i.e., of the difference between past and present evolution. This problem has received a general answer: climate change can be attributed to the effects of greenhouse gas produced by human activity. This general answer still needs to be made more precise in order to get quantitative understanding.

The difficulty of the detection and attribution problems is enhanced by the fact that the natural variability of the climate is large and its magnitude unknown. This basic problem has other worsening features: the observation data are recent, scarce, and heterogeneous.

Let us know briefly present the procedures that are carried out in order to overcome the above-mentioned difficulties. In order to solve the detection problem, the first task is to reconstruct the mean values of the most important variables, like temperature at different heights, for the last two millennia. Why "reconstruct"? Because we only have measurement results for the last ten decades or so, which means that previous values have to be calculated from the models' outputs. The available observation data are moreover partial and have themselves be completed and homogenized via statistical procedures. They are however valuable because they can be used to partially calibrate the models.

The next task is to scrutinize those huge sets of data in order to find out whether their evolution pattern has been recently modified. As mentioned before, the statisticians' way to formulate this task is to find out whether a signal indicating climate change can be detected against loud noise. There are two ways to fulfill this task. The first is by defining a null hypothesis and testing it against data (both observational and model-generated). This is the classical way. The null hypothesis is for example that there is no climate change currently occurring. This approach has been heavily criticized by the EB proponents (Berliner et al. 2000) The second way is the EB way. It amounts to answering the following question: do "initial conditions" (describing an old state of climate (Hasselmann 1998)) raise the probability of a positive signal when recent data are taken into account? Both ways are used for the detection and attribution tasks. The classical way was the first whereas the EB approach is more recent but quickly developing.

The main problem facing EB scientists is that they have to find out ways to impose constraints on the prior distributions because the observation data fail to do so. As a result, the choice of the prior distribution has an important impact on the computation of the posterior distribution. Whatever the importance of this difficulty, it does not outweigh the main advantage of EB methods, which is that they allow for an explicit description of the modellers' *state of knowledge*, the basis on which one can try to detect the signal of climate change. The evolution of this state of knowledge is thus taken into account as models progress.

3.2 The Study of Climate Sensitivity

Climate sensitivity is the degree to which climate responds to perturbation. In the context of the present focus on human-induced climate change, the relevant perturbation is the increase in the concentration of greenhouse gases within the atmosphere. This is the reason why the study of climate sensitivity is focused on the following question: How does climate change when atmospheric CO_2 concentration is doubled with respect to its value before the industrial revolution? Doubling of atmospheric CO_2 concentration is called "forcing". It might be asked why CO_2 concentration has been chosen in the definition of forcing whereas other gases have the same green-housing effect when their atmospheric concentrations increase. The answer is that the computation being already very heavy with one greenhouse gas, it has been judged unnecessary to make them more complicated, so that focusing on CO_2 concentration is a good proxy for studying climate sensitivity.

The main difficulty in trying to answer this question is that there exist retroactive mechanisms whose effect is to stabilize temperature increase due to forcing. It is thus extremely difficult to assess the hypothesis that climate reacts in such and such a way to forcing. The very formulation of this hypothesis calls for a Bayesian approach as the outputs of the models implementing forcing can readily be used to update the probability of the hypothesis. As the updating process can be iterated as often as permitted by available computational power, it looks as if the EB approach is likely to be successful for this task. As a matter of fact, the study of climate sensitivity is indeed dominated by EB approaches.

The main advantage of EB approaches in the study of climate sensitivity is that multiple and independent lines of evidence about climate sensitivity from, for example, analysis of climate change at different times, can be combined by using information from one line of evidence as prior information for the analysis of another line of evidence (Annan and Hargreaves 2006; Hegerl et al. 2006). However, it is still unclear to what extent the different lines of evidence provide complete information on the underlying physical mechanisms and feedbacks.

In order to take the effect of feedback mechanisms into account, it is necessary to consider the outputs of several models, because many individual models contain highly idealized representations of these mechanisms. The best way to overcome these poor representations is to try and compensate their defects by combining several models (for criticisms see Parker 2010). This approach consists in building up statistical ensembles of different models and averaging over their outcomes. Each model instantiates a specific scenario defined by a determined different distribution of the most uncertain parameters that are relevant to the study of sensitivity and feedback mechanisms. Traditional statistical estimators, like means, are then processed on the ensemble's outputs. The expected next step is to assess the likelihoods of these distributions against data. However, the available data are not good enough to provide sufficiently strong constraints on the distributions. The difficulty is thus to find out further constraints on the distributions. The EB approach is comparatively more efficient than the classical one because it better manages the relation between models' outputs and empirical data and allows for stricter control of this relation. EB approach to multi-models ensembles is thus rapidly developing.

3.3 *Regional Projections*

As a third and last illustration of how EB is used in climate statistics, I now turn to regional projections. "Projections" designates predictions relative to possible "scenarios". A scenario in this context is a possible future defined relative to a determined level of greenhouse gas emission. Modellers compute predictions in different cases, whose extremes are the case in which nothing is done in reducing greenhouse gas emissions and the case where the emissions are so reduced as to avoid increase of global mean temperature above 2°C. Besides the fact that the latter is definitely out of reach, an important feature of the notion of projection is that as the currently available models are different, they can compute different projections for the same scenario. Divergences are usually higher in more distant time. However, projections for the next few decades are very similar among most models.

The most publicized and debated projections are the global ones, concerning the entire planet. Until recently, these were the only computable ones. However, some models (and associated super computers) are now able to compute *regional* projections, that is, projections dealing with smaller scales, continental or even smaller. Reducing projection scale demands heavy computational power as much more details have to be put into the models in order to account for local peculiarities, like land cover. Such amount of computational power is only recently available.

EB methods are not used by all modellers computing regional projections, but there is one case in which these methods have been chosen as the best, unique tool allowing modellers for coherent and understandable projections: Great Britain. The British climate agency has indeed decided to used EB throughout, both in the computations and in the presentation of the results to decision makers (*UK Climate Projections science report: Climate change projection*, http://ukclimateprojections.metoffice.gov.uk/media.jsp?mediaid=87893&filetype=pdf).[1]

As the last point, communication to decision makers, is a major challenge for climate scientists, let me now emphasize the importance of the dual aspect of climate projections, which are both science-based and directed toward expertise. On the one hand, climate scientists try to better understand climate as a huge complex system governed by hydrodynamic and thermodynamical laws, sensitive to a large number of atmospheric chemical reactions, influenced by the local behaviour of the biomass, and, unfortunately, by industrial activity. This is complex system science, based on computer models, statistics, a lot of work devoted to articulating small-scale to larger-scale models, etc. On the other hand, most climate scientists are engaged, besides their scientific work, in an entirely original expertise strategy aimed at policy makers. Why is this expertise strategy original? Because it emerged as whistle-blowing whereas usual expertise work is an answer to a question asked by someone outside the scientific community. In the case of climate, the demand for expertise did not come, at the beginning, from outside the scientific community, but from inside, in the sense that climate scientists themselves were feeling that it was of utmost importance to make policy makers, and virtually everybody, know that human activity induced climate change was very likely happening and that public action was very desirable. Climate scientists also believed that much more scientific work had to be done, and thus funded, in order to know more about climate. As a result, IPCC has been created as an expertise providing group whose first aim is to provide policy makers, and virtually everybody, with the best available knowledge about climate, and to regularly update this information.

The upshot of urgency, the global dimension of the scientific problem, necessitating efforts from as many scientific teams as possible, and high resistance level from industrial players, is that most climate scientists are double-hatted and devote part of their work to expertise and communication. Why is this distinctive feature of climate science relevant to the use of EB? Because some EB proponents claim that climate scientists would be better off by adopting an overall Bayesian framework both for

---

[1]This report has triggered a harsh debate within the philosophy of science community: cf. Frigg et al. (2014) and Winsberg and Goodwin (2016).

their scientific work and when speaking as experts, out of coherence. One major form of this coherence argument is that *because* the Bayesian framework is more easily understandable by non scientists, especially when dealing with uncertainties, climate scientists would be better off with a unified approach to probability and uncertainty that allows for incorporation of imperfect information in the decision process. This is at least what the current leader of British climate statisticians, Jonathan Rougier, is claiming (2007, 2012).

3.4 Summing Up: Main Reasons to Adopt EB in Climate Science

Let us now recapitulate the main reasons why some statisticians support EB within the context of climate science. These reasons hold for the three above-presented domains where EB has been actually implemented but also on a more general basis. Some of them could be found in other scientific fields, but others are specific to climate science.

One of the reasons presented as soon as the 1990s by Bayesian statisticians is that the study of climate is so huge an enterprise that the practices associated with traditional statistical cause-effect analysis are very difficult or even impossible to implement. For instance, due to the singularity of the actual climate, it is impossible to have even remote analogues of controlled experiments comparing responses to treatments. An important aspect here is that even though climate simulations can be iterated, their results cannot replace measurements on the actual course of climate with respect to statistical cause-effect analysis, because assumptions about the cause-effect relations are built up within the model and thus cannot be tested by multiple iterations thereof.

A related reason with more technical content is that according to EB proponents, the notion of statistical significance is unable to formalize the notion of significance that is relevant to the climate case. When performing the detection task, classical statisticians are looking for statistically significant indicators of climate change whereas finding out the *practical* significance of these indicators would be much more relevant and useful. One might object that this would be an entirely different purpose in the first instance. In a way, this is exactly what EB proponents are claiming. They believe that traditional statistics does not fit the purpose of climate science as it has to be done in the emergency situation that we are currently facing.

However, EB proponents' arguments are not all along this practical line, as they claim that EB statistics are more efficient than traditional statistics in the resolution of certain specific problems, like the management of uncertainty, as already mentioned.

4 Empirical Bayes as a Tool

Now that the above general presentation of EB has been illustrated by examples from climate statistics, let me come back to the question What is EB? by arguing in favour of the hypothesis that EB is a *tool*. This will allow me to discuss Carrier and Lenhard's proposals in the introduction to this volume. I shall first present some comments on the notion of a tool as used in the context of describing and analyzing scientific activity. Then I shall use these comments to investigate into the senses in which EB can be said to be a tool.

4.1 Scientific Tools

Let me first emphasize that the notion of a tool is so large that many elements of scientific activity could be called tools, from measurement operations, data sets, theories, to models, templates, statistical procedures, computers, simulations, etc. However, there is a way to make the notion more precise; it is by distinguishing between two components of its meaning. The first component is that a particular tool is usually defined by a purpose it has to fulfill, like driving nails or screwing bolts. Some purposes may themselves be very large, like "computation", allowing computers to be called "tools" in this sense. However, purpose-directedness seems to be an important aspect of being a tool. The second component of the notion of a tool is that nothing would be called a tool without being useful or being actually used.

An important aspect of the first component above is that a tool in this purpose-directed sense is *neutral* with respect to what it is applied to in order to fulfill its purpose. Whatever the metal in which the nails are made, if sufficiently hard, the hammer will drive them equally. Whereas this kind of neutrality may have interesting consequences for material tools, like being equivalently efficient over a range of different materials, its importance in the epistemic domain is often overlooked. In order to better see this point, I will take the example of mathematics.

Among the reasons why mathematics may be called a tool for empirical scientists, its neutrality with respect to what it is applied to is a major one. Mathematics is as efficient in solving problems in (some parts of) biology as it is in physics. Here, neutrality is relative to topic. The tool-nature of mathematics is also revealed by its neutrality with respect to theory or model, as differential equations, for instance, are both used in celestial mechanics and in quantum mechanics. Thus, with respect to purpose-directedness, it seems that mathematics satisfies the associated neutrality constraint.

Focusing on mathematics as an example allowing us to better understand the notion of a tool in science, we can now realize that there may be a tension between the two components of the notion of a tool. It seems that in order to be actually useful, and used as a tool, mathematics has to be complemented with something else. It is indeed not mathematics as such that is used for fulfilling common scientific purposes, like predictions, but mathematics *plus* physical or biological content under the form of hypotheses, however general. On the one hand, purpose-directedness comes with topic neutrality, but on the other, usability cannot be obtained without complementing mathematics with empirical hypotheses. Moreover, it is also necessary to be provided with some sort of methodological guides or recipes in order to be able to use mathematics in the right way. Without relevant training, you could not find out the approximate solution of the Hamiltonian you have designed up to represent the quantum system you are interested in, for instance.

Taking into account the necessity of complementing mathematics with empirical hypotheses and methodological guidance amounts to situating oneself in the "perspective from within", according to van Fraassen's expression (2008). The perspective from within involves the scientific agent's point of view when trying to find out solutions to her problems. It opposes the perspective from above, or the philosopher's perspective who tries to categorize the items she is not interested in herself, but which she is describing from a detached point of view. To put it in a nutshell, considering mathematics, as separated from empirical hypotheses and methodological guidance, as a tool is to adopt the perspective from above, whereas considering usable tools by actual scientific agents implies adopting the perspective from within and avoids detaching mathematics from associated items.

The view of tools I have just presented is consistent with the common idea that *models* are tools. Now, models are usually considered as complex wholes the parts of which are not easily isolated from each other. In particular, the mathematical part cannot be easily detached from the remaining of the model in the sense that it is inert, so to speak, without the other components.

4.2 EB Approaches from a Methodological Point of View

As we have seen in Sect. 1, EB is difficult to characterize because of its heterogenous and rapidly evolving nature. Is it a scientific method by itself, or just a statistical method? Should we apply to it the even more general expression of "a new way to perform statistics", or is it a new technique for computing statistical results? Still another option is to link it with the philosophical foundations of older Bayesianism and describe it as a general, perhaps philosophical, theory of scientific reasoning.

Asking for the nature of EB involves reflecting upon the nature of statistics from the "tool" perspective. This task is known to be full of philosophical loopholes as it involves hypotheses about the relationships between logic and mathematics as well as the nature of inference. Therefore, I shall not endeavour to present any hypothesis about the nature of statistics, but rather focus on EB as it *used* in practice. I am aware that a complete argument about EB as a tool would involve taking position about statistics, inference, and the relationships between mathematics and logic, but the best I can do is to let such a position emerge from my analysis of EB-in-practice.

For sure, EB as it is used in practice is a powerful statistical method, but what is responsible for its success is not only the mathematics within it, but *also* the associated stochastic models *and* the way priors are computed. Maybe EB is more a toolbox that a tool, as suggested by the variety of tasks it is said by its proponents to fulfill. It is indeed described as a methodology both for estimation and inference, but also as a modelling technique, and as a approach to problem-solving aiming at analyzing data, finding regularities, establishing links between data and hypotheses, establishing causal claims.

Beyond EB's efficiency, there is still another aspect of this approach that relates it with the notion of a tool, namely, that it is mostly viewed as useful and not as any theoretically well-grounded methodology. In brief, the justification of why its use is rapidly developing does not rely on principles of rationality or reasoning but only on its usefulness. It is an instrument, no more no less. This would be fine in another context in which the tasks this instrument is fulfilling were not considered fundamental. The problem is that analyzing data, finding regularities, establishing links between data and hypotheses, and establishing causal claims are considered fundamental tasks whose fulfillment should rely on well-justified principles. This is the reason why the debate between classical and Bayesian statisticians was so hot and the philosophy of statistics such a battlefield. From this perspective, EB appears as *just* a tool without the status of a well-grounded method.

As a concluding remark, I should emphasize that the view of EB as *just* a tool partly dovetails with Carrier and Lenhard's theses about the tool-character of mathematics. According to Carrier and Lehnard, there is a growing tendency to use mathematics as tools without caring whether this use is theoretically well-justified. The need for producing well-grounded representations of natural of social phenomena has been fading in the same time as new mathematics, or a new use of mathematics, allowed for short-circuiting theoretical representations. As a result, the traditional hierarchy opposing well-grounded theoretical representations to merely useful ones can be dispensed with, so that it is now clear that mathematics can be used without (much) theory. By using mathematics without relying on theories, scientists are not only saving the phenomena but also producing good science.

My former analysis of EB as a toolbox agrees with Carrier and Lenhard's theses to the extent that, as Mayo strongly emphasizes, EB's foundations lack conceptual clarity, to say the least. However, as I have myself emphasized, EB is not only mathematics, it is also about modelling. I claim that the mathematical part within EB cannot be detached from the modelling part so that EB is no example of *mathematics* as a tool, but of modelling plus statistical inference as a tool. I also claim that the engine of EB's efficiency is probably not mathematics per se, but machine computation, which seems to me a somewhat different topic from mathematics.

5 Conclusion

In this paper, I have presented a new trend in statistics called "Empirical Bayes" and I have illustrated it by examples from climate statistics. I have tried to avoid technicalities in the presentation of EB in order to focus on its methodological aspects and its involving the construction of stochastic models. Throughout the paper, I have also kept EB separated from its old-Bayesian ancestors in order to focus on the efficiency of this approach, a feature old-Bayesian statistics usually lack. A different story could have been told on EB by looking more closely at its Bayesian origin. I have rather chosen to focus on its entrenchment in computational science.

In Sect. 3, I have made clear how my analysis of EB relates to Carrier and Lenhard's theses about mathematics as a tool. On the one hand, EB is a nice example of a tool whose use is not strongly based on theoretical justifications, but on the other hand, EB is not only a mathematical tool but has other constituent parts that go beyond mathematics.

References

Annan, J. D., & Hargreaves, J. C. (2006). Using multiple observationally-based constraints to estimate climate sensitivity. *Geophys Res Lett, 33*, L06704.

Berliner, L. M., Richard, L. M., Levine, A., & Shea, D. S. (2000). Bayesian climate change assessment. *Journal of Climate, 13*, 3805–3820.

Casella, G. (1992). Ilustrating empirical Bayes methods. *Chemometrics and Intelligent Laboratory Sytems, 16*, 107–125.

Frigg, R., Bradley, S., Du, H., & Smith, L. A. (2014). Model Error and Ensemble Forecasting: A Cautionary Tale. Iin: Guichun C. Guo and Chuang Liu (eds.) Scientific Explanation and Methodology of Science. *Singapore: World Scientific, 2014*, 58–66.

Hasselmann, K. (1998). Conventional and Bayesian approach to climate-change detection and attribution. *Quarterly Journal of the Royal Meteorological Society, 124*, 2541–2565.

Hegerl, G. C., Crowley, T. J., Hyde, W. T., & Frame, D. J. (2006). Climate sensitivity constrained by temperature reconstructions over the past seven centuries. *Nature, 440*, 1029–1032.

Kass, R. (2011). Statistical inference: The big picture. *Statistical Science, 26*, 1–20.

Kennedy, M. C., & O'Hagan, A. (2001). Bayesian calibration of computer models. *Journal of the Royal Statistical Society Series B, 3*, 425–464.

Mayo D. G. (2013). Discussion: Bayesian methods: Applied? Yes. Philosophical Defense? In Flux. *The American Statistician 67*(1): 11–15. (Commentary on A. Gelman and C. Robert "Not only defended but also applied: The perceived absurdity of Bayesian inference" (with discussion)

Parker, D. (2010). Whose probabilities? Predicting climate change with ensembles of models. *Philosophy of Science, 77*(5), 985–997.

Rober C. (2016). Des spécificités de l'approche bayésienne et de ses justifications en statistique inférentielle. In Drouet (Ed.), *Eclairages philosophiques sur les méthodes bayésiennes*, Editions Matériologiques, Paris, to appear .

Rougier, J. C. (2007). Probabilistic inference for future climate using an ensemble of climate model evaluations. *Climatic Change, 81*, 247–264.

Rougier, J. C., & Crucifix, M. (2012). Uncertainty in climate science and climate policy. In L. Lloyd, & E. Winsberg (Eds.), *Conceptual issues in climate modeling*. University of Chicago Press (forthcoming).

van Fraassen, B. (2008). *Scientific representation: Paradoxes of perspective*. Oxford: Clarendon Press, Oxford.

Winsberg, E., & Goodwin, W. M. (2016). The adventures of climate science in the sweet land of idle arguments. *Studies in History and Philosophy of Modern Physics, 54*, 9–17.

Part III
Reflections on the Tool Character

On the Epistemic and Social Foundations of Mathematics as Tool and Instrument in Observatories, 1793–1846

David Aubin

1 Introduction

> The astronomer is dependent on his tools; the observatory is but the receptacle of his tools, his tool-chest so to speak (Harrington 1883–1884, 249).

One night, in June 1782, the Astronomer Royal Nevil Maskelyne (1732–1811) suddenly felt "much out of love with his instrument."[1] William Herschel (1738–1822) had come to Greenwich to stargaze in his company. But to realize that all telescopes in the Royal Observatory were so much inferior to Herschel's new reflector was disheartening to Maskelyne. Observatory scientists indeed loved their instruments. In their publications, they devoted hundreds of pages to the description of telescopes. They drew them in exquisite details. They lobbied for years to obtain the requisite funds to buy the most precise instruments, only to wait even longer for the most reputable makers finally to provide them. Then, they spent hours and hours chasing and charting their inevitable defects. They discussed at great lengths the operations required for their perfect calibration, attended to their proper care, and improved them constantly (Carl 1863; Bell 1922; King 1955; Chapman 1996). Much more rarely, however, did tools command their attention. Astronomers, as we know, also loved mathematics. "Every part of the operations of an observatory is

[1]William Herschel to Caroline Herschel, June 5, 1782 (Lubbock 1933, 115).

D. Aubin (✉)
Institut de mathématiques de Jussieu-Paris rive gauche (CNRS, Paris-Diderot, UPMC), Sorbonne Universités, UPMC Univ Paris 6, 4 place Jussieu, F-75252, Paris, France
e-mail: david.aubin@upmc.fr

J. Lenhard, M. Carrier (eds.), *Mathematics as a Tool*, Boston Studies in the Philosophy and History of Science 327, DOI 10.1007/978-3-319-54469-4_10

mathematical," wrote one of Maskelyne's successors, George Biddell Airy (1801–1892), in a memo dated December 4, 1861.[2] In the practice of observatory scientists, mathematical theories and methods often were as central as their instruments. As the Königsberg Observatory director Friedrich Wilhelm Bessel (1784–1846) famously commented, mathematical corrections themselves improved the precision of telescopes:

> Every instrument is made twice, once in the workshop of the artisan, in brass and steel, but then again by the astronomer on paper, by means of the list of necessary corrections, which he determines in his investigations. (Bessel 1848, 432); trans. (Crowe 1994, 156).[3]

In view of examining how scientists use mathematics as a tool, we therefore see that the specific cultural and epistemological spaces delineated by astronomical observatories provide a promising terrain for a careful study of attitudes with respect to mathematics, tools, and instruments, and to their interactions. Observatory scientists themselves compared mathematics to their most cherished instruments. Logarithms, for example, were seen as an "admirable artifice that, by shortening computations, extends astronomers' lives [just as] the telescope ha[s] increased their sight" (Biot 1803, 26). The link between the telescope and the logarithm echoed Kepler's striking frontispiece to the *Rudolphine Tables* (1627) where "in the upper part, inventions that were the most useful to Astronomy are represented: Galileo's telescope, Napier's logarithms, and Kepler's ellipse" (Delambre 1821, 1:558) (see Fig. 1). In astronomical context, the comparison between a mathematical device, the logarithm, and a physical instrument, the telescope, was explicit.

Of course, one has to remember that the word "instrument" is ambiguous (Van Helden and Henkins 1994). In the *Novum Organum*, Francis Bacon (1561–1626) described the way in which "Man, Nature[']s Minister and Interpreter," was able to act and understand through experience and reason. In its original seventeenth-century English translation, Bacon's second aphorism read:

> Things are performed by instruments and helps, which the Understanding needs as much as the Hand. Now as Mechanick Instruments assist and govern the Hands motion, likewise the instruments of the Understanding prompt and advise it. (Bacon 1676, 1); quoted in (Taub 2011, 691).

In this sense, it is self-evident that telescopes, logarithms, and ellipses should be thought of as different types of instruments for the eye or for the mind.[4] But, going beyond this simple identification, can we situate more precisely the

[2]"Remarks on the neglect, by the Junior Assistants, of the course of education and scientific preparation recommended to them." Airy Papers, Cambridge University Library, RGO 6/43, 235. About this memo, see Aubin (2009), 273 and 276–277.

[3]Jedes Instrument wird auf diese Art zweimal gemacht, einmal in der Werkstatt des Künstlers von Messing und Stahl; zum zweitenmale aber von dem Astronomen aud seinem Papiere, durch die Register der nöthingen Verbesserungen, welche er durch seine Untersuchung erlangt.

[4]In this article, I shall not refer to the "mathematical instruments" tradition, which obviously played a role in shaping the conception of instruments in the observatory culture. By the late eighteenth century, astronomers were relying on highly skilled makers, such as Jesse Ramsey and Edward

Fig. 1 On the top of the allegorical kiosk pictured on the frontispiece of Johnannes Kepler's *Rudolphine Tables* (1627). Muses are holding some of the principal tools and instruments used by Kepler, from left to right: what looks like a representation of an eclipse; a telescope with line drawing used in optical theory inside; logarithms represented by Napier's computing rods and the number 6931492, a close approximation of log 2; geometry holding a compass and a set square in her hands with a representation of an ellipse; an unequal-arm balance, and magnetism represented by loath stone and compass. On this image, see (Jardine et al. 2014)

astronomer's mathematical practices with respect to their instrumental practices? In earlier publications, we have offered a synthetic discussion of "observatory techniques" over the long nineteenth century (Aubin et al. 2010; Aubin 2015). Characteristic of the observatory culture, such techniques involved instrumental technology and observation practices, the social organization of working practices, and cognitive tools such as mathematical theories and methods. Whether they originated or not in this culture, observatory techniques formed a coherent set of practices and technologies. A specific aspect of the observatory culture was the role played by numbers and their collection, manipulation, storage, and transformation. Mathematics, I argued elsewhere, was just another instrument in observatory scientists' panoply (Aubin 2009, 282).

The present chapter is intended as a contribution to the philosophical debate regarding the "unreasonable effectiveness," or applicability of mathematics to the natural sciences (Wigner 1960; Steiner 1998; Bangu 2012). If we take seriously the idea that mathematics was conceived *as* a tool or *as* an instrument—and, as I try to show, I believe that we must—then we need to pay attention to a few things. First, as Steiner has argued, this conception of the applicability of mathematics will slide the problem away from concepts, laws, and language to *practice*.[5] On what ground indeed could a telescope be applied to the understanding of astronomical

Troughton, fellows of the Royal Society whose names were, as we shall see, routinely attached to the high-precision instruments they produced (Bennett 2011; Chapman 1995).

[5]As expressed by Gottlob Frege in 1884, there is a simple solution to the applicability problem: "The laws of numbers, therefore, are not really applicable to external things; they are not laws of nature. They are, however, applicable to judgments holding good of things in the external world: they are laws of the laws of nature" (Frege 1980), quoted by (Withold 2006, 74). For another visions of mathematics as a language in this context, see Sarukkai (2005).

phenomena? This had very little to do with the way the instrument *expressed* scientific truths, but everything to do with the way it was devised, built, maintained, studied, and used. Second, to make the comparison with mathematics effective and culturally meaningful, I believe it is essential to pay a closer attention to the variety of material tools and instruments one encounters in relevant scientific environments. Documenting the case of the Paris Observatory in 1793, I am led to distinguish, for the sake of my argument, between scientific instruments and mere tools, and among the latter between simple tools and high-precision tools. Third, as I show, such distinctions necessarily involved social criteria regarding users and the conditions of their use of devices. There were therefore high stakes in terms of users' status in designating their equipment as tools or instruments. Fourth, these distinctions also applied to the way mathematics was used by various people as tool and as instrument. Like in the philosophy of mathematical practice (Mancosu 2008), this approach shifts the focus of our attention from mathematical foundations to a wider range of practices. Looking at the computers' daily routine at Greenwich in 1839, I exemplify the various meanings mathematics as a tool can have there. Finally, pursuing this approach raises new questions: How far can the analogy between mathematics and tools or instruments go? Have the notions of care, improvements, maintenance, and fixes corresponding meanings as far as mathematical instruments are concerned? Did astronomers develop with regards to their mathematical instruments the same kind of personal attachment, intimate knowledge about the ins and outs, and attention to the life history of their most prized material instruments? To provide extensive answers to such interrogation would be enough for a whole research program. With the perspective I adopt here, there are a number of cases in the history of mathematics that might be revisited with profit. Astronomers tinkered over and over again with the mathematical instruments that were handed down to them in order to increase their precision.[6] In the last part of this paper, I will illustrate the fruitfulness of this approach by considering the case of Bessel functions.

2 Tools vs. Instruments in the Paris Observatory, 1793

To try and make sense of the distinction between tools and instruments in the context of the observatory, one may look at inventories. Take the case of the "inventory of instruments of the National Observatory in Paris" from 1793.[7] This

[6]One may think, among other cases, of the Gauss–Laplace theory of errors (Sheynin 1973) or Poincaré's qualitative dynamics (Roque 2015).

[7]See "Pièce justificative N ° X: Inventaire des instrumens de l'Observatoire national de Paris en 1793," Archives nationales F17/1219; Archives de l'Observatoire D.5.38; repr. (Cassini 1810), 208–217). The date of this report is uncertain, but September 19, 1793 is a reasonable estimate (Wolf 1902, 349). Dated "19 of the first month of Year II" [which should be 19 vendémiaire, that is, October 10], the report was said to be completed when discussed by the Commission temporaire

report was drafted in rather dramatic circumstances. Following the abolition of the Academy of Sciences by the Republican Government, the traditional organization of the Paris Observatory under the Cassini dynasty was overturned. A decree stated that the Observatory was to be "republicanized" and that the former director Jean-Dominique Cassini, also known as Cassini IV (1748–1845), would loose all prerogatives over his former assistants. As a result Cassini resigned from his position and, on September 19, 1793, a delegation was sent to the Paris Observatory in order to draw a list of all its instruments.

With respect to the distinction between instruments and tools [*outils*], this inventory is highly revealing, in part because it was written, not by astronomers, but by the delegates appointed by the Revolutionary *Commission temporaire des arts*, set up by Government following the suppression of the Academy of Science in August 1793.[8] These delegates were the physicist Jacques Charles (1746–1823), and the instrument makers Étienne Lenoir (1744–1832) and Jean Fortin (1750–1831) who unlike scientists were not highly experienced as writers. Fully conscious of what was at stake, Cassini however oversaw the inventory with keen eyes, as well as Jean Perny de Villeneuve (1765–?), Cassini's former assistant and now temporary director of the Observatory, and the young Alexis Bouvard (1767–1843), who would stay at the observatory for the next fifty years.[9]

The inventory of instruments drafted by the commissioners was divided in several sections: clocks (17 items), refracting telescopes (13), achromatic objectives (4), simple objectives (26), reflectors (5), micrometers (13), generic instruments (26), and—interestingly—"tools and bits of machines" [*outils et débris de machines*] (24). As we can see, "tools" were listed here as a special kind of "instruments," but perhaps not as highly valued as the others, as can be inferred from their showing up pell-mell with other bits and pieces at the end of the inventory. Astronomers' knowledge about their instruments was rather intimate: their dimensions, origins, flaws, and present states were all precisely established. The most valuable ones had proper names attached to them and in fact often were the combination of several instruments. The first refracting telescope was described as such:

> N° 1. Achromatic telescope by Dollond, objective with three glass [pieces] of 42 lines of aperture, 3 feet and half of focal length; it has three oculars, one terrestrial and two celestial ones; it is mounted on a mahogany stand with cupper columns with all its motions; to this

des arts on September 26. According to his biographer, Cassini left the Observatory never to come back, on October 3rd (S.-Devic 1851, 205).

[8]Informations about the Commission and these commissioners can in particular be found in its proceedings (Tuetet 1912). On scientific instruments at this time, see also Daumas (1953). On Charles, see Blondel (2003).

[9]Under the Terror, Cassini was jailed in the English Benedictine Convent on February 14, 1794. He was freed after Robespierre's downfall, on August 5, 1794, but never returned to the Observatory. For more information on the history of the Paris Observatory during the French Revolution, I refer to Chapin (1990) and Aubin (2013); more complete accounts in French can be found in Cassini (1810), S.-Devic (1851), Wolf (1902).

> telescope an heliometer by Bouger can be fitted, simple objective, plus a wire micrometer by Hautpois (Cassini 1810, 209).[10]

Let us emphasize the individuality of such instruments. Cassini for example remembered having used Dollond's refractor to observe the phases of Saturn in 1774 (Cassini 1793, 153). Also known as the *lunette du prince de Conti* because the Prince had bought it after the Duke of Chaulnes's death in 1769, the Dollond refractor was rented to Cassini III on several occasions and sold to the Observatory in 1778 (Wolf 1902, 242–243); (Barty-King 1986, 92). Found to have serious defects, it had fallen in disrepair at the time of the inventory.[11]

"Tools" in contrast were more generic, even when they were better built. In fact, the tools mentioned in the inventory were not simple tools like hammers or screwdrivers, but rather high-precision special-purpose tools, like three long steel rulers, which had been "worked for an infinite [amount of] time and with infinite care to obtain a straight line," and drawing marble tables "polished in mirror-like manner" (Cassini 1810, 215–216). These tools, Cassini explained in a footnote added later, had been acquired to equip the workshop set up at the Observatory in 1784 in an ill-fated attempt at fostering the development of high-precision instrument making in Paris (Wolf 1902, 277–286; Daumas 1953, 358–360). Only the most notable tools therefore were deemed worthy of mention in the inventory, as they already were in Cassini's detailed account books, which indicated the rather steep price paid for the steel rulers and the marble tables (respectively, 480 and 368 *livres* Wolf 1902, 279–280). Although much higher, the cost of furnishing and equipping the workshop with common tools was significantly given in bulk in Cassini's accounts. While there was no entry for the word *instrument* in Diderot and D'Alembert's *Encyclopédie* a generic definition was given to *outil* by Louis de Jaucourt, one of the most prolific contributors to the encyclopedia. He provided a similar distinction between tools and instruments:

> TOOL, ... an instrument used by workmen and artisans to work on the different tasks of their professions, crafts, and trades; thus are hammers, compasses, planes, squares, braces, etc. ... Let us simply add that workmen distinguish somewhat between tools and instruments, and that not all instruments are tools (Diderot and D'Alembert 1751–1765, 11:718, orig. emphasis).[12]

Jaucourt's definition echoed the inventory made by Charles, Fortin, and Lenoir who agreed that tools were a specific genre of instruments, albeit less noble, not all

---

[10] N ° 1. Lunette achromatique de Dollond, objectif à trois verres de 42 lignes d'ouverture, 3 pieds et demi de foyer ; elle a trois oculaires, un terrestre et deux célestes ; elle est montée sur un pied d'acajou à colonne de cuivre avec tous ses mouvemens; à cette lunette s'adapte un héliomètre de Bouger, objectif simple, plus un micromètre filaire de Hautpois.

[11] About instruments in state of disrepair, see Schaffer (2011).

[12] OUTIL, (...) instrument dont les ouvriers & artisans se servent pour travailler aux différens ouvrages de leur profession, art & métier ; tels sont les marteaux, les compas, les rabots, les équerres, les villebrequins, &c. (...) Nous ajoutons seulement que les ouvriers mettent quelque différence entre les outils & les instrumens ; tout outil étant instrument, & tout instrument n'étant point outil.

instruments being tools. In a book by Louis Cotte (1740–1815), one finds more complex and more interesting webs of meaning around those terms.[13] Intent of showing in his *Vocabulaire portatif des mécaniques* that the beauty of God's work was no less present in the products of human art and industry than in Nature, Cotte explained the differences among the terms *machine*, *instrument*, *apparatus*, and *tool*:

> We understand by *machine* a combination of several simple machines, such as the lever, the winch, the pulley, etc. whose action is to stand in for man's strength and to produce great effects in little time and at little expanse in all mechanical operations in which they are employed. ... The *instrument* is a kind of machine, but [one] susceptible of great precision, to be employed in scientific operations which require accuracy, like astronomy, practical geometry, surgery, etc. The *apparatus* is a combination of different instruments whose combination contributes to the demonstration of physical, mathematical, and chemical truths. The *tool* is a simple instrument, often of the sort of the wedge, which is useful in the manual and common operations of the crafts and trades (Cotte 1801, viii–ix, orig. emphasis).[14]

In Cotte's view, tools were simple instruments, which themselves were a special sort of machines. The distinctions he established were based on two criteria: dichotomy between simplicity and precision, on the one hand, and the kinds of operations they were employed in, on the other. Instruments were characterized by precision and their use in operations geared at discovering new truths about nature, whereas tools were necessary simple and to be used by craftsmen in their daily and mundane occupations.[15]

Going back to the Paris Observatory inventory, we see that Cotte's criteria apply well, but not completely. In the inventory, the only explicit distinction was social. "Instruments" were high-precision objects used by savants in their scientific operations in order to uncover scientific truths, whereas "tools" were to be used by craftsmen and makers in the workshop. But there was nothing common or

[13]Father Cotte, a cleric, is known for his work on meteorology and on the popularization of natural history, physics, and astronomy (Pueyo 1994).

[14]On entend par *Machine* une combinaison de plusieurs machines simples, telles que le levier, le treuil, la poulie, etc. dont le résultat est de suppléer aux forces de l'homme et de produire de grands effets en peu de tems et avec peu de dépense dans toutes les *Opérations mécaniques* où elles sont employées (...). L'*Instrument* est bien aussi une espèce de machine, mais susceptible d'une très-grande précision, pour pouvoir être employée dans les *Opérations scientifiques* qui demandent de l'exactitude, comme l'astronomie, la géométrie pratique, la chirurgie, etc. L'*Appareil* est une combinaison de différens instrumens dont la réunion concourt à démontrer les *vérités physiques, mathématiques, chimiques, etc.* L'*Outil* est un instrument simple, le plus souvent de l'espèce du coin, qui sert dans les *Opérations manuelles et habituelles* des arts et des métiers.

[15]It is interesting to note that, in the definition of the word *instrument*, the *Oxford English Dictionary* today explains, similarly, that the distinction between tools, instruments, and machines, is based on social, rather than lexicographical, grounds: "Now usually distinguished from a tool, as being used for more delicate work or for artistic or scientific purposes; a workman or artisan has his tools, a draughtsman, surgeon, dentist, astronomical observer, his instruments. Distinguished from a machine, as being simpler, having less mechanism, and doing less work of itself; but the terms overlap." Note added to the definition of "Instrument" (www.oed.com).

mundane about the tools listed, and their precision seemed highly valued. At the Paris Observatory, *tools* were specialized instruments acquired to help artisans make the optical *instruments* required by astronomers. Instruments surely lay both literally and figuratively at the center of the observatory, but the place of tools was ambiguous in this cultural space. While simple tools went unmentioned, some of the high-precision tools found their place in the list side by side with instruments. Like instruments, such tools possessed their own history and individuality, but they were to be used primarily by different people, not astronomers.

There is a final twist we need to take into account. In a culture that praised inventiveness, to design one's own tools might however be good ground for claiming some degree of recognition from the astronomers' part. At the end of the seventeenth century, Abraham Sharp, John Flamsteed's assistant at Greenwich, thus was highly praised for his skill as an observer and as a mathematician, but also for his talent as a mechanic, having "made himself most of the tools used by joiners, clock-makers, opticians, mathematical-instrument-makers, &c." (Anon. 1781, 462). When the Dudley Observatory was built in Albany, New York, in 1856, the overseeing board duly noted the director Benjamin Apthorp Gould's technical ingenuity with his tools: "Great difficulties were encountered in boring or drilling the horizontal holes through the stone piers of the meridian circle, a difficulty attributed by Dr. Gould in great measure to the inefficiency of the tools. He changed the whole character of the drills, using cast iron instead of steel; and with much simpler appliances has accomplished the work successfully" (Henry et al. 1858, 23).[16] The use of tools was reserved to people with a lower status in the observatory, but inventively tinkering with them could help assistants and astronomers alike gain some degree of recognition.

3 The Eight-Hour Day: Mathematics as a Tool at Greenwich, 1839

In this context, whether observatory mathematics could be seen as a simple tool, a high-precision tool, or an instrument crucially hinged on the way it was put to use: Who would use it? How? And to what purpose? To examine this issue, let us focus on the Royal Observatory Greenwich under Airy's directorship. In the memo quoted above, Airy explicitly spelt out his view of the place of mathematics in an observatory.

> The action and faults of telescopes and microscopes require for their understanding a knowledge of Mathematical Optics. Every discussion and interpretation of the observations requires Mathematical Astronomy. The higher problems, such as the discovery of the ele-

[16]On the history of the Dudley Observatory, see Wise (2004).

> ments of a comet's orbit from observations, require the high Mathematics of Gravitational Astronomy.[17]

In short, mathematics was everywhere. Moreover, in Airy's understanding, formal hierarchies among personnel hinged on their level of mathematical knowledge much more than anything else.[18] At the bottom of the scale, were supernumerary computers. In addition to being able to "write a good hand and good figures" and "to write well from dictation, to spell correctly and to punctuate fairly", computers were to have rudimentary mathematical knowledge, essentially restricted to arithmetic, including vulgar and decimal fractions, extraction of square roots, use of logarithms, and the use of $\pm$. Next came the Assistant, first, second and third grades, whose competencies were mostly evaluated through their knowledge of mathematics, from simple Euclidean geometry to analytical mechanics "especially in reference to Gravitational Astronomy." In Airy's organization, computers were the mathematical artisans of the observatory, and, as such, I would like to say that they used mathematics as a tool, and not as an instrument. They used mathematics like apprentices in optical workshops handled the basic unspoken tools such as drills and bores, and not like astronomers manipulated telescopes in observatories. To illustrate this point, let us focus on a significant episode of the labor history of computing: the way in which Greenwich computers earned the eight-hour day.[19] It all started on Monday afternoon, January 21, 1839. Computers were chitchatting in the Royal Observatory's Octagon Room. Rain had fallen nearly all day and the Astronomer Royal had left for London.[20] One of them started to complain about their working conditions. His was a new face in the room: some H. W. Bowman who had just been hired. His age is unknown, but like most of the other computers there, he must have been very young, twenty years old at the most. Less than two weeks earlier, Airy had described what Bowman's new job would be about: "The work is almost entirely of calculation, and it is highly important that the computer should understand the use of the $+$ and $-$ signs in the various operations of additions, subtraction, multiplication and division."[21] Before a formal offer could be made to Bowman, it was agreed that the head of the computing bureau, J. W. Thomas, would check the value of his work. Now, Bowman seemed unhappy with the job. If he were to work so many hours, he lamented to his colleagues, he could not possibly live long. Typically, computers at Greenwich worked from 8 a.m. to 8 p.m.; they had a one-hour break for dinner,

[17] Airy, "Remarks on the neglect, by the Junior Assistants, of the course of education and scientific preparation recommended to them" (Dec. 4, 1861). RGO 6/43, 235.

[18] Several slightly different copies of this memo are extent in Airy's papers at the Cambridge University Library. A first draft was written on November 20, 1856 and a slightly revised version was adopted on May 10, 1857. In the following I quote from RGO 6/43, 170–175. For a more detailed analysis of this memo, see Aubin (2009, 277).

[19] On the division of computing labor at Greenwich, see Grier (2005).

[20] RGO 6/24/33: Airy's Diary

[21] Airy to Bowman, January 8, 1839. RGO 6/526, 86.

while Saturday afternoons and Sundays were free. Therefore, Bowman went on, "time must have some weight in making [him] an offer."[22]

Brought in the observatory as a computer, Bowman however seems to have had few skills worth bragging about. At first he had trouble with additions and subtractions: "He knew nothing of either," his superior Thomas reported. "The – he called a stroke and the + he called a cross and he considered the sign × equally the same." Asked to find the logarithm of the sine of $42°19'$ in Jean-François Callet's *Tables portatives* (published in 1783 on the basis of William Gardiner's), Bowman did not know where to begin. But despite foreboding beginnings, two weeks later Thomas was forced to aver that Bowman "has come pretty middling—he can add –, ×, and ÷ very well providing the numbers be of one denomination." All in all, depending on whether he was willing to study or not, the new recruit could "be valued equal Mr Richard Dunkin."[23] This comparison spoke highly in favor of Bowman. Richard Dunkin was the son of William Dunkin, himself a computer of considerable experience who had worked for the *Nautical Almanac* for more than twenty years. Hired at twenty-three, he was allowed to stay in Truro, Cornwall, working in his home. In 1832 when the Nautical Almanac Office was established in London by its new superintendent, Lieutenant William Samuel Stratford, to his regret Dunkin Sr. was forced to leave with his family his "semi-independent position at Truro," and accept a "daily sedentary confinement to an office-desk for a stated number of hours in the company of colleagues all junior to himself in age and habits" (Dunkin 1999, 45).[24] William Dunkin never ceased to regret his previous life. Like many in his cohort who reluctantly saw the industrial age forced upon them, Dunkin wished his sons to do better. Having worked as a miner in his youth, he remembered that: "the underground tributer's work was occasionally far more lucrative than the scientific work to which his after-life was devoted" (Dunkin 1999, 42–43). He thus took great care of his sons' education which he expected would help them get a foothold in business: "His great desire was that they should be educated for a mercantile life, and that they should not become computers. As a computer all his life from youth, he always, and perhaps truly, considered that it was not a profession that gave much prospect of advancement in social position" (Dunkin 1999, 46). In Camden Town, where the Dunkins had their house, the boys were schooled at Wellington House (where Charles Dickens had been a pupil in 1824–1827). Later, at least two of them were sent to a French boarding school, near Calais. But sorrows repeatedly stroke the Dunkins. William's first son died in

[22]Thomas to Airy, January 21, 1839. RGO 6/525, 29, orig. emphasis; (Dunkin 1999, 72).

[23]All quotations in the above paragraph from Thomas to Airy, January 21, 1839. RGO 6/524 File 10bis, 352, orig. emphasis. "Units of the same kind but of different magnitudes, as pounds, shillings, pence, and farthings, which are units of value ... are called units of different denominations" (De Morgan 1836, 25).

[24]Stratford (1791–1853) entered the Navy on February 10, 1806; he was the first secretary of the Royal Astronomical Society in 1826–1831 and superintendent of the *Nautical Almanac* from 1830 to his death. On the story of this institution up to that point, see Dunkin (1898) and Croarken (2003).

March 1832. Three years later, an abscess formed on one of his feet, which would not go away. The walk to the Nautical Almanac Office got harder and harder until William was allowed to compute at home again. Meanwhile, in less than five weeks in the spring of 1836 two more of Dunkin's children, a boy and a girl, had died. On July 3, 1838, William Dunkin also passed away, leaving his wife alone with two boys: Richard aged 15 and Edwin two years his elder. Lieutenant Stratford knew the Dunkin family was in duress when he recommended the boys to Airy. Richard and Edwin were recalled from their boarding school (Dunkin 1896, 197), and in August of that year, both joined Airy's team of computers, where the boys quickly proved to be excellent. On January 21, 1839, after having reported Bowman's complaints to his boss, Thomas nonetheless warned Airy that if working conditions remained too harsh, the Dunkins might lend receptive ears to offers they might receive from elsewhere—"and to lose these two would be a great loss."[25] Bowman indeed was not the only one to whine about the practical aspects of his new job. Earlier, a young computer named Thaddeus Foley also had had his conflict with Thomas: he "condemns the place. He styles it a beastly place, a slavery, and that no one but a half starved beggar would stop in it."[26] Consulted on the matter, Lieutenant Stratford could not hide his surprise at Airy's readiness to "binding persons" for so long every day.[27] How could the Astronomer Royal expect work to be done well in such conditions? "As to myself," Thomas the overseer went on, "I complain not," but like Stratford he supported his computers' complaints, if only for fear of injuring the work.

> Being constant at work a person becomes stupefied and although still working at the same rate now begins to commit blunders and the examiner will with the same state of mind run over them and mark them as correct—such has been the case and such will be the case. And I am confident that to continue to work 11 hours a day much longer will not answer the purpose.[28]

In the culture of the nineteenth observatory, precision was everything (Wise 1995). The paradox of the computer's work therefore was posed as such: how could one rest assured that a low-skilled employee with a small pay would be carrying out a tedious and repetitive task for several long hours every day without making any error; and when errors inevitably occurred, how could they be promptly detected? In the event, the Astronomer Royal was sensitive to Thomas's argument. He agreed to reduce working hours from 8 a.m. to 5 p.m. with an hour break, or from 8 p.m. to 4 p.m. if the computers were to take no break at all. Everyone agreed they preferred to finish at the earliest time possible. The Greenwich computers had earned the eight-hour day.

---

[25] All quotes are from Thomas to Airy, January 21, 1839. RGO 6/525, 29.

[26] Thomas to Airy, undated [1838]. RGO 6/525, 15, orig. emphasis. Esq. Mathematical Master at the Royal Naval School, Camberwell, Foley was later elected a Fellow of the Royal Astronomical Society; see *Monthly Notices of the Royal Astronomical Society* 6 (1844), 52.

[27] Stratford's opinions about computers are in Thomas to Airy, January 21, 1839. RGO 6/525, 29.

[28] Thomas to Airy, January 21, 1839. RGO 6/525, 29.

Social progress notwithstanding, this episode—perhaps more than the case of the personal equation to which we shall come back—illustrates the way in which Airy's observatory was transformed into a factory with a strict discipline (Schaffer 1988). The mathematics mobilized by computers consisted of simple tools indeed and their work was mostly manual. Mathematics was for sure used as a tool at Greenwich: it was a tool, and perhaps a machine-tool, used by computers to perform a task assigned to them under the constant supervision of their boss; it was a tool used by assistant to climb up the hierarchy; it was a tool used by the observatory director to assert his authority over his staff and the general public. But, as a simple tool, it hardly deserved special attention. These aspects of the use of mathematics in the observatory leaving few public records are therefore rather hard to study historically. There are however instances when mathematics ceased to be a simple tool and became one of the most powerful instruments at observatory scientists' disposal.

4 Improving Instruments with Mathematics and Mathematics as an Instrument in Königsberg, 1815–1824

In the observatory context, using mathematics as an instrument is a whole other matter than using it as a simple tool. Like all other instruments—or, rather, in combination with them—mathematics used as an instrument was put to the service of increasing precision. At the end of his life, Bessel was already terminally ill when on October 5, 1854 he wrote a touching letter to Airy.[29] Referring to the publication over which Airy's computers had toiled for years, Bessel expressed his great pleasure "in the evening of my scientific life to see a work completed on whose advancement I have bestowed a great part of the morning and noon of that life."[30] He praised the Astronomer Royal for having "brought through to an end the great and still extending labour" and recalled how at the beginning of his career he had set on the similar task that would be Airy's main inspiration:

> When, forty years ago, I entered upon my astronomical course, and found myself in the full possession of the bodily strength and activity which are indispensable to render a life useful to science, I found myself naturally obliged to consider carefully the state of astronomy, and to exert myself to make out clearly what must be done in order to establish it more firmly than it appeared to be established: in order to give it a form which should not withdraw itself from the ever-advancing improvement (founded on the very nature of science) in our knowledge of the heavenly movements, but should rather enable us to increase progressively the correctness of the earlier determination by the use of the later observations, so that they might asymptotically draw near to the truth which never can be reached.

[29]On Bessel, see Hammel (1984) and Ławrynowicz (1995), as well as Olesko (1991).

[30]Bessel to Airy, October 5, 1845. RGO 6/530, 75–76. I quote from one of the two (nearly identical) English translations in Airy's papers (ibid., 76–78 and 79–80).

To improve the correctness of observations made earlier led Bessel to study carefully every source of errors he could think of. Errors of astronomical observation, he wrote, formed two classes: those that "are dependent on innumerable accidental causes and therefore can be considered to follow the general propositions of the calculus of probability" and those "that are provoked by constantly acting causes and which are to be ascribed to the deviation of the instruments from their *mathematical* ideal or from their manner of treatment."[31] Skirting around the issue of the personal equation, Bessel had first believed that the second class of errors could be diminished "through the *insight* of the observer and rigour in the investigation of the instrument and in the method of observation" (Bessel 1819, 19, orig. emphasis). By carefully tracking down Bessel's role in the well-visited history of the personal equation, Christoph Hoffmann has reminded us that in this case the observer was not considered as a human laborer in need of discipline, but rather as a source of errors among others.[32]

For our purpose, Bessel's final treatment of the personal equation rather points toward the conclusion that mathematics served in the hand of the skilled astronomer as a high-precision tool that would improve the accuracy of observation, whether by investigating the instrument, the physical conditions in which it was used, or even physiological differences between observers. Bessel's *Tabulæ Regiomontanæ* (1830) were a masterwork of error analysis based on Greenwich observations of fundamental stars (including Maskelyne's) and gave the mean and apparent motions of 38 stars from 1750 to 1850. About these tables, John Herschel (son of William's) wrote: "It affords the first example of the complete and thorough reduction of great series of observations, grounded, in the first instance, on a rigorous investigation, from the observations themselves, of all the instrumental errors, and carried out on a uniform plan, neglecting no minutiæ which a refined analysis and a perfect system of computation could afford" (Herschel 1847, 203–204). Herschel also quoted the Danish astronomer Heinrich Christian Schumacher (1780–1850) according to whom: "One may almost assert that one exact and able calculator is capable of doing better service to astronomical science than two new observatories" [ibid., 203]. In this sense, mathematics was an instrument that truly replaced optical instruments.

In Bessel's practice, mathematics was thoroughly used as a specialized tool, like the expensive rulers and marble tables found at the Paris Observatory, and his innovative use of such tool drew attention to his work. In a letter to Airy about his method (November 9, 1833), Bessel commented that: "[i]t would be useless, to enter here into the particularities of the computations" (Bessel 1875–1876, 3:462). There was no need to discuss such simple tools as mere computation. To improve the precision of the observations by means of the mathematical treatment

---

[31](Bessel 1819), 19; trans. (Hoffmann 2007), 346, my emphasis: "die eine enthält die eigentlichen Beobachtungsfehler, die von unzähligen zufälligen Ursachen abhängen und deshalb den allgemeinen Sätzen der Wahrscheinlichkeitsrechnung folgend angesehen werden können; die andere begreift die von beständig einwirkenden Ursachen herrührenden, der Abweichung der Instrumente von ihrer mathematischen Idee, oder ihrer Behandlungsart zuzuschreibenden."

[32]See Hoffmann (2007). On the personal equation, see also Schaffer (1988) and Canales (2001).

of raw data, Bessel however emphasized some high-precision tool coming from practical astronomy, in particular the computation of *elements*, that is, the main constants (aberration, nutation, refraction) to take into account in the reduction of astronomical data. "Every new inquiry, increasing the weight of the result, issueing [sic] from the combination of this and former inquiries, the remaining error, probably will diminish continually; but this error, never vanishing entirely, it will (generally speaking) be necessary, to exhibit the result of a computation, depending upon some assumed values of certain Elements in a form open to further corrections" (Bessel 1875–1876, 3:463). In other words, as a high-precision tool, mathematics could be used to reduce the errors, but remained dependent on a set of theoretical and observational assumptions.

Bessel went further and also explored various uses of mathematics that take us even closer to the astronomer's idea of instrument. When he set up his observatory in Königsberg, Bessel received a Dollond telescope of 4 feet of focal length and 2.7 inches of aperture (Bessel 1815, iii). As he wrote to Carl Friedrich Gauss on 30 December 1813, this was "one of the best instruments in existence," believing he would be able to observe with it angles with a precision of $1''$ to $2''$ (Gauss and Bessel 1975, 181). However, it was not perfect. In his letter to Gauss, Bessel tellingly described in the same breath a modest shade he had devised to protect his telescope from sun warmth and new methods in mathematical analysis he had come up with to improve the precision of his instrument by analyzing its small defects. Having undertaken the microscopic investigation of the small ellipticity and eccentricity of the pivot, Bessel explained that he used what he called an application of Gauss's least-square method, which he had come upon on another occasion and which he felt was "very elegant." In his later publication, Bessel explained: "Before giving the detailed investigation of this and other errors of the circle, I allow myself a digression about the solution of a class of equations frequently occurring in practical astronomy, which will find their application here."[33]

In my view, Bessel's "digression" represents the moment when he ceased to view mathematics as a tool, even a high-precision one mobilized to improve his Dollond refractor, and started to consider mathematics itself as an instrument. Mathematics as an instrument should also be submitted to the same inquiry as physical instruments and observers. In doing so, mathematics itself would be perfected not for its own sake, but for the sake of improving the precision of astronomical observations. Let us briefly say what Bessel's digression was about. It concerns Bessel's own approach at determining the coefficient of a finite trigonometric series representing a function, developed independently from Joseph Fourier's (Dutka 1995, 127). On July 2, 1818, Bessel read a paper at the Berlin Academy of Sciences where he extended his earlier digression to infinite trigonometric series, therefore providing a

[33]Bessel (1819), ix; Bessel (1875–1876), 2:24: "Ehe ich die nähere Untersuchung dieses und der übrigen Fehler des Kreises mittheile, erlaube ich mir eine Abschweifung über die Auflösung einer häufig in der praktischen Astronomie vorkommenden Classe von Gleichungen, die ihre Anwendung finden wird."

first formal treatment of what came to be called Fourier series. When a function of U is expanded in the series (Bessel 1819); (Bessel 1875–1876, 1:18):

$$U = A' \sin u + A'' \sin 2u + \cdots + A^{(i)} \sin iu + \ldots$$
$$+B' \cos u + B'' \cos 2u + \cdots + B^{(i)} \cos iu + \ldots$$

then, in general, the coefficients could be expressed as

$$A^{(i)} = \frac{1}{\pi} \int_0^{2\pi} U \sin iu \, du,$$
$$B^{(i)} = \frac{1}{\pi} \int_0^{2\pi} U \cos iu \, du$$

Pointing out that this type of series expansion was useful in a variety of astronomical questions, he went on to describe the way these developments could be used to give an analytic solution of what he called the "Keplerian problem," that is, the computation of the true variation (or anomaly) of an elliptic orbit perturbed by another body. Bessel explained that the difference between the true and mean anomalies could be expressed as a trigonometric series of the mean anomaly, whose coefficient would be expressed as a power series of the eccentricity.

Pursuing his studies on that problem, Bessel presented a paper titled "Study of the Part of Planetary Perturbations Arising from the Motion of the Sun" at the Berlin Academy of Science on January 29, 1824. In this remarkable communication (Bessel 1824), he considered that when the motion of a planet around the sun was perturbed by another planet, this perturbation had two parts: the direct action of the perturbing planet and the indirect action of the sun whose trajectory was also perturbed. To tackle this problem, Bessel followed Laplace's method and expressed the problem in polar coordinates. Extending his earlier approach to this case, he found that results could be given as a function of two integrals he noted I_k^h and J_k^h. Exploring further mathematical properties of these integrals, he established several of their mathematical properties, for example the recursive property:

$$kI_k^{h-1} - 2hI_k^h + kI_k^{h+1} = 0.$$

Typically, for an observatory scientist, Bessel also provided several pages of tables giving numerical values of the functions I_k^0 and I_k^1 for different values of k. These expressions came to be called Bessel functions. Of the mathematical tools developed by astronomers in the early nineteenth century, they certainly are among the most prominent.[34] But if Bessel functions are a high-precision tool for mathematical physicists and engineers, I want to argue that, in the observatory culture, they had more similarities with instruments than with tools. Indeed, Bessel did not

[34]See, e.g., Watson (1966) and Hochstadt (1986), ch. 8.

develop this analysis by tinkering with existing mathematics in order to improve the efficiency of an existing instrument but rather invented new mathematical theories to test Newton's theory of universal gravitation. Bessel indeed saw the finality of his investigations as addressing the question "whether astronomical theories are everywhere in such great agreement with the observations so as to reject all doubts regarding the truth of Newton's hypothesis."[35] Just like telescopes, mathematical analysis itself was in Bessel's hands an instrument that could be perfected in order to probe the very foundations of the Newtonian theory.

5 Conclusion

As Bessel himself had intuited, mathematics would soon literally become an instrument for the discovery of new celestial bodies (Ławrynowicz 1995, 219). On September 23, 1846, the planet Neptune was discovered at the place computed by Urbain Le Verrier (1811–1877), who would soon declare that "[t]he *great instrument* with the help of which all these [astronomical] questions will be solved will be none other than the study and computations of perturbations."[36] Struck by this discovery, Alexander von Humboldt (1769–1859) was curious to know what great development of the human thought, what new "organ" (like the telescope, algebra, or analysis) to use his own term, had made this discovery possible and asked his friend the mathematician Carl Jacobi (1804–1851) for his opinion.[37] In a "thunderous letter," Jacobi replied: "Good God!" This was not the result of any "deep thought, but [that of] a nimble hand." No deep mathematical thinking was involved in Le Verrier's discovery. The only reason why mathematicians did not themselves take care of all astronomy was because this science relied on numerical results: it was "boring to have to escort any thin idea with 10,000 logarithms."[38]

This exchange underscores discrepant conceptions of mathematics as a tool and as an instrument. It nicely illustrates the slippage that could ensue from too

[35]Bessel (1824); repr. Bessel (1875–1876), 1:86: "Ob aber die astronomischen Theorien allenthalben in so grosser übereinstimmung mit den Beobachtungen sind, dass dadurch jeder Zweifel an der Wahrheit der Newton'schen Annahme zurückgewiesen wird, dieses ist eine Frage, welche wohl Niemand bejahen wird, deren genaue Erörterung jedoch sehr wichtig ist und die grössten Fortschritte der Wissenschaft verheisst." For a study of Bessel's cosmological understanding of Newton's law, see Merleau-Ponty (1983), 119–122.

[36]Excerpt from Le Verrier inaugural lecture at the Sorbonne; quoted in *Revue scientifique et industrielle* 28 (1847), 131.

[37]Humboldt to Jacobi, December 22, 1846 (Humboldt and Jacobi 1987, 103).

[38]Jacobi to Humboldt, December 26, 1846; (Humboldt and Jacobi 1987, 104): "Du lieber Gott! Hier heißt es nicht, Gedanken tief, sondern Hand flink." Below, Jacobi added: "dieser Sachen kriegen erst durch die wirkliche numerische Ausfürung Werth, und es ist langweilich, jeder dünnen Gedanken sogleich mit 10,000 Logaritmen escortiren so müssen" (Humboldt and Jacobi 1987, 105–106). Humboldt himself called Jacobi's a "donnerdend Brief" (Humboldt and Jacobi 1987, 109).

close an association with material devices. While in the hands of Le Verrier and Bessel, mathematical analysis and computation had become powerful instruments for astronomical research, others like Jacobi relegated this to mere manual work of deskilled computers. By emphasizing the boredom of computational work in the observatory (Donnelly 2014), Jacobi was degrading astronomers' status to that of menial workers, and the use of mathematics to the mere handling of a tool—a high-precision tool to be sure, but just a tool. "[I]f Leverrier has seen a new planet with the eye of mathematics," Jacobi added, 'I have myself given a new eye to mathematics."[39] The mathematician found that toying with the instrument itself was more fun than using it. The design, care, and maintenance of such tools, he seemed to say, was of a higher purpose. Only this work ensured that mathematics remained applicable to astronomy and able to provide the required level of precision. A new conception of mathematics as *Instrumentenkunde* (Carl 1863) was emerging, according to which mathematics could and should be developed autonomously from the needs of astronomical workers, or any other user of mathematics, for that matter. As Jacobi famously wrote about Fourier, a question about numbers was worth as much as a question about the system of the world and mathematics ought solely to be pursued "for the honor of human spirit."[40]

References

Anonymous. (1781). Biographical memoirs of Mr. Sharp. *Gentlemen's Magazine, 51*,461–463.

Aubin, D. (2009). Observatory mathematics in the nineteenth century. In E. Robson & J. A. Steddal (Eds.), *Oxford handbook for the history of mathematics* (pp. 273–298). Oxford: Oxford University Press.

Aubin, D. (2013). On the cosmopolitics of astronomy in nineteenth-century Paris. In S. Neef, D. Boschung, & H. Sussman (Eds.), *Astroculture: Figurations of cosmology in media and arts* (pp. 61–84). Paderborn: Wilhelm Fink.

Aubin, D. (2015). L'observatoire: régimes de spatialité et délocalisation du savoir, 1769–1917. In D. Pestre (Ed.), *Histoire des sciences et des savoirs de la Renaissance à nos jours* (Vol. 2, pp. 54–71). Paris: Le Seuil.

Aubin, D., Bigg, C., & Sibum, H. O. (Eds.). (2010). *The heavens on earth: Observatories and astronomy in nineteenth-century science and culture*. Rayleigh: Duke University Press.

Bacon, F. (1676). *The Novum Organum of Sir Francis Bacon, Baron of Verulam, Viscount St. Albans, Epitomiz'd, for a Clearer Understanding of His Natural History, Translated and Taken Out of the Latine by M.D.*. London: Thomas Lee.

[39]Jacobi to Humboldt, December 21, 1846 (Humboldt and Jacobi 1987, 100): "wenn Leverrier mit dem Auge der Mathematik einen neuen Planeten gesehen hat, habe ich damals der Mathematik ein neues Auge eingesetzt." Jacobi was referring to his theory of elliptic functions (1829) which was used by Le Verrier in his work.

[40]Jacobi to Legendre, July 2, 1830 (Jacobi 1881–1884, 1:453): "le but unique de la science, c'est l'honneur de l'esprit humain, et (...) sous ce titre, une question de nombres vaut autant qu'une question de système du monde." In a complementary approach, one might also consider Bessel acting as a research-technologist in the sense put forward in (Shinn 2008).

Bangu, S. (2012). *The applicability of mathematics in science: Indispensability and ontology*. Basingstoke: Palgrave Macmillan.
Barty-King, H. (1986). *Eyes right: The story of Dollond & Aitchison opticians, 1750–1985*. London: Quiller Press.
Bell, L. (1922). *The telescope*. New York: MacGraw-Hill.
Bennett, J. (2011). Early modern mathematical instruments. *Isis, 102*, 697–705.
Bessel, F. W. (1815). Beschreibung und Untersuchung der Instrumente. *Astronomische Beobachtungen auf der Königlischen Universitäts-Sternwarte in Königsberg* (1813–1814) 1:iii–xxiv. Repr.: Ueber das Dollond'sche Mittagsfernrohr und den Cary'schen Kreis. In Bessel (1875–1876).
Bessel, F. W. (1819). Analytische Auflösung der Kepler'schen Aufgabe. *Abhandlungen der Berliner Akademie Akademie der Wissenschaften in Berlin. Mathematische Classe* (1816–1817): 49–55. Repr. In Bessel (1875–1876).
Bessel, F. W. (1824). Untersuchung des Theils der planetaruschen Störungen, welcher aux der Bewegung der Sonne entsteht. Repr. In Bessel (1875–1876).
Bessel, F. W. (1848). *Populäre Vorlesungen über wissenschaftliche Gegenstände* (H. C. Schumacher, Ed.). Hamburg: Perthes-Besser & Mauke.
Bessel, F. W. (1875–1876). *Abhandlungen* (3 Vols) (R. Engelmann, Ed.). Leipzig: Wilhelm Engelmann.
Biot, J.-B. (1803). *Essai sur l'histoire générale des sciences pendant la Révolution française*. Paris: Duprat & Fuchs.
Blondel, C.-J. (2003). Un enfant illustre de Beaugency: Le physicien et aéronaute Jacques Charles (1746–1823). *Les Publications de l'Académie d'Orléans : Agriculture, Sciences, Belles-Lettres et Arts*, no. 4 (December), 1–81.
Canales, J. (2001). The single eye: Reevaluating ancien régime science. *History of Science, 39*, 71–94.
Carl, P. (1863). *Die Principien der astronomischen Instrumentenkunde*. Leipzig: Voigt & Günther.
Cassini, J.-D. (1793). Des principales observations faites depuis 1671 jusqu'en 1789, sur les phases de l'anneau de Saturne. *Mémoires de l'Académie des sciences pour 1789*, 142–153.
Cassini, J.-D. (1810). *Mémoires pour servir à l'histoire des sciences et à celle de l'Observatoire royal de Paris, suivis de la vie de J.-D. Cassini écrite par lui-même et des éloges de plusieurs académiciens morts pendant la Révolution*. Paris: Bleuet.
Chapin, S. L. (1990). The vicissitudes of a scientific institution: A decade of change at the Paris Observatory. *Journal for the History of Astronomy, 21*, 235–274.
Chapman, A. (1995). *Dividing the circle: The development of critical angular measurement in astronomy, 1500–1850* (2nd ed.). Chichester: John Wiley.
Chapman, A. (1996). *Astronomical instruments and their users: Tycho Brahe to William Lassel*. Aldershot: Variorum.
Cotte, L. (1801). *Vocabulaire portatif des mécaniques, ou définition, description abrégée et usage des machines, instrumens et outils employés dans les sciences, les arts et les métiers, avec l'indication des ouvrages où se trouve leur description plus détaillée*. Paris: Delalain.
Croarken, M. (2003). Tabulating the heavens: First computers of the Nautical Almanac. *IEEE Annals of the History of Computing, 2*, 48–61.
Crowe, M. J. (1994). *Modern theories of the universe: From Hershel to Hubble*. New York: Dover.
Daumas, M. (1953). *Les Instruments scientifiques aux XVIIe et XVIIIe siècles*. Paris: Presses universitaires de France.
Delambre, J.-B. (1821). *Histoire de l'astronomie moderne* (2 Vols.). Paris: Veuve Courcier.
De Morgan, A. (1836). Arithmetic and Algebra. In: *Library of useful knowledge, mathematics*. London: Baldwin and Cradock.
Diderot, D., & D'Alembert, J. Le R. (Eds.). (1751–1765). *Encyclopédie, , ou dictionnaire raisonné des sciences, des arts et des métiers* (17 Vols.). Paris & Neuchâtel.
Donnelly, K. (2014). On the boredom of science: Positional astronomy in the nineteenth century. *British Journal for the History of Science, 47*,479–503.

Dunkin, E. (1896). Richard Dunkin [obituary notice]. *Monthly Notices of the Royal Astronomical Society, 56*, 197–199.

Dunkin, E. (1898). Notes on some points connected with the early history of the "Nautical Almanac." *The Observatory, 21*, 49–53, 123–127.

Dunkin, E. (1999). *A far off vision: A Cornishman at Greenwich observatory, autobiographical notes by Edwin Dunkin, F.R.S., F.R.A.S. (1821–1898) with notes on the lives & work of his father, brother and son* (P. D. Hingley & T. C. Daniel, Eds.). Truro, Cornwall: Royal Institution of Cornwall.

Dutka, J. (1995). On the early history of Bessel's functions. *Archives for History of Exact Sciences, 49*, 105–134.

Frege, G. (1980). *The foundations of arithmetic* (J. L. Austin, Trans.). Oxford: Blackwell. (Originally published 1884)

Gauss, C. F., & Bessel, F. W. (1975). *Briefwechsel*. [= Carl Friedrich Gauss, *Werke. Ergänzungreihe*, vol. 1] Hildeheim: Gorg Olms. Repr. from *Briefwechsel zwischen Gauss und Bessel*. Leipzig: Wilhelm Engelmann, 1880.

Grier, D. A. (2005). *When computers were human*. Princeton: Princeton University Press.

Hammel, J. (1984). *Friedrich Wilhem Bessel*. Leipzig: Teubner.

Harrington, M. W. (1883–1884). The tools of the astronomer. *The Sidereal Messenger, 2*, 248–251, 261–268, 297–305.

Henry, J., Bache, A. D., & Pierce, B. (1858). *Defense of Dr. Gould by the scientific council of the Dudley Observatory* (3rd ed.). Albany: The Scientific Council.

Herschel, J. F. W. (1847). A brief notice of the life, researches, and discoveries of Friedrich Wilhelm Bessel. *Notices of the Royal Astronomical Society, 7*, 200–214.

Hochstadt, H. (1986). *The functions of mathematical physics*. New York: Dover.

Hoffmann, C. (2007). Constant differences: Friedrich Wilhelm Bessel, the concept of the observer in early nineteenth-century practical astronomy, and the history of the personal equation. *British Journal for the History of Science, 40*, 333–365.

von Humboldt, A., & Jacobi, C. G. J. (1987). *Briefwechsel* (H. Pieper, Ed.). Berlin: Akademie.

Jacobi, C. G. J. (1881–1884). *Gesammelte Werke* (8 Vols). Berlin: G. Reimer.

Jardine, N., Leedham-Green, E., & Lewis, C. (2014). Johann Baptist Hebenstreit's Idyll on the Temple of Urania, the Frontispiece Image of Kepler's Rudolphine Tables, Part 1: Context and Significance. *Journal for the History of Astronomy, 45*, 1–19.

King, H. C. (1955). *The history of the telescope*. London: Charles Griffin.

Ławrynowicz, K. (1995). *Friedrich Wilhelm Bessel 1784–1846* (K. Hansen-Matyssek & H. Matyssek, Trans.). Berlin: Birkhaüser.

Lubbock, C. A. (1933). *The Herschel chronicle*. Cambridge: Cambridge University Press.

Mancosu, P. (Ed.). (2008). *The philosophy of mathematical practice*. Oxford: Oxford University Press.

Merleau-Ponty, J. (1983). *Les Sciences de l'univers à l'âge du positivisme: étude sur les origines de la cosmologie contemporaine*. Paris: Vrin.

Olesko, K. (1991). *Physics as a calling: Discipline and practice in the Königsberg seminar for physics*. Ithaca: Cornell University Press.

Pueyo, G. (1994). Les deux vocations de Louis Cotte, prêtre et météorologiste (1740–1815). *Bulletin des Académie et Société lorraines des sciences, 33*, 205–212.

Roque, T. (2015). L'originalité de Poincaré en mécanique céleste: pratique des solutions périodiques dans un réseau de textes. *Revue d'Histoire des Mathématiques, 21*, 37–105.

S.-Devic, M. J. F. (1851). *Histoire de la vie et des travaux scientifiques et littéraires de J[ean]-D[ominique] Cassini IV*. Clermont, Oise: Alexandre Daix.

Sarukkai, S. (2005). Revisiting the unreasonable effectiveness of mathematics. *Current Science, 88*, 415–423.

Schaffer, S. (1988). Astronomers mark time. *Science in Context, 2*, 115–145.

Schaffer, S. (2011). Easily cracked: Scientific instruments in states of disrepair. *Isis, 102*, 706–717.

Sheynin, O. B. (1973). Mathematical treatment of astronomical observations (a historical essay). *Archives for the History of Exact Sciences, 11*, 97–126.

Shinn, T. (2008). *Research-technology and cultural change: Instrumentation, genericity, transversality*. Oxford: Bardwell Press.
Steiner, M. (1998). *The applicability of mathematics as a philosophical problem*. Cambridge, MA: Harvard University Press.
Taub, L. (2011). Reengaging with instruments. *Isis, 102*, 689–696.
Tuetet, L. (Ed.). (1912). *Procès-Verbaux de la Commission temporaire des arts* (2 Vols). Paris: Imprimerie nationale.
Van Helden, A., & Henkins, T. L. (Eds.). (1994). *Instruments in the history of science*. Special issue of *Osiris* 9.
Watson, G. N. (1966). *A treatise of the theory of Bessel functions* (2nd ed.). Cambridge: Cambridge University Press.
Wigner, E. P. (1960). The unreasonable effectiveness of mathematics in the natural sciences. *Communications on Pure and Applied Mathematics, 13*, 1–14.
Wise, G. (2004). *Civic astronomy: Albany's Dudley Observatory, 1852–2002*. Astrophysics and Space Science Library, Dordrecht: Kluwer.
Wise, M. N. (Ed.). (1995). *The values of precision*. Princeton: Princeton University Press.
Withold, T. (2006). Lost on the way from Frege to Carnap: How the philosophy of science forgot the applicability problem. *Grazer Philosophische Studien, 73*, 62–82.
Wolf, C. (1902). *Histoire de l'Observatoire de Paris de sa foundation à 1793*. Paris: Gauthier-Villars.

Approaching Reality by Idealization: How Fluid Resistance Was Studied by Ideal Flow Theory

Michael Eckert

Abbreviations

DMA Deutsches Museum, Archiv, München
GOAR Archiv des Deutschen Zentrum für Luft- und Raumfahrt, Göttingen

The ideal and the real appear most often as antagonists. The former transforms the target of an investigation into an object that is amenable to the laws of physics and further mathematical analysis; the latter abstains from undue simplification and aims at a close correspondence between theory and observation. Ideal gas theory, for example, approaches reality by replacing a real gas with a hypothetical counterpart described by a simplified gas law. Real gases require more complex descriptions for a precise agreement between theoretical and observed behavior. Ideal flow theory is another example. By the neglect of fluid resistance it dismisses a crucial feature of real flows of water, air or other fluid media. According to ideal flow theory a body moving at constant speed through a fluid does not experience a resistance (D'Alembert's Paradox). Therefore, such an idealization is inappropriate for investigating the drag of a body in the real flow of water or air. At least this seems to be the obvious lesson from the failure to cope with real fluids in terms of a theory that regards fluids as inviscid.

Yet ideal flow theory, together with the associated mathematical tools of potential theory, paved the way for the understanding of fluid resistance. In this chapter I trace the history of this paradoxical juxtaposition of idealization versus practice in fluid mechanics. Already the birth of ideal flow theory in the eighteenth century was deeply rooted in practical concerns. Leonhard Euler's conception of ideal flow theory (Euler's equations) was preceded by a practical request: the analysis of

M. Eckert (✉)
Deutsches Museum, Museumsinsel 1, D-80538, München, Germany
e-mail: m.eckert@deutsches-museum.de

J. Lenhard, M. Carrier (eds.), *Mathematics as a Tool*, Boston Studies in the Philosophy and History of Science 327, DOI 10.1007/978-3-319-54469-4_11

pipe flow for fountains at Frederick the Great's summer palace Sanssouci. Another idealization was introduced in the nineteenth century by the Scottish engineer William Rankine who developed the notion of streamlines – originally perceived as "ship-lines" in an attempt to determine the optimal shape of ship hulls with minimal resistance. Rankine's idealization involved the so-called "method of sources and sinks," a powerful tool for optimising streamlined bodies. In the twentieth century this method was found useful for the design of Pitot tubes (instruments to measure flow velocities in terms of pressures) and for the comparison of theoretical and experimental results concerning the flow around airship models in a wind tunnel. These investigations became instrumental for discriminating between various kinds of resistance (form drag, skin friction, lift-induced drag).

In this chapter I analyse idealization in fluid mechanics as a tool for practical ends. Although the proverbial gulf between hydrodynamics, perceived as pure physics, and hydraulics, its engineering counterpart, is often regarded as the result of idealization, I will emphasize here the uses of idealization for practice. In Sect. 1 I briefly illustrate how real flow problems in the eighteenth and nineteenth centuries were idealized in order to obtain mathematical tools for practical applications. Section 2 is dedicated to Ludwig Prandtl and the birth of boundary layer theory widely perceived as a bridge between theory and practice in fluid mechanics. Section 3 is focused on a device for measuring flow speed, the Pitot tube, which was optimized by applying the theory of ideal flow. Section 4 deals with the shedding of vortices in the wake of an obstacle in a flow, where ideal flow theory was instrumental for explaining the stable arrangement of alternating vortices ("Kármán vortex street"). Section 5 is concerned with the aerodynamics of wings where ideal flow theory sheds light on the phenomenon of "induced drag" – a resistance caused by vortices trailing away from the wing tips. All these cases illustrate how Ludwig Prandtl and his school of applied mechanics (Heinrich Blasius, Georg Fuhrmann, Theodore von Kármán) developed useful theories for hydraulic and aeronautical engineering from idealized concepts.

1 The Practical Roots of Ideal Flow Theory

In the eighteenth century the term hydraulics was not yet distinct from hydrodynamics. Both denoted the science of moving water in its broadest sense – including mathematical analysis as well as practical application (Zedler 1735; Chambers 1738; Rees 1819). Bernoulli's law, for example, today regarded as the law of energy conservation in flows, was discovered in the course of practical investigations of pipe flow. It is named after Johann Bernoulli (1667–1748) and Daniel Bernoulli (1700–1782), father and son. The father, Johann, introduced it in his treatise *Hydraulica*; the son, Daniel, in his *Hydrodynamica*. In an attempt to claim priority over his son, Johann antedated his *Hydraulica* to "Anno 1732" in the printed title, while Daniel's *Hydrodynamica* appeared in 1738. But *Hydraulica* was published only five years after Daniel's *Hydrodynamica* (Rouse 1968; Mikhailov 2002; Szabo

1987; Calero 1996). Disregarding their priority dispute, neither the father nor the son aimed at a general law of energy conservation. Bernoulli's law, as it appears first in Daniel's *Hydrodynamica*, refers to the outflow of water from a container through an attached pipe with an orifice of given size; the problem was to find the pressure against the wall of the pipe as a function of the height of water in the container and the size of the orifice. Problems like these belonged to the "art of raising water" including "the several machines employed for that purpose, as siphons, pumps, syringes, fountains, jets d'eau, fire-engines, etc," as a contemporary encyclopedia categorized them (Chambers 1738).

It was left to Leonhard Euler (1707–1783) to give Bernoulli's law its now familiar form. In 1755, he derived the equations of motion for ideal (i.e. frictionless) fluids, the Euler equations (Truesdell 1954). Although his treatise appears unrelated to practice, this impression is misleading. Euler's work on hydrodynamics was deeply rooted in the same eighteenth century water art problems that gave rise to the treatises of the Bernoullis.

Yet modern scientists and historians ridiculed Euler's idealization as the origin for the schism between theory and practice. "Unfortunately, he omitted the effects of friction, with embarrassing practical consequences. When Euler applied his equations to design a fountain for Frederick the Great of Prussia, it failed to work," a physicist argued about a famous mishap at the summer palace of the Prussian King at Sanssouci (Perkovitz 1999). I have analysed Euler's alleged failure elsewhere Eckert (2002, 2008). Here it may suffice to contrast the widespread slander of Euler as a mathematician who was "letting his mathematics run away with his sense of reality" (Bell 1937, pp. 168–169) with the opposite view that Euler revealed in his work "a highly perceptive engineering mentality that illustrates the depths of his technical knowledge" (Steele 2006, p. 290). It is true that the hydraulic design for the pumps and pipes at Sanssouci was doomed to failure, but Euler was not responsible for this flawed design; he was only asked to analyse the cause of its failure. In the course of this analysis he established what finally entered the textbooks of hydraulic engineering as the non-stationary pipe flow theory. Euler correctly explained why the Sanssouci water art system failed. The design for pumping water to a reservoir involved a great distance between the pumps and the reservoir which necessitated the application of high pressures in order to set the water in the pipes in motion. The inertia of the water in the pipes–not friction–was the culprit. By approaching the problem from an idealized perspective Euler arrived at a basic understanding of the nature of the problem. The neglect of friction allowed him to focus on the inertial motion of water flow in pipes which enabled him to discern the major deficiencies of the design. Euler derived from this ideal pipe flow theory practical suggestions that could have been used to avoid further failure. But his advice was ignored.

The Sanssouci affair illustrates how Euler used idealization in order to cope with practical problems. By the same token he had solved other problems in naval architecture, ballistics, and hydraulic machinery, before he arrived at the general equations of fluid motion that established ideal flow theory.

Another contribution to ideal flow theory is due to the Scottish engineer William John Macquorn Rankine (1820–1872): the "method of sources and sinks" for constructing streamlines around bodies in ideal flows. It originated from Rankine's interests in naval architecture. Rankine attempted to construct "the paths in which the particles of a liquid move past a solid body," as he explained to the British Association for Advancement of Science in 1864. Originally he had considered these paths "as figures for the horizontal or nearly horizontal water-lines of ships," but in the course of his mathematical analysis he preferred the notion of "Stream-Lines" for this graphical representation of flow patterns "as being a more general term" (Rankine 1864, p. 282).

In a subsequent "Mathematical Theory of Stream-Lines" (Rankine 1871, p. 287) he explicitly referred to George Green's famous Essay from the year 1828 in which the methods of potential theory had been developed and applied to electricity and magnetism (Cannell and Lord 1993). Rankine was not the first to apply the mathematical apparatus of potential theory in hydrodynamics, as earlier studies of ideal flow problems by Hermann von Helmholtz and others demonstrate (Darrigol 2005), but Rankine's use of this tool resulted in a display of stream-lines that was mathematically constructed from the superposition of a uniform flow with a given distribution of point-like flow sources and sinks ("foci"). If the total flow ejected from the sources equaled that sucked up by the sinks, the resulting flow resembled that around a closed body. The shape of this body was dependent on the distribution of sources and sinks. The method was applicable "not only to bifocal, quadrifocal, and other stream-line surfaces having foci situated in one axis, but to all stream-line surfaces which can be generated by combining a uniform current with disturbances generated by pairs of foci arranged in any manner whatsoever, or having, instead of detached focal points, focal spaces" (Rankine 1871, p. 291). In other words, the method could be used to compare the ideal flow generated by an appropriate distribution of sources and sinks with actual flows around bodies with corresponding shapes.

Potential theory limited the use of Rankine's method to ideal flow, but this did not prevent him from considerations on friction. He closed his theory of stream-lines with "Remarks on the Skin-resistance" in which he referred to the "well known" observations among engineers in naval architecture, "that the friction between a ship and the water acts by producing a great number of very small eddies in a thin layer of water close to the skin of the vessel." Thus he anticipated a boundary layer as the site of skin friction. He also coined the notion of "eddy resistance" for what was later named vortex drag or form drag. However, he was aware that such a resistance was beyond the scope of his method (Rankine 1871, p. 291) (see also Darrigol 2005, pp. 274–277).

2 Prandtl's Boundary Layer Concept

By the end of the nineteenth century hydrodynamics had developed into a sophisticated mathematical theory. The equations of motion had been extended from Euler's equations to the Navier-Stokes equations by adding a term that was supposed to account for friction. However, actual flow phenomena often eluded the efforts of mathematical theory to such an extent "that technology has adopted its own procedure to deal with hydrodynamical problems, which is usually called hydraulics," as the author of a textbook on hydrodynamics observed in 1900. "This latter specialty, however, lacks so much of a strict method, in its foundations as well as in its conclusions, that most of its results do not deserve a higher value than that of empirical formulae with a very limited range of validity" (Wien 1900, p. III).

The gap between hydraulics and hydrodynamics–by now perceived as the epitome for the gulf between down-to-earth engineering and ivory-tower science–was most pronounced when fluid resistance in pipes was analysed with regard to the flow velocity and the diameter of the pipe. "Physical theory predicts a frictional resistance proportional to the velocity and inversely proportional to the square of the diameter, according to the technical theory it is proportional to the square of the velocity and inversely proportional to the diameter." Thus Arnold Sommerfeld (1868–1951) alluded to the gap between hydraulics and hydrodynamics in a public talk at the Technical University Aachen in 1900. "The physical theory agrees splendidly in capillary tubes; but if one calculates the frictional losses for a water pipeline one finds in certain circumstances values which are wrong by a factor of 100" (Sommerfeld 1900).

Although the discrepancy was not resolved for at least another decade, the concept by which the solution would finally be found took shape only four years later. "I have posed myself the task to do a systematic research about the laws of motion for a fluid in such cases when the friction is assumed to be very small," Ludwig Prandtl (1875–1953) introduced in 1904 a paper which became famous as the starting point for the boundary layer concept (Prandtl 1905, p. 485). The occasion for this presentation was the Third International Congress of Mathematics, held in August 1904 in Heidelberg, but Prandtl's performance was unusual with regard to the audience at this event. The paper contained little mathematics. Prandtl was then a young professor of mechanics at the Technical University Hanover. His primary interest in fluid mechanics was engineering, not mathematical. At the time of his Heidelberg presentation he had just accepted a call to Göttingen University, where the mathematician Felix Klein (1849–1925) considered him an ideal candidate to add technical mechanics to the academic fields of a university. Klein's effort was motivated by the contemporary tension between academic and technical universities (Manegold 1970). Prandtl was at first hesitant whether he should accept the Göttingen offer because of his affiliation with engineering. "The gravest doubt emerged from my sense of belonging to technology," he wrote to Klein in May 1904, but the prospect of "the beautiful scientific Göttingen intercourse" helped to overcome his doubts (quoted in Rotta 1990, p. 297).

Prandtl's motivation for his Heidelberg talk, therefore, was rooted in practice rather than theory. When he had finished his study of engineering at the Technical University Munich in 1900, his first employment (before he was called to the Technical University Hanover in 1901) was as engineer in a machine factory where he had to design an exhaust system for wood shavings and swarf. "I had arranged a tapered tube in order to retrieve pressure," Prandtl recalled many years later, "instead of a pressure retrieval, however, the flow of air became detached from the walls" (Prandtl 1948). When Prandtl was called to Hanover he made this phenomenon the subject of further study. Why would a flow become detached from the wall? He constructed a water canal with a camera for visual observation of flow detachment. At the Heidelberg Congress he presented photographs of flow detachment obtained from this canal–quite an unusual performance for the attending mathematicians. Prandtl sketched the central idea behind his boundary layer concept as follows (Prandtl 1905, p. 486):

> By far the most important part of the problem concerns the behaviour of the fluid at the walls of solid bodies. The physical processes in the boundary layer between the fluid and the solid body is addressed in a sufficient manner if one assumes that the fluid does not slip at the walls, so that the velocity there is zero or equal to the velocity of the solid body. If the friction is very small and the path of the fluid along the wall not very long the velocity will attain its free stream value already at a very close distance from the wall. Although friction is small, within the small transitional layer, the abrupt changes of velocity result in considerable effects.

If the influence of friction is limited to a thin layer, the flow beyond this layer at greater distance from the wall can be assumed as ideal flow–thus the theory of inviscid fluids could still be used outside the boundary layer. For the flow in the boundary layer along a flat plate Prandtl derived approximate equations (by cancelling terms from the Navier-Stokes equations) from which he obtained a numerical result for the skin friction coefficient. However, he did not disclose what mathematical method he had invoked to arrive at his result, nor did he discuss the problems involved with the derivation of his boundary layer equations from the Navier-Stokes equations. From a mathematical vantage point Prandtl's boundary layer approach was justified only many years later by the method of matched asymptotic expansions (see, e.g. Schlichting and Gersten 2000, chapter 14.2).

Disregarding qualms about the mathematical justification of his approach, Prandtl asked his first doctoral student, Heinrich Blasius (1883–1970), to elaborate the (laminar) flow in the boundary layer along a flat plate in more detail (Blasius 1907; Hager 2003). Blasius reduced Prandtl's boundary layer equations by a suitable transformation of variables to an ordinary differential equation that could be solved in terms of a power series. Thus he was able to derive the velocity profile in the laminar boundary layer along a flat plate ("Blasius flow") and to improve Prandtl's first estimate for the ensuing friction coefficient. Albeit limited to laminar flow only, this was the first successful computation of skin friction based on rational mechanics combined with empirical observation.

3 How to Shape Pitot Tubes and Air-Ships?

The boundary layer concept provided a rehabilitation of ideal flow theory for many practical cases. For fluids with small viscosity like water and air, the thickness of the laminar boundary layer is extremely thin in many practical applications. Therefore, the ideal flow regime reaches close to the actual surface of a body that moves through such a fluid. In other words, the tools of ideal flow theory, such as Rankine's method of sources and sinks, would be applicable in order to compute the flow velocity and pressure in the vicinity of bodies constructed by this method.

It is therefore not accidental that the first computation of this kind is due to Blasius. After his doctoral work under Prandtl he was employed as a physicist at the Berlin Testing Facility for Hydraulics and Naval Architecture (Versuchsanstalt für Wasserbau und Schiffbau). One of his first tasks was to improve so-called Pitot tubes–devices for determining the flow velocity by measuring the pressure difference between the "stagnation point" in front of the tube and at a site where the flow is tangential to the opening. But Pitot-tubes with different shapes yielded different results. "The Pitot-tubes are here discredited very much," Blasius wrote in a letter to Prandtl shortly after he had started his new job in Berlin (Blasius to Prandtl, 29 July 1908, GOAR 3684).

In his Göttingen laboratory, Prandtl used a similar device, combined with a high-precision liquid column manometer, for measuring the airspeed in a wind tunnel. "Designed with the appropriate shape, the difference of pressure [measured at the openings at the front and on the sides] corresponds to the velocity head," Prandtl explained the principle of this instrument (Prandtl 1909). Thus he expressed the well-known result from ideal flow theory, $h = v^2/2g$, where g is the gravity constant and the "velocity head" h is the height of a water column generated by the pressure difference between the stagnation point at the front opening (where the flow velocity is zero) and the free stream velocity v measured at the opening on the side. The Göttingen device was later called "Prandtl tube" (Prandtlsches Staurohr, see Rotta 1990, p. 48). The velocity head measured with different Pitot tubes differed by a "gauge factor" from the ideal value. As Blasius confided to Prandtl, the instrument originally used in the Berlin laboratory had a gauge factor of 1.82. In order to bring this factor close to 1, the appropriate shape was crucial.

When Blasius began to analyse different shapes of Pitot-tubes with regard to their gauge factor, the "Göttingen shape" served him as a role model (Blasius to Prandtl, 13 August 1908, GOAR 3684). It was close to a tube, for which Blasius measured a gauge factor of 1.08 and which he named after the French hydraulic engineer Henry Darcy (Brown 2003). "We obtain a mathematical expression for the distribution of velocities in this flow by placing a source in parallel flow," Blasius resorted to Rankine's method, "the water of the source is carried downstream and we get the stream-line S shown in the figure which we can consider as the boundary of a Darcy tube" (Blasius 1909). Thus he started a detailed analysis of the "Rankine half-body" (Fig. 1), as this shape became known in the textbooks on fluid mechanics.

By the same time, Prandtl asked Georg Fuhrmann (1883–1914), who had studied engineering at the Technical University Hanover, to analyse various airship models in the Göttingen wind tunnel in order to find a shape with minimal air resistance. In order to compare experiment with theory, Fuhrmann used Rankine's method for the construction of theoretical shapes by an appropriate distribution of sources and sinks. As he wrote in a letter to Prandtl in September 1909, he combined sources and sinks in such a way that the resulting stream-lines enclosed bodies with equal

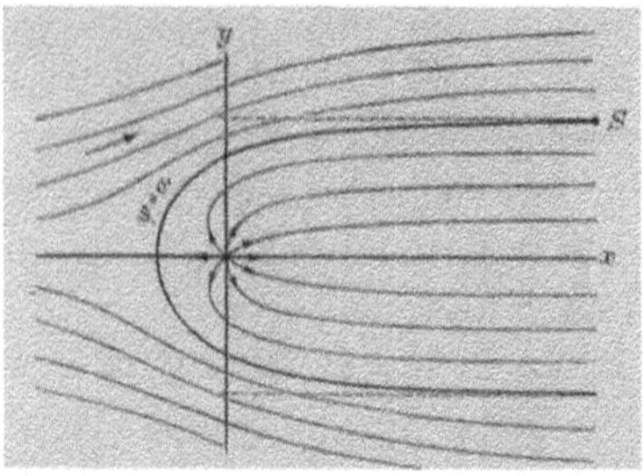

Fig. 1 Sketch of a "Rankine half-body" as the ideal shape for a Pitot tube (Blasius 1909, Fig. 3)

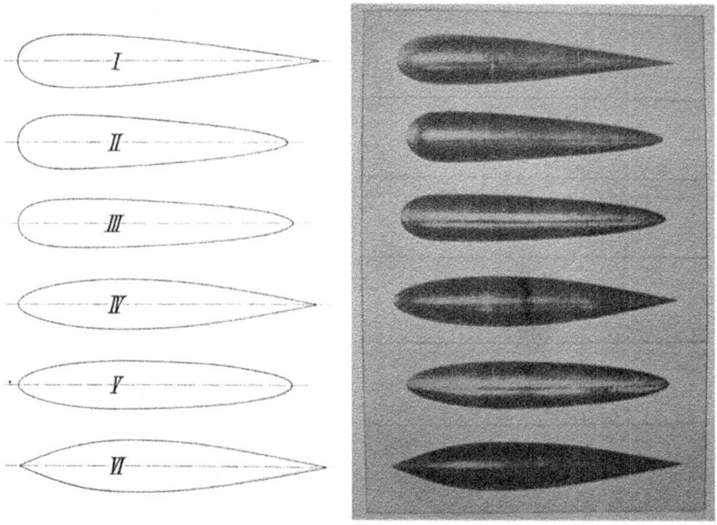

Fig. 2 Fuhrmann designed airship shapes by Rankine's method of sources and sinks (*left*). The airflow around models of the same shapes (*right*) was measured in a wind tunnel; the model with minimal drag was the third from below

diameter and equal volume (Fuhrmann to Prandtl, 19 September 1909, GOAR 2628). "The shape of the models was determined by a mathematical procedure that allowed us to compute by a special approach the flow and particularly the distribution of pressure under the assumption of a frictionless fluid in order to arrive at a comparison between these hydrodynamic methods and the measurements," Prandtl reported about Fuhrmann's use of Rankine's method. "It may be added that the agreement between theory and experiment is very satisfactory" (Prandtl 1911, p. 44).

In his final treatise on this subject Fuhrmann compared six different airship-models (Fig. 2) with respect to the distribution of pressure and velocity according to Rankine's method with measurements in the wind tunnel. Around the front part of the models there was close agreement between theory and experiment, but at the stern there were more or less severe differences dependent on the respective shape. "According to the theory of flow detachment by Prof. Prandtl this is completely understandable," Fuhrmann explained these discrepancies, "because in the rear part, where the flow is slowed down by friction and enters a region of higher pressure, the conditions for flow detachment and vortex formation are fulfilled; these vortices correspond to what is called backwash at ships." (Fuhrmann 1912, p. 105)

Thus two different sources of flow resistance were discerned: In the front part of the models, the distribution of flow velocity and pressure outside the boundary layer obeys the laws of ideal flow theory, i.e. the flow resistance is only due to skin friction exerted inside the boundary layer. When the flow becomes detached from the surface, the resistance is due to vortex shedding. Unlike skin friction, this resistance depends largely on the shape of the model; therefore it is called form drag.

4 Vortex Streets

Another incentive to apply ideal flow theory in combination with the boundary layer concept resulted from the doctoral work of Karl Hiemenz (1885–1973) whom Prandtl had asked to investigate the flow around a cylinder placed perpendicularly in a uniform flow of water (Hiemenz 1911). Hiemenz built a canal where he measured the pressure around the cylinder and in its wake. He used Prandtl's high-precision Pitot-tube for measuring the pressure and designed the canal so that he could control the uniformity of the water flow, but "the vortex tail which forms in the wake of the immersed obstacle is a rather unstable fabric. On the side of the tail vortices are separating which, as they move away from the body, drift to the side and, by reaction from the close walls, cause an irregular oscillation of the entire tail." (Hiemenz 1911, p. 12)

Prandtl's assistant, Theodore von Kármán (1881–1963), became curious in this wake oscillation and analysed the stability of rows of vortices detached from opposite sites at the cylinder's perimeter. He used ideal flow theory so that he could apply the mathematical tools of potential theory for his analysis. In this idealized view the vortices were moving away from the cylinder with an opposite sense of rotation along parallel rows (Fig. 3). Kármán compared an arrangement where the vortices of one row were situated exactly opposite to the vortices of the other row, with one where they were staggered to another in both rows. He found that only the latter configuration is stable with regard to a small disturbance. Although the result was obtained with ideal flow theory, it implied an important consequence about

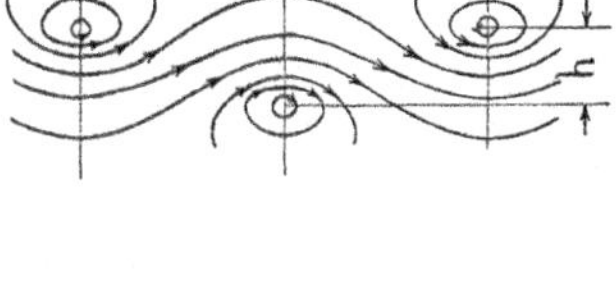

Fig. 3 Kármán vortices in the wake of an obstacle (Kármán 1911, p. 513)

the fluid resistance: The trailing vortices transport a momentum that allows one to compute the form drag of the cylinder (Kármán 1911, 1912).

Experiments performed in Prandtl's laboratory largely confirmed Kármán's theory which predicted a universal vortex arrangement with a ratio $h/l = 0.283$, where h is the distance between the vortex rows and l the separation of vortices in each row (Fig. 3); the experimentally observed value was 0.305. Furthermore, the resistance by vortex shedding derived for a flat plate (placed perpendicularly in a stream) and a cylinder resulted in friction coefficients which roughly agreed with experimentally observed values (Kármán and Rubach 1912).

Kármán, however, was not the first who investigated these vortex rows. Henri Bénard (1874–1939), famous for his discovery of regular patterns in horizontal layers of fluid heated from below (Bénard cells), studied the rows of alternating vortices in the wake of obstacles by means of sophisticated optical methods since 1904. He even filmed this phenomenon and described it in several papers published after 1908. No wonder he was annoyed about the "Kármán vortex street". He considered Kármán's analysis as too idealistic. In 1926 he published a paper with the title "About the Inaccuracy for Fluids in Real Conditions of the Kármán's Theoretical Laws Regarding Alternated Vortex Stability" in which he blamed Kármán for inappropriate comparisons between theory and experiment. When they met in the same year at a congress, Kármán, according to a later recollection of Bénard, admitted that the Göttingen experiments from the year 1912 could not rival with Bénard's own experiments (Wesfreid 2006, pp. 23–27). In his own recollection, Kármán dated their meeting four years later and ridiculed the priority dispute. "All right," he recalled his response to Bénard, "I do not object if in London this is called Kármán Vortex Street. In Berlin let us call it Kármánsche Wirbelstrasse, and in Paris, Boulevard d'Henri Bénard." (Kármán and Edson 1967)

5 The Induced Drag of Airfoils

It is remarkable that ideal flow theory–even if only under certain conditions and with input from experimental measurements, such as the frequency of vortex shedding–could be used as a tool for the determination of a fluid resistance due to vortex shedding. Kármán's theory could not explain how vortices are detached from an obstacle in a flow, but it could account for the stable arrangement of these vortices and the momentum transferred to the obstacle by vortex shedding.

Another remarkable application of ideal flow theory concerns the lift and drag of airfoils. By 1912 Prandtl's research began to focus on airplanes, in particular the determination of the lift and drag of airfoils (Prandtl 1912). For airfoils with infinite span, ideal flow theory appeared as a powerful tool in order to compute the lift, as Wilhelm Martin Kutta (1867–1944) and Nikolai Joukowsky (1847–1921) had shown in 1902 and 1906, respectively, by what became known as the circulation theory. The flow around the profile of an airfoil of infinite span could be perceived as the superposition of a uniform and a circulatory flow, with the

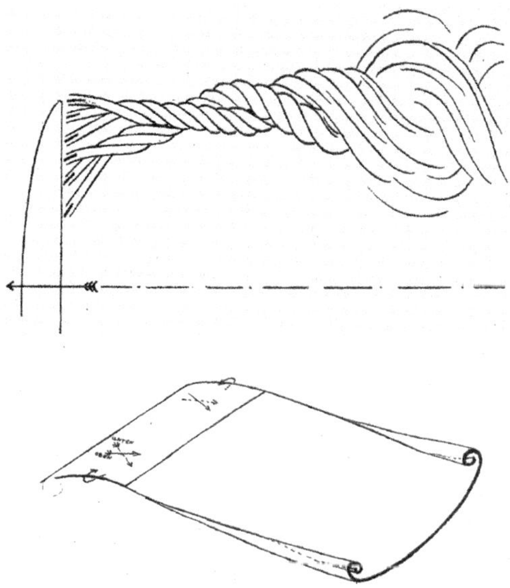

Fig. 4 Lanchester perceived the airflow at the wingtips like drilled ropes (*top*, from (Lanchester 1907, p. 127)); Blasius imagined (*below*) that this flow is "rolling up like a bag" (GOAR 3684)

circulation responsible for the lift. In other words, lift could be imagined as the result of a vortex traveling with and circulating around the airfoil.

Prandtl and others attempted to extend this concept to wings of finite span (Bloor 2011). The British engineer Frederick Lanchester (1868–1946) had speculated in 1907 that the vortex around the airfoil gives rise to vortical pigtails trailing away from the wingtips (Lanchester 1907, 127). Lanchester's book was translated into German in 1909 and stirred vivid debates among Prandtl and his disciples. "The question about the constitution of the vortical pigtails (Wirbelschwänze), which are drawn in Lanchester's fig. 86 like drilled ropes, suggested to me the following," Blasius alluded to this debate in a letter to Prandtl in November 1910. Instead of drilled ropes he perceived the flow at the wingtips as "rolling up like a bag" ("tütenförmig aufrollen") (Fig. 4). He recalled that Prandtl had already mentioned earlier that "such a spiral vortex" was trailing away from a wingtip. But he regarded a mathematical analysis of such wingtip vortices as hopeless ("Durchrechnung hoffnungslos") (Blasius to Prandtl, 17 November 1910, GOAR 3684).

Prandtl had presented a similar sketch presumably already in a lecture in the summer of 1909 (Rotta 1990, pp. 188–193). In November 1911 he made a step forward in order to render the mathematical analysis less hopeless. He idealized the wingtip vortices as a pair of vortices whose vortex threads could be perceived as parallel straight lines originating at the wingtips and extending against the direction of flight to infinity (Fig. 5). "Their distance is equal to the span of the wing, their strength equal to the circulation of the flow around the wing" (Prandtl 1912, pp. 34–35).

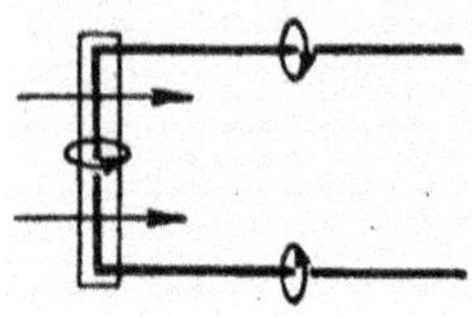

Fig. 5 Prandtl's idealization of the airflow around a wing as superposition of a uniform flow with a horseshoe vortex (Prandtl 1912)

Thus the analysis was reduced to the computation of the strength of horseshoe-like vortices–one vortex fixed to the wing and a pair of parallel vortices left in the air by the wingtips. Viscosity played no role in this idealized scheme. In 1914, a collaborator of Prandtl explained the V-shaped formation of bird-flight in terms of the horseshoe vortices of this scheme. He argued that outside the downwash area of the horseshoe vortex of a leading bird the wingtip vortices on the left and on the right generate an upwind for the succeeding bird. If each succeeding bird moved in the upwind of his predecessor, the whole group of birds would fly in a V-shaped formation. The computation of the flow of air in the vicinity of horseshoe-shaped vortex threads was analogous to the computation of the strength of the magnetic field in the vicinity of an electric current (Wieselsberger 1914).

Although this was only a first step towards a full-fledged airfoil theory (Bloor 2011; Epple 2002; Eckert 2006), the scheme of the horseshoe vortex involved an important consequence which was recognized long before the theory was ripe for publication in 1918. The vortices trailing off from the wingtips exert a resistance on the airfoil! Unlike the lift in the Kutta-Joukowsky theory for an airfoil of infinite span, the lift of an airfoil with finite span is accompanied by a drag that is unrelated to the viscosity of the air and depends on the wing span. After Fuhrmann's comparison of ideal and real flow around airship models and Kármán's theory of drag due to vortex shedding, the drag induced by the lift of a wing was another example for the rehabilitation of ideal flow theory for practical goals. Further elaboration of airfoil theory yielded a formula how this lift-induced drag depends on the wing span and how the coefficients of lift and drag for wings of different planform and span are related to another. Wind tunnel tests confirmed this formula and thus provided confidence that the airfoil theory indeed turned out practical results. When Hugo Junkers, one of Germany's leading manufacturers of airplanes in the First World War, learned about these results in April 1918, he was "very surprised" and exclaimed: "Had we known them earlier we could have spared all our test runs" (DMA, NL 123, no. 11). Nothing could have demonstrated more clearly to what extent ideal flow theory had become a practical tool that was useful even for the design of airplanes–despite its limitation on inviscid fluids. But it should be emphasized that this practical use was confined to basic insights concerning the dependence of induced drag on the wing span. Like Euler's pipe flow theory or the construction of Rankine's bodies, the airfoil theory was "useful" in a specific sense–not as a blueprint for engineering design but as a guide for the experimental and theoretical investigation of otherwise inaccessible phenomena.

6 Conclusion

The lift-induced drag of airfoils belongs to the category of form drag (like the resistance due to the shedding of Kármán vortices). In order to account for the total drag of an airfoil, one has to add the drag due to skin friction. Without the idealization of the theory of inviscid fluids with its mathematical techniques from potential theory, it would not be possible to understand the underlying physical causes of different kinds of fluid resistance. Ideal flow theory, combined with the boundary layer concept, was instrumental in order to discern the realms of viscous from inertial forces.

Of course, ideal flow theory by itself fails to account for fluid resistance. Prandtl left no doubt that "the hydrodynamics of the frictionless fluid leads to completely useless results regarding the resistance problem" (Prandtl and Tietjens 1934, Vol. II, p. XV). But contrary to the popular myth of the antagonism between the ideal and the real, ideal flow theory was–and still is–an important tool for practical applications.

This lesson is particularly pertinent for the education of engineers. No textbook about engineering fluid mechanics may afford to ignore ideal flow theory and the associated mathematical tool box of potential theory and complex analysis. From Prandtl's lectures in the early twentieth century to modern engineering courses in the twenty first century it has been regarded as essential to devote a considerable part of the courses to ideal flow theory. A contemporary textbook on *Applied Hydrodynamics*, for example, devotes half of its content to ideal flow theory because it provides the techniques to study "two-dimensional flows in regions where the effects of boundary friction are negligible: e.g., outside of boundary layers. The outcomes include the entire flow properties (velocity magnitude and direction, pressure) at any point. Although no ideal fluid actually exists, many real fluids have small viscosity and the effects of compressibility may be negligible. For fluids of low viscosity the viscosity effects are appreciable only in a narrow region surrounding the fluid boundaries." Therefore, the theory "is applicable with slight modification to the motion of an aircraft through the air, of a submarine through the oceans, flow through the passages of a pump or compressor, or over the crest of a dam, and some geophysical flows." (Chanson 2009, p. 4)

At least in fluid mechanics, therefore, the ideal and the real are no antagonists. The study of ideal flow is not only not opposed to the analysis of real flow, but to the contrary crucial for understanding the latter.

Acknowledgements This chapter is based on research for a biography of Ludwig Prandtl funded by the Deutsche Forschungsgemeinschaft. Further thanks go to the organizers and participants of the workshop on "Historical Perspectives on Mathematics as a Tool" at the Centre for Interdisciplinary Research in Bielefeld, held from 13 to 15 November 2013, for the opportunity to present a preliminary version. I also appreciate discussions with my colleagues at the research institute of the Deutsches Museum.

References

Bell, E. T. (1937). *Men of mathematics*. London: The Scientific Book Club.

Blasius, H. (1907). *Grenzschichten in Flüssigkeiten mit kleiner Reibung*. Dissertation, Universität Göttingen.

Blasius, H. (1909). Über verschiedene Formen Pitotscher Röhren. *Zentralblatt für Bauverwaltung, 29*, 549–552.

Bloor, D. (2011). *The enigma of the aerofoil: Rival theories in aerodynamics, 1909–1930*. Chicago: Chicago University Press.

Brown, G. O. (2003). Henry Darcy's perfection of the pitot tube. In O. B. Glenn, D. G. Jürgen, & H. H. Willi (Eds.), *Henry P. G. Darcy and Other Pioneers in Hydraulics: Contributions in Celebration of the 200th Birthday of Henry Philibert Gaspard Darcy* (pp. 14–23). Reston, VA: American Society of Civil Engineers.

Calero, J. S. (1996). *La genesis de la mecanica de fluidos*. Madrid: Universidad Nacional de Educacion a Distancia.

Cannell, D. M., & Lord, N. J. (1993). George Green, Mathematician and Physicist 1793–1841. *The Mathematical Gazette, 77*(478), 26–51.

Chambers, E. (1738). *Cyclopaedia, or, a universal dictionary of arts and sciences*. London: Ephraim Chambers.

Chanson, H. (2009). *Applied hydrodynamics: An introduction to ideal and real fluid flows*. London/New York/Leiden: CRC Press.

Darrigol, O. (2005). *Worlds of flow*. Oxford: Oxford University Press.

Eckert, M. (2002). Euler and the fountains of Sanssouci. *Archive for History of Exact Sciences, 56*, 451–468.

Eckert, M. (2006). *The dawn of fluid dynamics. A discipline between science and engineering*. Berlin/Weinheim: Wiley-VCH.

Eckert, M. (2008). Water-art problems at Sanssouci. Euler's involvement in practical hydrodynamics on the eve of ideal flow theory. *Physica D, 237*, 1870–1877.

Epple, M. (2002). Präzision versus Exaktheit: Konfligierende Ideale der angewandten mathematischen Forschung: Das Beispiel der Tragflügeltheorie. *Berichte zur Wissenschaftsgeschichte, 25*, 171–193.

Fuhrmann, G. (1912). *Theoretische und experimentelle Untersuchungen an Ballonmodellen*. Dissertation, Universität Göttingen.

Hager, W. H. (2003). Blasius: A life in research and education. *Experiments in Fluids, 34*, 566–571.

Hiemenz, K. (1911). *Die Grenzschicht an einem in den gleichförmigen Flüssigkeitsstrom eingetauchten geraden Kreiszylinder*. Dissertation, Universität Göttingen.

von Kármán, T. (1911). Über den Mechanismus des Widerstandes, den ein bewegter Körper in einer Flüssigkeit erfährt. *Nachrichten der Königl. Gesellschaft der Wissenschaften, mathematisch-physikalische Klasse* (pp. 509–517).

von Kármán, T. (1912). Über den Mechanismus des Widerstandes, den ein bewegter Körper in einer Flüssigkeit erfährt. *Nachrichten der Königl. Gesellschaft der Wissenschaften, mathematisch-physikalische Klasse* (pp. 547–556).

von Kármán, T., & Edson, L. (1967). *The wind and beyond*. Boston: Little Brown.

von Kármán, T., & Rubach, H. (1912). Über den Mechanismus des Flüssigkeits- und Luftwiderstandes. *Physikalische Zeitschrift, 13*, 49–59.

Lanchester, F. (1907). *Aerodynamics, constituting the first volume of a complete work on aerial flight*. London: Constable and Co.

Manegold, K.-H. (1970). *Universität, Technische Hochschule und Industrie: Ein Beitrag zur Emanzipation der Technik im 19. Jahrhundert unter besonderer Berücksichtigung der Bestrebungen Felix Kleins*. Berlin: Duncker und Humblot.

Mikhailov, G. K. (Ed.). (2002). *Die Werke von Daniel Bernoulli*. Basel: Birkhäuser.

Perkovitz, S. (1999). The rarest element. *The Sciences, 39*, 34–38.

Prandtl, L. (1905). Über Flüssigkeitsbewegung bei sehr kleiner Reibung, *Verhandlungen des III. Internationalen Mathematiker-Kongresses* (pp. 484–491). Heidelberg/Leipzig: Teubner.
Prandtl, L. (1909). Die Bedeutung von Modellversuchen für die Luftschiffahrt und Flugtechnik und die Einrichtungen für solche Versuche in Göttingen. *Zeitschrift des Vereins Deutscher Ingenieure, 53*, 1711–1719.
Prandtl, L. (1911). Bericht über die Tätigkeit der Göttinger Modellversuchsanstalt. *Jahrbuch der Motorluftschiff-Studiengesellschaft, 4*, 43–50.
Prandtl, L. (1912). Ergebnisse und Ziele der Göttinger Modellversuchsanstalt (vorgetragen am 4. November 1911 in der Versammlung von Vertretern der Flugwissenschaften in Göttingen). *Zeitschrift für Flugtechnik und Motorluftschiffahrt, 3*, 33–36.
Prandtl, L. (1948). Mein Weg zu hydrodynamischen Theorien. *Physikalische Blätter, 4*, 89–92.
Prandtl, L., & Tietjens, O. (1934). *Hydro- and aeromechanics. Vol. I: Fundamentals of hydro- and aeromechanics; Vol. II: Applied hydro- and aeromechanics*. New York: Dover Publications.
Rankine, W. J. M. (1864). Summary of the properties of certain stream-lines. *Philosophical Magazine, Series 4, 28*, 282–288.
Rankine, W. J. M. (1871). On the mathematical theory of stream-lines, especially those with four foci and upwards. *Philosophical Transactions of the Royal Society of London, 161*, 267–306.
Rees, A. (1819). *Universal dictionary of arts, sciences, and literature*. London: Longman.
Rotta, J. C. (1990). *Die Aerodynamische Versuchsanstalt in Göttingen, ein Werk Ludwig Prandtls. Ihre Geschichte von den Anfängen bis 1925*. Göttingen: Vandenhoeck und Ruprecht.
Rouse, H. (Ed.). (1968). *Hydrodynamics by Daniel Bernoulli and hydraulics by Johann Bernoulli*. New York: Dover.
Schlichting, H., & Gersten, K. (2000). *Boundary layer theory*. Berlin/New York: Springer.
Sommerfeld, A. (1900). Neuere Untersuchungen zur Hydraulik. In: *Verhandlungen der Gesellschaft deutscher Naturforscher und Ärzte*. 72:56.
Steele, B. D. (2006). Rational mechanics as enlightenment engineering: Leonhard euler and interior ballistics. In J. B. Brenda (Ed.), *Gunpowder, explosives and the state* (pp. 281–302). Aldershot: Ashgate.
Szabo, I. (1987). *Geschichte der mechanischen Prinzipien*. Basel/Boston/Stuttgart: Birkhäuser.
Truesdell, C. A. (1954). Rational fluid mechanics, 1687–1765. In A. T. Clifford (Ed.), *Leonhardi euleri opera omnia, II, 12* (pp. 9–125). Lausanne: Orell Füssli.
Wesfreid, J. E. (2006). Scientific biography of Henri Benard (1874–1939). In *Dynamics of Spatio-temporal cellular structures: Henri Benard centenary review* (Springer tracts in modern physics, Vol. 207, pp. 9–40). New York: Springer.
Wien, W. (1900). *Lehrbuch der Hydrodynamik*. Leipzig: Hirzel.
Wieselsberger, C. (1914). Beitrag zur Erklärung des Winkelfluges einiger Zugvögel. *Zeitschrift für Flugtechnik und Motorluftschiffahrt, 5*, 225–229.
Zedler, J. H. (1735). *Grosses vollständiges Universal-Lexicon aller Wissenschaften und Künste*. Leipzig/Halle: Zedler.

Idealizations in Empirical Modeling

Julie Jebeile

1 Introduction

In empirical modeling, mathematics has an important utility in transforming descriptive representations of target system(s) into calculation devices, thus creating useful scientific models. Descriptive representations in this context are pieces of knowledge about the properties and the behavior of target system(s) which are not yet expressed in mathematical terms. They are one form of models that, when mathematized, become inferentially useful. Mathematics thus allows models to fulfill their inferential role which can be to predict, explain or design experiments. In other words, if they are not mathematized, models would be partially descriptive representations of little impact and use.

Because mathematics here transforms descriptive representations into calculation devices, the transformation may be considered as the action of a tool or, more exactly, of several tools. In this paper, I assume that the idealizations involved in the transformation going from descriptive representations to a useful model could be such tools. I then examine whether these idealizations have the usual expected properties of tools, i.e., whether they are being adapted to the objects to which they are applied, and whether they are to some extent generic.

Ordinary tools—such as pliers, screwdrivers and wrenches—transform the objects on which they are used for a given purpose. They do so by virtue of having physical properties which are adapted to the objects.[1] That said, tools cannot

[1]Mutual adaptedness seems to be a specific property of some ordinary tools. Here, not only the tool is adapted to the object on which it is used, but also the object is adapted to the tool. This

J. Jebeile (✉)
IRFU/Service d'Astrophysique, CEA Paris-Saclay, 91191, Gif-sur-Yvette, France
e-mail: julie.jebeile@gmail.com

J. Lenhard, M. Carrier (eds.), *Mathematics as a Tool*, Boston Studies in the Philosophy and History of Science 327, DOI 10.1007/978-3-319-54469-4_12

be idiosyncratic (or they would lose their utility) and must rather apply to a certain range of objects. For example, an Allen key is a tool of hexagonal cross-section that is used to drive a range of bolts and screws as soon as they have a hexagonal socket in the screw head. What about idealizations in empirical modeling?

One might hope that, by contrast, mathematics applies identically to descriptive representations whatever they may be. In this way, it would be a very efficient tool in the modeling field because it would be universal. I will put forward the claim that this cannot be true and assess to which extent the transformation of models into mathematical terms is actually adapted to the specific empirical nature of the target system(s).

In this paper, I will first describe how models are built, focusing on their transformation into mathematical terms. This transformation has two phases: the first is about expressing the initially representational content of the model in mathematical equations. The second is about making these equations tractable. As we will see, idealizations are involved in both phases of transformation.

I will then argue that, for the model to be useful, its transformation into mathematical terms has to be adapted to the target system(s). Thus, the idealizations involved in the process must be constrained by the specificity of the target(s). Not only are these idealizations designed for models to be inferential, but they must also be chosen in such a way that models preserve at least a minimum amount of relevant accurate information about the systems under study. For my argument to be general enough, I will consider transformations that result in both analytically solvable and numerically solvable models.

That said, I will further argue that adaptedness here does not mean restriction to the sole systems under study. I will show that the idealizations involved in a transformation may well be suited for modeling other empirical systems as well. In that sense they function as mathematical tools that have a certain scope of application, as one would expect from tools in the ordinary sense of the word.

2 Model Building

In this section, I will describe the process of building models and will place special emphasis on the transformation into mathematical terms. My aim is to show that idealizations play a central role in this transformation.

I shall specify that my way of describing model building differs dramatically from the mapping accounts although they are important contributions in the discussions about the role of mathematics in empirical modeling (e.g., Batterman 2010, Bueno and Colyvan 2011, Pincock 2007). As the proponents of these accounts

is a case of screwdrivers and screws, Allen keys and bolts, or hammers and nails. Some ordinary tools do not share this property, however (e.g., rakes, scissors and shovels). Because this property is specific, it will not be considered in this paper.

identify something called the mathematical structure of models, they treat as one the formal feature and the representational content of models. It may be a relevant philosophical approach to study in some way the mathematical role in scientific models. Here, however, I want to emphasize and to characterize the action of mathematics as a tool on a descriptive representation to make it a calculation device. Thus, a diachronic presentation of model building is preferred instead.

In this presentation, I offer a conceptual account of the stages in empirical modeling. It is inspired by Cartwright' account of theory entry proceeding (1983, Essay 7) but also differs from her account in that hers is more descriptive (mine is more idealized). According to Cartwright, scientists start with an unprepared description which contains everything they know about the system under study, including information they think relevant in whatever available form; information can be theoretical components, observations or measurements. From this a prepared description is established in preparation for mathematically formalizing the physical phenomenon under study. It contains theoretical principles applied to idealized cases, as well as boundary conditions and ad hoc terms used to match the empirical data.

The account of empirical modeling, as proposed here (see Fig. 1), is a revised and idealized version of Cartwright's conception in that it re-scales the transformation into mathematical terms within empirical modeling. In this account, I deliberately magnify the mathematical process so that it becomes clear how mathematics actually transforms initially descriptive representations into models.

In my account, I suggest that, at the very beginning, model building consists of an initial description of the target system(s) involved in the phenomenon under study. The situation is likely to be the same as what Cartwright describes as the phase of the unprepared description, except that, here, the description is supposed to still not have a mathematical form.

Then, scientists have to make a choice among all the information within the initial description, and select the relevant aspects about the phenomenon under study to be included in the model that is to be built. In other words, they have to make some

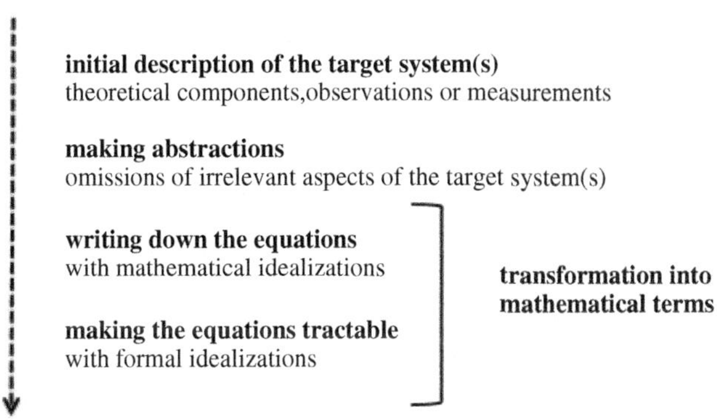

Fig. 1 Conceptual account of empirical modeling

abstractions.[2] Without this selection, modeling would not be possible. For Godfrey-Smith (2009), abstracting is required whatever the nature of the description may be (scientific or literary, for example) because there is no useful description which would exhaustively contain all the aspects of a system. For instance, there is no point in specifying the Moon phase for describing the motion of a body, or, of including the presence of oxygen or the average temperature for describing the trajectory of planets.[3]

The description, once cleared of the less relevant details, contains all the relevant representational aspects but does not yet have the desired inferential power (of predicting, explaining or designing experiments).[4] For the model to be a calculation device, this description must be transformed into a useable set of mathematical equations. This corresponds to what I have earlier called the transformation of models into mathematical terms.

The transformation (which could be also called mathematization) consists in two phases: the first is about expressing the initially representational content of the model in a first set of mathematical equations, and the second is about making this set of equations tractable. Both phases are necessary for the model to become a calculation device. The first phase aims at creating a preliminary version of the calculation device, and the second aims at making this device useable for inferential tasks. That said, building a mathematical model often starts from another mathematical model. This situation can be expressed as feedback loops in my schema of modeling phases.

The transformation involves, in its two phases, idealizations, which are simplifying assumptions expressed mathematically in the equations[5] (see e.g., Cartwright 1983; McMullin 1985; Hacking 1983; Laymon 1980, 1985, 1989a, b, 1995 for discussions on idealizations):

In the first phase, the expression of equations consists in translating the initially representational content into a mathematical language, i.e., in terms of mathematical symbols (e.g., numbers, variables, constants, vectors, matrix), operations (e.g.,

[2]Abstractions differ from idealizations in that they are omissions of some aspects in the target system which are not relevant for the problem being studied (e.g., to neglect gravitational force in subatomic phenomena), whereas idealizations are distortions (Jones 2005; Godfrey-Smith 2009). Idealizations can be omissions but, in this case, these omissions distort the representation in that they are omissions of relevant aspects.

[3]Abstracting can sometimes later be part of mathematization. For instance, difficulties in formulating equations might occur and lead to a different abstraction. This is, however, an additional aspect that I do not treat in this paper.

[4]What is relevant might actually depend on the final success in constructing a useful model, and therefore be identified as such at later stages of modeling. In such cases, there might be some back and forth in the process of model building.

[5]I shall stress that this way of defining idealization differs from the view on which model as a whole is an idealization. It is considered here that idealizations are only parts of a model. Unlike models, idealizations have no inferential power on their own. For instance, the Ising model will not be considered as an idealization but as being composed of idealizations. A mass point is an idealization, but is not a model in that, alone, it has no inferential power.

addition, integration, derivation) and functions (e.g., cos, sin, log). This translation is required because mathematical language enables to perform afterwards the expected inferential tasks. Mathematical language is generic in that the symbols, the operations and the functions may well be used to describe any empirical system. It therefore constitutes the formal part of the model that is clearly not specific to the system(s) under study.

That said, the linguistic translation of the initially representational content requires to make mathematical idealizations which fit this content to the constraining adopted mathematical language. Mathematical idealizations can be about taking some of the spatio-temporal properties out of the target system(s), thus abstracting away the "imperfections" of matter from the target system(s), for the final description of these systems to match ideal geometrical forms. For example, a wooden wheel or a balloon may be represented as a perfect circle.

The equations obtained in the first phase may be intractable—either analytically or numerically. In this case, one must proceed, in the second phase, to formal idealizations.[6] This is done either by considering the physical problem under study, or by considering the mathematical form of the equations.

In the former case, one tries to simplify the physical problem by omitting or by distorting some aspects of the system(s) in order to get results from calculation: "Complicated features of the real object(s) are deliberately simplified in order to make theoretical laws easier to infer, in order to get the process of explanation under way" (McMullin 1985, p. 258). Examples of formal idealizations are results from considering a body as being a mass point, an infinite surface or a frictionless surface, from replacing non-linear interactions by linear interactions, or from assuming an oscillation as being small. In some cases, considering the physical problem and trying to simplify its description are, however, not enough.

Because the equations remain intractable or complicated to solve, additional formal idealizations—which are sometimes called "approximations" in this context (Ramsey 1990, 1992; Laymon 1989a, 1989b)—are conceived based on the mathematical form of the equations. Here, formal idealizations do not always correspond with familiar idealized physical situations. Let us consider the example of a body of unit mass falling in a weakly resisting medium.[7] It is described by $dv/dt = g - kv$, with g the acceleration due to gravity and k a friction coefficient. Let us assume that the speed is zero when the body starts to fall ($t = 0$). Then $v(t) = (g/k)(1 - \exp(-kt))$ $= gt - gkt^2/2 + gk^2t^3/6 - \ldots$. When the friction resistance is insignificant (i.e., k is small), the terms with k can be removed and the speed is approximately described by the first term in the power series $v(t) = gt$. This formal idealization corresponds to a conceptually well-identified distortion—i.e., the absence of air—that is associated with a familiar idealized physical situation—i.e., a free fall in vacuum. That said, a formal idealization may consist in removing the terms after the second or the third order (or any superior order) of the power series. Here, the idealization would hardly

[6] I borrow the distinction mathematical vs. formal idealizations from McMullin (1985).

[7] I take this example from Norton (2012) for a different purpose though.

correspond with a familiar idealized physical situation. In our language and division of the world into concepts, we lack an expression to identify an aerial space which exerts on a body a resistance such that its speed equals exactly $gt - gkt^2/2 + gk^2t^3/6$, for example.

In a nutshell, mathematical idealizations make it possible to express the model in terms of mathematical equations. Formal idealizations are required to make these equations tractable and thereby useable. Without mathematical and formal idealizations, the model would not be a calculation device but merely a partial description of the system being studied.

I have shown how important idealizations are in the transformation of models into mathematical terms. It follows that the transformation is adapted to the empirical nature of the systems under study if the choice of the involved idealizations depends on the specificity of the systems. In the next section, I will contend that these idealizations should actually be constrained by the nature of the systems so that the transformation should be adapted to it.

3 Adaptedness of Idealizations

I will now argue that idealizations are tool-related in that they make the model useable. I will further argue that they are adapted to the kind of phenomena to which they are applied if the model is expected to be not merely useable, but useful.

Idealizations are tool-related. Following Carrier and Lenhard, this means that "they result from the properties of the tool and make sure that the tool can be used in the first place" (cf. the general introduction). Let me briefly clarify in what sense idealizations are tool-related. In the transformation of models into mathematical terms, idealizations are used to shape initially representational components into a mathematical form, so that models become tractable and therefore usable. In constraining the representation by a mathematical language, idealizations give models a new property, i.e., inferential power. In other words, idealizations make it possible to create a device from which calculations can be done.

I suggest that their capacity to give models inferential power results from the fact that they *essentialize* the features of the target(s) that they denote. They essentialize in the sense that they reduce the features to something formally and representationally essential. It is formally essential in that it is a mathematical object making tractable the equation that contains it. It is representationally essential in that it represents a feature of the target that is relevant for building the representation one needs.

One might think that, since idealizations are tool-related, they must meet mere mathematical constraints but no representational constraint. In the first phase, mathematical constraints are due to the mathematical language that the model must fit. In the second phase, mathematical constraints are due to the available mathematical means of solving equations (i.e., whether we know how to solve them). This would be true if the model is expected to be useable. A model is useable in that it is mathematically tractable and can be used to make calculations and to provide results. That said, in empirical science, a model is expected to be more than useable; it should be useful. A useful model has to be useable and has to provide reliable results for answering the questions the scientist is asking. My aim is to show that, for a model to be useful, the choice of the idealizations involved in the transformation must depend on the nature of the target systems.

Idealizations transform the initially representational content from the unprepared description into a calculation device by distorting this representational content. Thus, models become calculation devices to the detriment of their representational function. That is why a lot of philosophical attention has been paid on whether, once idealized, models can teach us something about the world (e.g., McMullin 1985; Laymon 1989a, b; Bokulich 2008, 2009; Strevens 2007; Weisberg 2007). Thus, idealizations are internally stressful concepts. Employing idealizations involves a trade-off between making a model useable and making a model useful. Therefore, the scientist has to find a compromise when choosing the idealizations to be included in the model.

For a model to be useful, I suggest that idealizations must be selected in a way that makes the model sufficiently accurate for the purpose at hand. I thus follow van Fraassen when he writes that "A representation is made with a purpose or goal in mind, governed by criteria of adequacy pertaining to that goal, which guide its means, medium, and selectivity" (2008, p. 7). He further claims that a "representation useful for particular purposes will involve selective distortion, and representation is closely involved with useful misrepresentation. Even when likeness is crucial to the purpose, we must look for likeness only in respects that serve the representation's purpose—and only to the extent that they do so" (2008, p. 87). Hence, idealizations that are the distortions in a model are allowed to the extent that they are selected adequately for the purpose at hand.

I will show that the choice of idealizations relates to issues of representation, and choosing an adequate idealization hinges on considerations about the specific empirical nature of the target system(s). As I will elaborate with the example of Prandtl's model, the choice of adequate idealizations is constrained by the specificity of the targets. Idealizations must not make the model deviate too much representationally from the actual description. Thus, I will claim that not only are idealizations in the transformation tool-related but they are also object-related. Object-related idealizations "create a simpler version of the relevant objects and their relationships so that mathematical models of them become more tractable" (cf. the general introduction) and, I shall add, so that mathematical models become of real utility in empirical science.

Let me illustrate this with the story of Prandtl's model. The study of a flow past an obstacle has long been an important industrial issue, dating back to the rise of steamboat navigation during the Victorian times. It was imperative that British engineers know how to assess the resistance of water on a boat in order to design optimal hull shapes. However, such a study is mathematically difficult. D'Alembert (1782), as well as other physicists and engineers studying fluid mechanics such as Poncelet, Saint-Venant, Boussinesq, Ranquine or Froude, suggested, tested, and tried to improve models of fluid resistance. But all of them faced the same thorny problem: one of the required idealizations leads to unacceptable aberrations.

Two important formal idealizations were introduced in the models. The first was an incompressible flow condition, which leads to the assumption that the influences of pressure and temperature on mass density are negligible. This idealization was unproblematic. The second idealization neglected all effects of viscosity (i.e., internal friction in the fluid). This idealization seems prima facie justified for applications involving fluids like air and water, which have a low viscosity. It results in equations with explicit solutions, which are known as the Euler equations. Thus, the idealization helps to provide a useable model. Nevertheless, the Euler equations lead to absurd results, at least if interpreted physically. One such result, discussed by d'Alembert, is the cancellation of the drag of a moving body. In other words, there would be no force exerted by the fluid on the body, which contradicts our most familiar observations: air slows down a balloon that was initially thrown horizontally, for example. The Euler equations raise other difficulties since they cannot provide explanations of phenomena such as pressure loss in a pipe or separating flow past an obstacle. These phenomena are the result, directly or indirectly, of fluid viscosity. The no-viscosity assumption is therefore too strong, even when modeling low viscosity fluids. It is not a harmless idealization in the sense of Elgin and Sober (2002); it jeopardizes the representational adequacy of a model that contains it.

If the idealizations in the transformation were only constrained by mathematics, such an issue would not appear. This example illustrates that, for a model to be useful, idealizations need to be sensitive to the nature of the target system(s). Thus, other idealizations are required which take into account viscosity in low viscosity flows. What are they?

With the objective of solving d'Alembert's paradox in mind, Navier and Stokes both contributed in establishing the fundamental equations of Newtonian fluid mechanics in a continuous medium in which they both introduced terms of fluid viscosity. In 1845 the final version of the Navier-Stokes equations was established. These equations derive from balances of mass, momentum, total energy and entropy applied on a fixed or mobile volume of fluid. Unfortunately, they form a complex system of equations with non-linear partial derivatives whose analytical resolution is still today a real challenge for mathematicians. Their resolution is one of the seven Millennium Problems raised by the Clay Mathematics Institute.

In order to get a useful model that is both inferential and representationally good enough, idealizations must be found which are more subtle than the no-viscosity assumption. Prandtl's model illustrates this achievement. Before Prandtl's model, successfully applying the Navier-Stokes equations was practically impossible. During the 19th century, engineers and fluid mechanists were forced to establish phenomenological laws in order to solve their problems (Darrigol 2005). More than 150 years after the first work of d'Alembert (1782) who discussed the forces exerted by a fluid on a solid body, Prandtl made a significant contribution in the use of these equations. His model is explicitly presented as a means of applying the Navier-Stokes equations to the concrete problems of a flow past a solid (Heidelberger 2006).

Ludwig Prandtl himself introduced the model in 1904 in these terms at the third international congress of mathematicians at Heidelberg (Darrigol 2005, p. 283).

In his model, Prandtl assumed the influence of forces induced by low viscosity as confined in the wake behind the obstacle as well as inside a layer, i.e., the boundary layer. Low viscosity fluid around an obstacle is conceived of as the interaction between two components. The first component is the boundary layer which is around the obstacle and whose viscosity is not zero and whose velocity profile evolves linearly. The velocity profile starts from zero at the interface fluid/obstacle—this is called the no slip condition—and reaches a maximal value equal to the viscosity in the second component. The second component is the wake. Within the wake, the viscosity is zero and the mean velocity remains constant. For this component alone, the Euler equations apply. With such an idealization, it has been possible to describe the trajectories of bodies in a flow—e.g., plane flight in the atmospheric air—or the fluid flow past an obstacle—e.g., flow of a river against a boat—, and thus to study hydrodynamic behaviors such as the onset of vortices.

The idealization conceived by Prandtl is more subtle than the simple omission of the terms of viscosity in the Euler equations. Idealizing for building Prandtl's model is not about automatically omitting or modifying certain terms in the Navier-Stokes equations that do not meet mathematical expectations. Because, even if the boundary layer is thin in ordinary fluids such as water and air, it has significant effect on flow in virtue of the high gradient of velocity that it implies. In this case, the right idealization is constrained by the specificity of the system being studied. Here the specificity is that the viscosity, albeit low, plays a crucial role to describe drag, pressure loss or flow separation. Understanding the relevance of this physical aspect is more important than merely considering its low numerical value, for instance. In other words, for a model to be useful and not just useable, it seems that idealizations cannot be made without considering the physical problem.

Another aspect of the story reinforces the idea that idealizing properly requires the consideration of the specificity of the physical problem. The concept of boundary layer was obtained by Prandtl through observations, i.e., visualizations of experiments that he conducted in a water tunnel. Morrison claims that since the boundary layer concept is the product of direct observations, it is what she calls a phenomenological abstraction. She writes indeed that "The model is phenomenological not because there is no theory from which to draw but because it is motivated solely by the phenomenology of the physics; in fact, once the model is developed theory does play an important role in its application" (Morrison 1999, p. 54). It seems that Prandtl even developed a certain intuition through the visualization of the phenomena before even setting the equations of his model. On that subject, Prandtl says: "Herr Heisenberg has [. . .] alleged that I had the ability to see without calculation what solutions the equations have. In reality I do not have this ability, but I strive to form the most penetrating intuition [*Anschauung*] I can of the things that make the basis of the problem, and I try to understand the processes. The equations come later, when I think I have understood the matter" (quotation taken from Darrigol, 2005, p. 287).

The difficulty scientists may have in finding appropriate idealizations has been overlooked by the philosophical discussions about idealizations. It is rather often assumed that scientists know well in advance the empirical consequences of the idealizations contained in a model (exceptions are notably Laymon and Ramsey). A famous example is a freely falling object. It is often suggested that scientists only use the assumption of zero air resistance for predicting the velocities of heavy bodies because they know that the assumption becomes inadequate when studying low-mass objects for which air resistance cannot be neglected. (They also know that, in that case, they need to provide an approximate phenomenological term for the air resistance force.) It may be sometimes true that scientists know the scope of idealizations especially when they use them for a long time. Prandtl's model, however, exemplifies a case where it is not true. The model also shows that there is no unique recipe that may help one to automatically make the following inference: "the fluid viscosity has a low value" then "the term of viscosity can be omitted in the equations." Such a point is even made in textbooks on fluid mechanics: "Approximation is an art, and famous names are usually associated with successful approximations: Prandtl wing theory, Kármán-Tsien method for airfoils in subsonic flow, Prandtl-Glauert approximation for subsonic flow, Janzen-Rayleigh expansion for subsonic flow, Stokes and Oseen approximations for visas flow, Prandtl boundary-layer theory, Kármán-Opohlhausen boundary-layer approximation, Newton-Busemann theory of hypersonic flow" (Van Dyke 1975, p. 2).

In a nutshell, not only are idealizations designed for models to be inferential, but they also must be chosen in such a way that models preserve at least a minimum amount of relevant accurate information about the systems under study. Hence, this indicates that the idealizations in the transformation have to be both tool-related and object-related. Thus, the transformation is adapted to the nature of the target(s).

In the next section, I will generalize this claim in studying different transformations of the representation of a same phenomenon. I shall emphasize that distinct transformations are related to distinct possible tasks.

4 Plurality of Transformations

The computational turn, which is the historical conjunction between the development of numerical analysis and the advent of computers, offers new means of writing and solving model equations, i.e., new transformations. It is possible to use numerical methods for expressing and solving equations. In this section, I will consider the case where different transformations are used to build a model of a same target system. In this descriptive part, I will emphasize, through the study of a flow past a cylinder, that each transformation involves its own idealizations, thus being adapted in a unique way to the target system. It will follow that, since idealizations involved in a transformation are both tool-related and object-related, each transformation transforms the initially descriptive representation into a model that can be more or less adequate for the purpose at hand.

I will further investigate the example of low viscosity and incompressible flow past a cylinder. It can be studied under analytic method—e.g., Prandtl's model—

or numerical methods applied on computer—i.e., discretization-based method, computational method of molecular dynamics, Monte Carlo method or cellular automata. All the models that are created based on these methods have in common that they express in their own way the fundamental principles of fluid mechanics, i.e., conservations of momentum, mass and energy (except nevertheless the Monte Carlo method which violates conservation of momentum as we will see). In these models, the problem is considered as two-dimensional. A first difference, however, is that some are based on a macroscopic description of the fluid while others are based on a microscopic description. The fluid is therefore idealized as a continuous medium for some models and as a set of discrete entities for the others. Let me further present the five transformations in the following.

4.1 Transformation 1: Prandtl's Model

When they describe the fluid macroscopically, hydrodynamics models necessarily contain approximate versions of the Navier-Stokes equations. Since these equations describe the behavior of the fluid idealized as a continuum, as said before, they derive from balances of mass, momentum, total energy and entropy applied on a fixed or mobile volume of fluid. Thus, they contain the fundamental principles of fluid mechanics. Two additional idealizations are often used. First, a boundary condition near the cylinder is established, which is the no slip condition: at the interface with the obstacle, the fluid velocity is supposed to be zero. Second, the fluid is considered as incompressible.

The system of equations thus obtained—composed of constraint equation, boundary condition and incompressible flow assumption—cannot be solved as such. Indeed, the system contains non-linear terms. A first analytical approach allows one to solve it nevertheless. This is the Prandtl model. As seen earlier, in this model, the fluid is supposed to have two interacting components. The first component is the boundary layer which is around the obstacle and whose viscosity is not zero and whose velocity profile evolves linearly. Within the second component, i.e., the wake, the viscosity is zero and the viscosity is constant.

4.2 Transformation 2: Discretization of the Navier-Stokes Equations

A second transformation allows one to avoid the two idealizations made by Prandtl, thus making the model more accurate. It is the numerical resolution of the Navier-Stokes equations by a discretization-based method. In this computational model, the numerical scheme is based on the integration of equations by finite element method. First, finite element method consists in discretizing the physical domain into finite elements. Then the partial differentials of the equation variables are replaced by the formal idealizations obtained with the values of the variable at the nodes of each finite element. Lastly, the obtained equations are integrated on each finite element of the meshing, and for each time step when the module of temporal dependence is required. This approach is less idealized than Prandtl's model and can therefore provide more precise results.

4.3 Transformation 3: Molecular Dynamics Model

The macroscopic representation of fluid—on which the Navier-Stokes equations are based—is not valid if the assumption of fluid as continuum does not hold. This situation corresponds to a Knudsen number superior to 0.01.[8] Here a microscopic description of fluid is required. This is the case of trajectories of spatial vehicles (Meiburg 1986). Because the fluid in which these vehicles move has three different regimes: continuum fluid regime, transitional flow regime and collisionless flow regime. Yet, in the two latter regimes, the fluid cannot be considered as a continuum. Consequently, it must be represented as a discrete set of entities. In this case it is required to use the Boltzmann equation in order to describe the average behavior of fluid particles. Furthermore, it is expected that conservation of linear momentum, conservation of kinetic energy and conservation of angular momentum are satisfied. From this, two methods can be used, namely, the molecular dynamics model and Monte Carlo method.

In the molecular dynamics model it is generally assumed that there are several thousands of particles. The initial distribution of these particles is randomly determined in the space or is explicitly set. It is the same process with their initial velocities. The particles then evolve with their own velocities. They can also interact between each other following the 'potential well' model. In this model, the collision is represented as if a particle fell in a potential well, i.e., in a local minimum of potential energy. At each new time step the instant of the next collision is calculated. This is done by examining all the pairs of fluid particles. Then all the particles are moved forward at the same time in accordance with the laws of classical physics, and the new velocities of the particles involved in a collision are calculated (Meiburg 1986, p. 3108).

4.4 Transformation 4: Monte Carlo Method

Like the molecular dynamics model, the numerical approach based on a Monte Carlo method proceeds on the calculation of the trajectories of particles. In this

[8]Knudsen number is a dimensionless parameter. It indicates the flow regime depending on the fluid continuity (while Reynolds number indicates the flow regime depending on the turbulence).

transformation, the fluid is represented as a lattice of cells. At the initial step, particles are randomly located on the cells or determined beforehand. Like in the molecular dynamics models, at each time step, the new positions of the particles are calculated in the space depending on their respective velocities. The Monte Carlo method differs nevertheless in the way the interactions between particles are handled. The position and the instant of a collision are here not determined by the calculation of the trajectories of the particles, but meet merely statistical considerations. In each cell, among all the particles, two particles only are selected randomly, independently of their positions. A collision between two particles is then considered: the new positions and velocities of the two particles are calculated; they are supposed to be rigid spheres rather than potential wells as they are in the case of the molecular dynamics model.

The statistical assumption that concerns the assessment of collisions makes it possible to greatly facilitate the calculations. This is the reason why, compared to the molecular dynamics model, the Monte Carlo method allows for a much higher speed of calculation running on computer. The side effect, however, is the violation of the conservation of angular momentum for the interactions between particles (see Meiburg 1986, p. 3109). In the model, recall that the calculation of the collisions is done independently of the position of the particles. One can assume that all the directions are equiprobable for the velocity of the two particles after collision. In the calculation, the direction of this velocity is therefore randomly set in selecting an azimuthal angle and a polar angle. Consequently, two components of the velocity after collision are determined by the chosen Monte Carlo method. They therefore do not remain possible variables. Moreover, the four other variables of the problem—namely, the coordinate of the velocity after collision and the three coordinates of the velocity before collision—are set when the conservations of linear momentum and kinetic energy are satisfied. Consequently, they are not available in order to meet the conservation of angular momentum.

4.5 *Transformation 5: Cellular Automata-Based Model*

Another transformation is based on cellular automata (Rothman and Zaleski 2004; D'Humières and Lallemand 1986).[9] The principle in cellular automata consists in representing the fluid as a lattice. At each node of the lattice stands a site. The state of each site received a value among a finite number of possible states. At time t_{n+1}, it depends on the value of the states of the neighboring sites at time t_n with which

[9]The development of cellular automata is more recent than the development of differential equations since it started in the 1940s with the work of Ulam and von Neumann at Los Alamos. For a general philosophical discussion on cellular automata, see e.g., Fox Keller 2003 and Rohrlich 1990. There are also different hydrodynamics models based on cellular automata. For an exhaustive presentation of these models, see the forthcoming paper of Barberousse, Franceschelli and Imbert entitled "Cellular Automata, Modeling, and Computation" and Barberousse and Imbert (2013)."

the site is connected. In the representation of the fluid, the fictive particles of the fluid possess the same mass and the same velocity. They only differ to each other in the direction of their velocity; the velocity can receive only six possible values (for a hexagonal lattice) (see Rothman and Zaleski 2004, chapter 1, p. 1–2). At each time step, the particles move from one site of the lattice to another site of the lattice following the direction of their velocity. They can collide if two particles or more arrive at the same site at the same time. Some collisions can produce the scattering of the particles. In this case, the velocity vector of the particles is modified. The total number of the particles and the sum of the velocity vectors do not change. This means that the mass and the momentum are conserved.

4.6 *Idealizations for the Purpose at Hand*

Each of the presented methods implies a specific transformation which, as I want to emphasize, involves its own idealizations:

First, the macroscopic representations—i.e., Prandtl's model and the discretized version of the Navier-Stokes equations—are in some way more idealized than the microscopic representations—i.e., the molecular dynamics model, Monte Carlo method and cellular automata-based model—in that they include the assumption of fluid as continuum.

Among the macroscopic representations, Prandtl's model is a more simplified version than the discretized version of the Navier-Stokes equations insofar as the latter does not need to assume somewhere that the viscosity is zero.

Among the microscopic representations, the Monte Carlo method includes a statistical assumption about collisions. This assumption is an idealization that the molecular dynamics model does not need to contain. Furthermore, while the molecular dynamics model and the cellular automata-based model are derived from the same fundamental physical principles of mechanics—i.e., conservations of momentum, number of particles and energy—, these two models differ from each other in their degree of idealization. In the molecular dynamics model, the fluid particles are conceived as rigid spheres and their positions and velocities can take a high number of possible values. In the cellular automata-based model, the fluid particles are considered as points; the possible values of their positions and their velocities are forced to evolve within a discrete hexagonal lattice. Therefore, in a certain sense, the cellular automata-based model is an idealized version of the molecular dynamics model in which differences of mass are canceled and possible directions of velocity are limited.

The analysis of the different methods has shown that each method applies its own transformation with its own set of idealizations. Let me now illustrate that each transformation transforms the initially descriptive representation into a model that can be more or less adequate for the purpose at hand.

First, the macroscopic representations are not suitable for situations where Knudsen number is superior to 0,01, but can be sufficiently accurate for other cases.

Second, unlike the discretized version of the Navier-Stokes equations, Prandtl's model is not sufficiently accurate for contemporary highly computerized studies in aeronautics or aerospace, but may be used to get an analytic understanding of how vortices appear at a rear of an obstacle.

Third, among microscopic representations, the Monte Carlo method enables one to make more rapid calculations, which can be sometimes required, but is not appropriate when it is about studying the onset of vortices. The statistics assumption that the Monte Carlo method involves has no major consequence when it is about modeling flows of Rayleigh-Stokes type (Meiburg 1986), but, as it fails to meet conservation of angular momentum and as angular momentum plays a central role for describing vortices, it may not be adapted for reproducing the onset of vortices at the rear of the obstacle. Here, molecular dynamics simulations are more adequate in order to reproduce the distributions of vortices (Meiburg 1986).

Fourth, cellular automata-based model is not adapted in cases where differences of mass within the fluid matter or where all directions of velocity must be taken into account. It may, however, be helpful in other cases.

Until now, I have argued that, for a model to be useful, the transformation has to be adapted to the nature of the target systems, and that, depending on the purpose at hand, it may be more adequate or less adequate. The question then arises: does this make the transformation idiosyncratic? I will offer an answer to this question in the final section.

5 Scope of Application

Another usually expected property of tools is that they are to some extent generic. In other words, a tool is supposed to apply to some range of objects (rather than to a unique object). And yet if the transformation of models is idiosyncratic, i.e., if it is excessively adapted to the target system, it cannot be considered as the action of a tool. Note that the question is about the starting point, i.e., the systems to which it is applied, not the endpoint of the tool, i.e., into what it transforms. In this final section, I will argue that, even though a similar transformation does not strictly apply identically to all systems, the idealizations involved in the process may well be suited for a range of empirical systems. In that sense, idealizations may function as mathematical tools that have a certain scope of application.

I have suggested that idealizations give models inferential power in that they *essentialize* the features of the target(s) that they denote. They essentialize in the sense that they reduce the features to something formally and representationally essential. Therefore, an idealization that has been used in a model can be used in another model as soon as the two models share similar features to which the idealization is adapted.

This is the case with models used to study similar systems. In the previous example, the boundary layer applies to a class of hydrodynamics systems that share a certain physical feature, i.e., physical singularity. Examples are flow past a circle, flow over airfoil and flow over flat plate.

The boundary layer was initially used in a particular case as an idealization in order to make the Navier-Stokes equations analytically solvable. Today, the scope to which it adequately applies is much larger. It survived Prandtl's model and is now commonly used to describe concrete cases in fluid mechanics. It actually now belongs to the standard vocabulary of fluid mechanists and is a single-handedly research topic (see e.g., Khujadze et al. 2010).

The boundary layer has thus been mathematically defined as "a narrow region where the solution of a differential equation changes rapidly. By definition, the thickness of a boundary layer must approach 0 as [the perturbing parameter] $\varepsilon \longrightarrow 0$" (Bender and Orszag 1978, p. 419). Therefore, as soon as a system has such a narrow region, the boundary layer may be an adequate idealization depending on the purpose at hand. This has also led to the development of the boundary layer theory which is "a collection of perturbation methods for solving differential equations whose solutions exhibit boundary-layer structure" (Bender and Orszag 1978, p. 420). The boundary layer is conceived more generally as leading to the following mathematical simplifications:

> There are two standard approximations that one makes in boundary layer theory. In the outer region (away from a boundary layer) y(x) is slowly varying, so it is valid to neglect any derivatives of y(x) which are multiplied by ε. Inside a boundary layer the derivatives of y(x) are large, but the boundary layer is so narrow that we may approximate the coefficient functions of the differential equation by constants. Thus, we can replace a single differential equation by a sequence of much simpler approximate equations in each of several inner and outer regions. In every region the solution of the approximate equation will contain one or more unknown constants of integration. These constants are then determined from the boundary or initial conditions using the technique of asymptotic matching [...] (Bender and Orszag 1978, p. 421).

The boundary layer is here defined in a general way which includes the definition given by Prandtl in his model. This shows that the boundary layer is certainly not idiosyncratic but actually applies to a range of systems which share a common physical feature, i.e., discontinuity, that can be expressed mathematically in the previous terms.

On the same grounds, the scope of application of the boundary layer may actually extend beyond the range of hydrodynamics systems. It may also be a relevant idealization to describe the skin effect in electromagnetism, since this effect displays high variability. The skin effect is produced by an alternating electric current which has a high density within a conductor and is largest near the surface, while it decreases with greater depths in the conductor. The boundary layer can be associated with the skin depth in which the electric current flows.[10]

The scope of idealizations can more generally extend in case of formal analogy. In such a case, idealizations contained in a model A are transposed in a model B

[10]The fact that idealizations have a certain scope of application may explain why some models are repeatedly used within and across scientific domains, e.g., the harmonic oscillator, the Ising model, a few Hamiltonians in quantum mechanics, the Poisson equation, or the Lokta-Volterra equations (see Barberousse and Imbert (2014) for an analysis of such recurring models).

in that the equations in A are algebraically identical to the equations in B. This is a case of the analogies between waves of light, sound and water (Hesse 1966, p. 10–12) where the same equation $y = a \sin(2\pi f x)$ can be employed in the three cases,[11] between the atom and the solar system, between nuclear fission and the division of a liquid drop, or between electrostatic attraction and the conduction of heat (see Bailer-Jones 2009).

Therefore, idealizations function as tools in that they are adapted to typical empirical features of the investigated phenomena, which may be redundant in nature (e.g., oscillation, stochastic feature, discontinuity, etc.). They can thus be transposed to other cases which share a certain representational similarity. This should come as no surprise since, as said before, they reduce features of the phenomena to something formally and representationally essential. In their "toolbox," scientists may choose such or such a mathematical tool, ready to be used, depending on the system(s) under study.

In arguing that idealizations have a scope of application because they essentialize features of the target system, I do not want to suggest that they are parts of some mathematical structure of the empirical world or to support a platonic conception of idealizations. Rather, I want to claim that they can be used as tools because they have a story in the building of models and, as such, are recognized by scientists as being adequate for such or such modeling. A model is never built from scratch, but based on what is known to work in other models. For example, idealizing fluid as a continuum (rather than a discrete set of molecules) is a very common assumption in fluid mechanics models, as it is for other idealizations (e.g., incompressible flow assumption), since it has proven to be an adequate one. In other words, idealizations could be seen as tools in the scientific toolkit. Depending on the kind of target system, scientists could choose such or such idealization that is known to be adequate for modeling the system.

De-idealization may sometimes be required, however, in order to make the model sufficiently accurate (and therefore useful). It consists in adding features of the target (that were originally left out) back into the models and/or correcting idealizations that originally appear in the models (McMullin 1985; Laymon 1995). When de-idealization is required, it means that an additional aspect of the system is relevant for the purpose at hand that was not captured (or not properly captured) by the idealized model. This is the purpose at hand that determines which components are relevant to include in the model. Let me give an example.

In order to derive the ideal gas law ($PV = nRT$) from a molecular model, gases are assumed to be perfectly elastic spherical molecules. These molecules exert no force and their volume is negligible in comparison with the volume occupied by the gas. These assumptions limit the application domain of the law which applies only

[11]In the case of waves of water, the equation describes the height of the water at the point x, with a being the maximum height or amplitude of the ripples, and f their frequency. In the sound model, it describes the amplitude of a sound wave at the point x, with a being the loudness and f the pitch. In the light model, it describes the amplitude of a light wave, with a being the brightness and f the color.

to ranges of normal temperature and pressure. Thus, it is not possible to predict the properties of biphasic systems or monophasic systems that evolve towards a biphasic state (state transitions) with this law, which instead requires the use of van der Waals' equation. Van der Waals' equation, which is given by, $P + (a/V2)\ (V-b) = RT$ (where a and b are associated with the intermolecular forces), is viewed as an improvement upon the ideal gas law because it takes into account attractive and repulsive intermolecular forces (see the chapter "Boon and Bane: On the Role of Adjustable Parameters in Simulation Models" by Hasse and Lenhard which discusses the ideal gas law and the role of adjustable parameters in great detail). By adding the intermolecular forces, it yields more accurate results at high temperatures and low pressures than the ideal gas law, and therefore it applies to a wider domain than does the ideal gas law. The introduction of intermolecular forces into the new model is a genuine de-idealization (McMullin 1985).[12]

To conclude, scientists may choose such or such idealization, depending on the system(s) under study and the purpose at hand. They can also decide to de-idealize in case the model is not sufficiently accurate.

6 Conclusion

I have first described how models are built, with special emphasis on the transformation. I have then argued that the transformation is always adapted to the target systems. The reason is that the idealizations involved in the process are both tool-related and object-related for the model to be useful. That said, I have further argued that adaptedness does not mean restriction to the sole systems under study. I have shown that the idealizations involved in a mathematical transformation may well be suited for other empirical systems as well and are in that sense mathematical tools that have a certain scope of application.

References

Bailer-Jones, D. M. (2009). *Scientific models in philosophy of science*. Pittsburgh: University of Pittsburgh Press.

Barberousse, A., & Imbert, C. (2013). New mathematics for old physics: the case of lattice fluids. *Studies in History and Philosophy of Science Part B: Studies in History and Philosophy of Modern Physics, Elsevier, 44*(3), 231–241.

[12]Some idealizations sometimes denote relevant aspects of the target model that de-idealization would fail to capture. These idealizations are called ineliminable (or essential) (Batterman 2005, 2009; Sklar 2000). They cannot be removed without losing the explanation of the phenomenon that is studied. This is the case with the thermodynamic limit, according to which the number of atoms in the system is infinite, which is necessary for explaining phase transitions, and in particular, the phase transition of a magnet at a certain critical temperature (Batterman 2005).

Barberousse, A., & Imbert, C. (2014). Recurring models and sensitivity to computational constraints. *The Monist, 97*(3), 259–279.

Batterman, R. W. (2005). Critical phenomena and breaking drops: Infinite idealizations in physics. *Studies in History and Philosophy of Science Part B, 36*(2), 225–244.

Batterman, R. W. (2009). Idealization and modeling. *Synthese, 169*(3), 427–446.

Batterman, B. (2010). On the explanatory role of mathematics in empirical science. *British Journal for the Philosophy of Science, 61*, 1–25.

Bender, C. M., & Orszag, S. A. (1978). *Advanced mathematical methods for scientists and engineers*. New York: McGraw-Hill Book Company.

Bokulich, A. (2008). Can classical structures explain quantum phenomena? *British Journal for the Philosophy of Science, 59*(2), 217–235.

Bokulich, A. (2009). Explanatory fictions. In M. Suárez (Ed.), *Fictions in science: Philosophical essays on modeling and idealization* (pp. 91–109). Londres: Routledge.

Bueno, O., & Colyvan, M. (2011). An inferential conception of the application of mathematics. *Noûs, 45*(2), 345–374.

Cartwright, N. (1983). *How the laws of physics lie*. Oxford: Oxford University Press.

D'Alembert, J. (1782). *Essai d'une nouvelle théorie de la résistance des fluides*. Paris.

D'Humières, D., & Lallemand, P. (1986). Lattice gas automata for fluid me. *Physica, 140A*, 326–335.

Darrigol, O. (2005). *Worlds of flow: A history of hydrodynamics from the Bernoullis to Prandtl*. Oxford: Oxford University Press.

Godfrey-Smith, P. (2009). Models and fictions in science. *Philosophical Studies, 143*(1), 101–116.

Hacking, I. (1983). *Representing and intervening: Introductory topics in the philosophy of natural science*. Cambridge: Cambridge University Press.

Heidelberger, M. (2006). Applying models in fluid dynamics. *International Studies in the Philosophy of Science, 20*(1), 49–67.

Hesse, M. B. (1966). *Models and analogies in science*. Notre Dame: University of Notre Dame Press.

Jones, M. R. (2005). Idealization and abstraction: a framework. In N. Cartwright & M. R. Jones (Eds.), *Idealization XII: Correcting the model : Idealization and abstraction in the sciences* (pp. 173–217). Amsterdam: Rodopi.

Keller, E. F. (2003). Models, simulation, and 'computer experiments. In H. Radder (Ed.), *The philosophy of scientific experimentation* (pp. 198–215). Pittsburgh: University of Pittsburgh Press.

Khujadze, G., van yen Nguyen, R., Schneider, K., Oberlack, M., & Farge, M. (2010). Coherent vorticity extraction in turbulent boundary layers using orthogonal wavelets. *Bulletin of the American Physical Society, 55*(16), 206.

Laymon, R.. (1980). Idealization, explanation, and confirmation. *PSA: Proceedings of the Biennial Meeting of the Philosophy of Science Association, 1980*, 336–350.

Laymon, R. (1985). Idealizations and the testing of theories by experimentation. In P. Achinstein, & O. Hannaway (Eds.), *Observation experiment and hypothesis in modern physical science* (pp. 147–173). Cambridge, MA: M.I.T. Press.

Laymon, R. (1989a). Applying idealized scientific theories to engineering. *Synthese, 81*(3), 353–371.

Laymon, R. (1989b). Cartwright and the lying laws of physics. *Journal of Philosophy, 86*(7), 353–372.

Laymon, R. (1995). Experimentation and the legitimacy of idealization. *Philosophical Studies, 77*(2-3), 353–375.

McMullin, E. (1985). Galilean idealization. *Studies in History and Philosophy of Science Part A, 16*(3), 247–273.

Meiburg, E. (1986). Comparison of the molecular dynamics method and the direct simulation Monte Carlo technique for flows around simple geometries. *Physics of Fluids, 29*(10).

Morrison, M. (1999). Models as autonomous agents. In M. S. Morgan & M. Morrison (Eds.), *Models as mediators* (pp. 38–65). Cambridge, UK: Cambridge University Press.

Norton, J. D. (2012). Approximation and idealization: Why the difference matters. *Philosophy of Science, 79*(2), 207–232.
Pincock, C. (2007). Mathematical idealization. *Philosophy of Science, 74*, 957–967.
Ramsey, J. L. (1990). Beyond numerical and causal accuracy: Expanding the set of justificational criteria. *Proceedings of the Biennial Meeting of the Philosophy of Science Association, 1*, 485–499.
Ramsey, J. L. (1992). Towards an expanded epistemology for approximations. *Proceedings of the Biennial Meeting of the Philosophy of Science Association, 1*, 154–164.
Rohrlich, F. (1990). Computer Simulation in the Physical Sciences. PSA 1990: *Proceedings of the Biennial Meeting of the Philosophy of Science Association*, 2:507–518.
Rothman, D., & Zaleski, S. (2004). *Lattice-gas cellular automata. Simple models of complex hydrodynamics*. Cambridge: Cambridge University Press.
Sklar, L. (2000). Theory and truth: Philosophical critique within foundational science. Oxford University Press.
Strevens, M. (2007). Why explanations lie: Idealization in explanation. New York: New York University. Unpublished manuscript. http://www.strevens.org/research/expln/Idealization.pdf.
Van Dyke, M. (1975). *Perturbation methods in fluid mechanics*. Stanford: The Parabolic Press.
Weisberg, M. (2007). Three kinds of idealization. *Journal of Philosophy, 104*(12), 639–659.

Forcing Optimality and Brandt's Principle

Domenico Napoletani, Marco Panza, and Daniele C. Struppa

1 Introduction

In a series of previous papers (Napoletani et al. 2011, 2013a,b, 2014) we described what we called the 'microarray paradigm' and we showed that there are specific methodological motifs that structure the approach of data analysis to scientific problems. By 'microarray paradigm' we referred to the belief that sufficiently large data collected from a phenomenon allows answering any question about the phenomenon itself. Answers are then found through a process of automatic fitting of the data to models that do not carry any structural understanding beyond the actual solution of the problem. This is a practice we suggested to label 'agnostic science'.

We argued as well in Napoletani et al. (2011) that, in data analysis, mathematics is "forced" onto the data. By this we mean that there are techniques expected to be useful, even when the assumptions under which these techniques should be applied do not appear to hold for the phenomenon under study. This process, which we called 'forcing', can be viewed as a direct, coarse, and willful attempt to bridge the gap between data and mathematics. This agnostic approach displays a role of mathematics in science that is essentially different from the traditional one:

D. Napoletani (✉)
University Honors Program and Institute for Quantum Studies, Chapman University, 92866, Orange, CA, USA
e-mail: napoleta@chapman.edu

M. Panza
CNRS, IHPST (UMR 8590 of CNRS, University of Paris 1 Panthéon-Sorbonne), Paris, France

Chapman University, 92866, Orange, CA, USA
e-mail: marco.panza@univ-paris1.fr

D.C. Struppa
Schmid College of Science and Technology, Chapman University, 92866, Orange, CA, USA
e-mail: struppa@chapman.edu

J. Lenhard, M. Carrier (eds.), *Mathematics as a Tool*, Boston Studies in the Philosophy and History of Science 327, DOI 10.1007/978-3-319-54469-4_13

rather than offering a structured interface between our problems and the raw data, mathematics now provides a rich repository of techniques for forcing. The extent to which forcing is possible, and therefore the relation between specific classes of phenomena and specific mathematical techniques to force onto them, suggests a shift in the focus of our understanding: from the phenomenon itself and its inner structure, to the structure of the algorithmic processes that are privileged in data analysis. The consequence is that the link between reasonable heuristic assumptions and successful problem solving through data analysis can be broken without impact on the effectiveness of the problem solving itself.

Here, we will show the implications of this shift by dissecting one of its most pervasive aspects: the search for optimization and the belief that optimization as such is always useful and necessary. For our purposes, to solve an optimization problem means to find the minimum of a fitness function $F(x)$ on a given domain $\mathscr{S}$ (usually a subset of $\mathbb{R}^n$ for some $n > 0$). We explore how forcing appears in the ways the fitness function $F(x)$ is built starting from a specific empirical problem, and in the methods used to approach the optimization problem itself.

We also propose that the way optimization techniques are used in practice often hints at a more basic methodological principle in data analysis, what we call 'Brandt's Principle', which can be expected to have wide applicability in problems from life and social sciences. The articulation of this suggestion goes through three sections. In Sect. 2 we explore how optimization can be seen as forcing. In Sect. 3, we draw the extreme consequence of regarding optimization as forcing; we suggest that, quite often, the local behavior of optimization algorithms is more important than its ultimate convergence to any optimal solution to a problem, and we introduce Brandt's principle, an operative principle of this approach. In Sect. 4, we speculate on the reasons of the effectiveness of this principle, particularly when applied to solve problems about what we suggest to call 'historical phenomena', i.e. phenomena significantly constrained by their past development. To this end, we first analyze a recently proposed principle of morphogenesis in developmental biology (Minelli 2011), and then explore its strong analogy with Brandt's principle.

2 Forcing Optimality

When using optimization to solve an empirical problem, there are two levels at which we may have forcing.

At one level, it is possible that the optimization method is forced upon the empirical problem: though the fitness function may or may not encode significant information on the problem and the underlying phenomenon, the method used to solve the optimization problem does not actually find the global extrema of the function, but nevertheless the original empirical problem is solved by the (non-optimal) output of the algorithm. To show an example of this situation, we analyze in Sect. 2.1 a popular method of optimization in data analysis, particle swarms optimization (PSO). Specifically, we show that its effectiveness is not bound to the actual ability of this algorithm to find a globally optimal solution to a problem, rather, PSO can be seen as a collection of local explorations of the solution space.

At another level, the fitness function is forced upon the empirical problem, even when poorly related to it. The more complex the empirical problem, the more the associated fitness function will be ad hoc, and many fitness functions, not all leading to the same optimum solutions, are possible. And yet, despite the ad hoc nature of the fitness function, the empirical problem is successfully solved. To explore this second level of forcing, we show in Sect. 2.2 how important problems, such as image processing, lead to entire families of fitness functions with different choices of parameters.

In Sect. 2.2 we also show that the complexity of these problems and associated fitness functions is such that the whole variety of optimization methods used to solve the corresponding optimization problems often reduce to a local, point by point search for solutions, with no hope in general for the identification of the global optimum, similarly to what we argued in Sect. 2.1 for the restricted case of PSO methods. We can imply therefore that forcing the fitness function usually leads to forcing the optimization method itself.

We conclude that, in data analysis, what is generally relevant in the use of optimization is not a special significance of global optimum, but rather its capacity to give mathematical form to the problem itself, and its effectiveness in generating local search algorithms acting on the space of potential solutions to the problem. It is in light of this conclusion that we claim that in data analysis, the following is typically true:

Using optimization methods to solve empirical problems is a form of forcing.

In the remaining part of the present section, we will support this claim through examples. This will allow us to distinguish the two relevant forms of forcing optimality most frequently used in data analysis. We are then left to understand in Sects. 3 and 4 the reasons of the effectiveness of forcing optimality.

2.1 *Forcing Optimization Methods*

In one of the early examples of computational methods inspired by biological systems, Kennedy and Eberhart motivated an innovative optimization method in Kennedy and Eberhart (1995) by assuming that birds move in a flock to optimize their search for food. Modifying some preexisting models of birds flocking behavior (Reynolds 1987), they built virtual particles that try to stay cohesive, while individually searching for optimality of a fitness function. The resulting searching method, called particle swarm optimization (PSO), turned out to be extremely popular in solving optimization problems, partly because of the attractiveness of a technique, such as PSO, that does not require much effort to be adapted to specific problems.

Here we want to show how the structure of this technique relies heavily on heuristic rules, to the point that PSO loses its original goal of finding a global optimum of a fitness function. PSO methods starts by setting (possibly randomly) position and velocity of multiple virtual particles, each of them associated with the evaluation of the fitness function at its position. The position and velocity of each particle is then slightly modified, taking into consideration both the location of its best evaluation up to that moment, and the location of the best evaluation of a set of neighboring particles. This updating process is justified by the hope that, as the particles move in the domain of the fitness function, and communicate their respective evaluations with each other, at least one of them will eventually settle its position to the sought for location of the global minimum. To which extent this assumption is true will be discussed shortly, after giving some details on the actual implementation of PSO methods.

Consider a swarm composed by N distinct particles, each of which is defined by a position $x_i = x_i(t) \in \mathbb{R}^n$ and a velocity $v_i(t) \in \mathbb{R}^n$, $(i = 1, \ldots, N)$, depending on the value of a temporal parameter t. Let $f(x)$, with $x \in \mathbb{R}^n$, be a fitness function that we wish to minimize on a domain $\mathscr{S} \subset \mathbb{R}^n$, and let $x_i(0) \in \mathbb{R}^n$ and $v_i(0) \in \mathbb{R}^n$ be the initial position and the initial velocity vector of the i-th particle, respectively.

We define $b_i(t)$ as the best position that the i-th particle achieved up to time t, which means that we have $f(b_i(t)) \leq f(x_i(t'))$ for all $t' \leq t$ and $b_i(t) = x_i(T)$ for some $T \leq t$. Assuming that each particle has a well defined neighborhood $\mathscr{N}_i$ (either a small set of particles within a given distance, or a set of particles that are a priori considered to be close to each other), we also define $l_i(t)$ as the best position within the neighborhood, which means that $f(l_i(t)) \leq f(b_j(t))$ for all $j \in \mathscr{N}_i$ and $l_i(t) = x_{j'}(T)$ for some $T \leq t$ and $j' \in \mathscr{N}_i$. We then update the velocity and position of each particle according to the following rules (Kennedy and Eberhart 1995; Dorigo et al. 2008):

$$v_i(t+1) = wv_i(t) + \varphi_1 U_1(t)(b_i(t) - x_i(t)) + \varphi_2 U_2(t)(l_i(t) - x_i(t))$$
$$x_i(t+1) = x_i(t) + v_i(t+1)$$

where $w > 0$ is a parameter called 'inertia weight' which gives a measure of the tendency of a particle to keep its direction of motion, φ_1, φ_2 are also positive and are called acceleration coefficients, and U_1, U_2 are random diagonal matrices with values in the interval $[0, 1)$ regenerated at each iteration. The choice of w, φ_1, and φ_2 is crucial to ensure that the algorithm output does not diverge. The whole term $\varphi_1 U_1(t)(b_i(t) - x_i(t))$ is referred to as 'cognitive component', as it quantifies a sort of memory of the particle, that keeps track of the location of its own best past or current evaluation of the fitness function. The effect of this term is to contribute to the velocity vector $v_i(t+1)$ a component that always points to $b_i(t)$, and that, the further $x_i(t)$ is from $b_i(t)$ the larger it is (not withstanding the variability of $U_1(t)$).

The term $\varphi_2 U_2(t)(l_i(t) - x_i(t))$ is referred to as 'social component', since it involves a local comparison within a set of neighboring particles, and it gives the particle a tendency to stay close to the location of the best past or current evaluation of the entire set of neighbors. Like the previous term, it contributes to the velocity vector a component that is always pointing in the direction of the best location ever found by the neighbors.

The purpose of local optimization is encoded in the structure of the PSO algorithm through these two components that decide the direction of the movement of the particle. The halting of the algorithm updating the positions and velocities of the particles is usually done either by setting a very large number of iterations before running the algorithm, or by terminating the iterative updating when the velocities of all or most particles get close to zero (presumably because they are close to the global minimum). In both these cases the estimate of the global minimum will be the minimum among the values of the fitness function at the location of the particles. The use of several particles allows to explore widely the fitness function on the domain $\mathscr{S}$, while maintaining a certain coherence among their evolution. However, there is no evidence in general that running the PSO algorithm will indeed lead to some global optimum in $\mathscr{S}$.

While there are many variants of PSO methods, they broadly respect the structure sketched above, and they turn out to be effective on a large variety of optimization problems, including many in image and signal processing, and in control theory (Poli 2008). Here we broadly summarize an image processing application that approaches the issue of image denoising using PSO methods (Bhutada et al. 2012). This is a particularly interesting example because it will introduce a way to build fitness functions that we will further explore in Sect. 2.2.

Let I be a discrete image, given as a real valued function defined on a grid of N discrete points. It is often more useful to represent I as a weighted sum of simpler images, that encode some specific characteristics we may expect in the image, for example, such simpler images could be oscillating patterns with fixed frequency, or they could be sharp spikes that are nonzero only over a few points on the grid. Let us call such collection of images a dictionary $\mathscr{G} = \{g_1, \ldots g_M\}$, and assume that $I = \sum_{m=1}^{M} c_m g_m$ for some coefficients c_m, so that, in a given dictionary $\mathscr{G}$, our image can be represented by the list of coefficients $c = \{c_1, \ldots c_M\}$. We note that in general we need more dictionary elements than the size of the image, i.e. $M > N$ (see Mallat 2008, chapter 5). In practice the image could be contaminated by noise, and in an idealized scenario we can assume that such noise is additive, that is, we have access only to the noisy image $\tilde{I} = I + W$, where W is a set of values drawn from some probability distribution. Under these conditions, the dictionary coefficients that we have access to are those of the noisy image $\tilde{I}$, i.e. $\tilde{c} = \{\tilde{c}_1, \ldots \tilde{c}_M\}$.

How can we reconstruct the original image from these noisy coefficients? It turns out that, if the dictionary is well chosen, we can set to zero or shrink the smallest coefficients (threshold them) without losing the most important characteristics of the original image, for example its edges and texture. Wavelet representations are among the dictionaries that allow this thresholding operation (Mallat 2008).

Without going into the details of their structure, we can say that they entail a way to decompose the image that encodes information at large scales (say, the rough contours of the objects in the image) and information at much finer scales (information on local textures and noise). Then it is possible to build a threshold function $T(\tilde{c}_m)$ that, according to the magnitude of each $\tilde{c}_m$, decides whether that coefficient needs to be replaced by zero, or, possibly, by some other value smaller than $\tilde{c}_m$. For example, early in the development of this field, it was proposed the threshold function defined by $T(\tilde{c}_m) = 0$ if $|\tilde{c}_m| < C$, $T(\tilde{c}_m) = \tilde{c}_m$ if $|\tilde{c}_m| > C$, for a carefully chosen value of C (Donoho and Johnstone 1994, 1995). This turned out to be too crude in some applications where more information is available on the type of images of interest, and more complex threshold functions, depending non linearly on several extra parameters, were suggested. In particular one recent type, developed in Nasri and Pour (2009) and used in Bhutada et al. (2012), depends from three parameters p, q, r. We denote it here by $T(\tilde{c}_m, p, q, r)$. This threshold function can be seen, for each choice of parameters p, q, r, as a piecewise smooth function very close to zero in a symmetric neighborhood close to the origin, and almost linear outside that neighborhood.

The parameters in $T(\tilde{c}_m, p, q, r)$ are chosen to keep the new reconstructed image as close as possible to the original one, and one way to do that is to assume that the error (fitness) function $E(p, q, r) = \|I - \sum_{m=1}^{M} T(\tilde{c}_m, p, q, r) g_m\|_2^2$, is minimized, where we denote by $\|*\|_2^2$ the square of the L_2 norm of a discrete image (i.e. the sum of the squares of the values of the image at each of the N points on which it is defined). Alternatively, under some restrictions on the type of dictionary used, one could minimize the simpler error function $E(p, q, r) = \sum_{m=1}^{M} |T(\tilde{c}_m, p, q, r) - c_m|^2$, as it is done in Bhutada et al. (2012). Assuming we know one representative image with and without noise, it is possible to minimize the function $E(p, q, r)$ above, with the hope that, once optimal parameters p, q, r are found, the threshold $T(\tilde{c}_m, p, q, r)$ with that choice of parameters will effectively denoise other noisy images.

Now, the complex dependance of the threshold function from its parameters causes a proliferation of local minima for the error function E. Having several local minima makes it very difficult to have a good, single initial estimate for the approximate location of the minimum, that is, a selection of values p_0, q_0, r_0 that are close enough to the globally optimal values. Such good estimate would be necessary if the global minimum on the domain $\mathscr{S}$ is searched with more traditional gradient descent methods (Arfken and Weber 2005, Sect. 7.3), which essentially modify the location of an evaluation of a fitness function in the direction of the steepest negative change of the function. What is more troubling, these methods depend on an estimation of the derivative of the error fitness function, which, in turn, depends from the coefficients c_m of the original image (with potentially very wide variations) making a numerical estimate of the derivative difficult.

PSO methods can overcome these difficulties thanks to the following two properties: they do not need a good initial estimate for the global minimum, since an entire population of initial evaluations is used instead; they do not need the computation of the derivatives of the error fitness function, and therefore their performance is robust with respect to variations in the image. In fact, in Bhutada et al. (2012) it is shown that PSO methods can be used successfully and quickly to minimize the error function E, either in the basic setting of uniform threshold across all the dictionary, or even the more computationally intensive sub-band setting, where the parameters are chosen independently for subgroups of the dictionary elements. The threshold parameters found by PSO are then shown to give effective image reconstructions, when confronted with other denoising techniques.

The preponderance of applications of PSO methods in image processing is not casual, as optimization problems derived in this setting often display the complex dependance from their parameters argued in the image denoising example (see Sect. 2.2). But despite the wide applicability of these methods, they are known to perform poorly when applied to combinatorial optimization problems (Poli 2008) where the true optimum is one of a finite number of discrete possibilities that, possibly, grows exponentially with respect to the number of the variables involved in the problem. However, these optimization problems are exactly those that are hardest to solve (many combinatorial optimization problems are NP-complete Wegener 2005), and for PSO methods to be a significant advance in optimization we would expect them to perform well when applied to these problems. As noted in the conclusion of (Poli 2008):

> most people will not care as to whether their new tool is guaranteed to give the absolute best performance on problem. What they want is something simple and reliable. Finally, probably the PSO has, at the moment, in the mind of many people the sort of magical black box flavor that attracted so many researchers to other area of artificial/computational intelligence before.

This black box belief was briefly discussed in Napoletani et al. (2013a) when we suggested that the microarray paradigm has a counterpart, in the use of computational processes, in the belief that a sufficiently complex machine should be able to solve any scientific problem.

The work in Pedersen and Chipperfield (2010) suggests that PSO methods can be conceptually and effectively simplified by recognizing the heuristic nature of the social and cognitive components in their implementation. Under this interpretation, they are methods of space exploration whose implementation is guided by a fitness function (to be optimized). However, they do not use in an essential way the analytical properties of the fitness function (for example its derivatives) and they are not expected to eventually converge to a global minimum. It is argued instead that only the actual performance of PSO methods on benchmark problems, where there is a consensus of what constitutes a solution, can discriminate those variants that are faster and more robust.

This means that the effectiveness of PSO methods does not depend on their having been conceived as optimization techniques. In fact, what counts here is not the optimization, but the capacity of the algorithm to appropriately modify the relevant values (position and velocity of each particle). Even if a fitness function f needs to be appealed to, for the algorithm to work, once this is done, optimization is no more a central concern of the resulting recursive algorithmic procedure. A PSO algorithm is accepted as suitable simply as long as, by following it, some solution to the original empirical problem is found, in a way that does not exceed some predetermined time and resources constraints, rather than in the best possible way.

It is true that a choice of a fitness function to be used by PSO methods might not be effective for a specific problem, and this would require it to be replaced by another function, until an appropriate fitness function is chosen. Still, this does not imply that we have a trial and error procedure, or some sort of hypothetico-deductive approach. PSO methods are shown to be effective, in a very specific algorithmic form, for entire classes of very different empirical problems even though they may not truly optimize sufficiently complex fitness functions. What happens is that a fixed mathematical technique, or better, a well specified algorithm, is forced onto the problem; if trials and errors occur, it is at the level of the fitness function, and always within the framework of PSO methods, which is mostly fixed under the passage from each trial to another. This is not to say that the fitness function itself is never forced onto the problem. Rather the contrary is often true, both in the application of PSO, and in other optimization methods used in data analysis: the specific choice of the fitness function (and its corresponding extrema) does not hinge on a structured heuristic insight into the empirical problem, which prefigures another sort of forcing. This is what we shall see in the next section.

2.2 Forcing Fitness Functions

The way image denoising is phrased as an optimization problem in Sect. 2.1 is exemplary of the way in which an optimization framework is used to phrase other complex problems about natural phenomena, irrespective of the specific method used to solve the resulting optimization problem. The emphasis in Sect. 2.1 was on how to solve an optimization problem, already derived from an empirical problem. Here instead we explore, in the context of more sophisticated image processing problems, the difficulties in the preliminary process of translating the empirical problem into an optimization one by choosing a fitness function.

A first consideration in image processing is that a natural image has important features that are relevant to its recognition and classification, for example overall contours, or different uniform textures in different regions, or different objects superimposed to each other. Capturing this structure involves first of all having a compact representation for the image. For dictionaries that efficiently encode edge information (Mallat 2008, chapter 9), this requirement can be approximately satisfied by requiring the minimization of $E_1 = \sum_{m=1}^{M} |T(\tilde{c}_m)|$. By $T(\tilde{c}_m)$ we indicate here symbolically the coefficients of the reconstructed image (without being concerned with the analytical shape of the function T), and we enforce that $\tilde{I} = \sum_{m=1}^{M} T(\tilde{c}_m) g_m$. Minimizing the function E_1 has been shown to lead to a sparse, compact, representation for the reconstructed image where only a few coefficients are nonzero (Donoho and Johnstone 1994, 1995). Suppose now that we wish to have a sparse image reconstruction and a denoising of the image at the same time. A way to proceed would be to look at the minimization of the overall fitness function

$E = \lambda E_1 + E_2$, where $E_2 = \|\tilde{I} - \sum_{m=1}^{M} T(\tilde{c}_m)g_m\|_2^2$ (a modification of the fitness function seen in Sect. 2.1 with respect to the noisy image $\tilde{I}$) and $\lambda > 0$ determines the relative strength of the two fitness terms in E (Donoho et al. 2006). What is important for our discourse, is that there is now an entire family of fitness functions, parameterized by λ. While under certain restrictive conditions on the dictionary it is possible to select heuristically a specific good value of λ (Chen et al. 2001), this is not the case in general, and we are left with the problem of choosing which, if any, of the fitness functions is truly the most suitable for the specific noisy image to represent sparsely.

More sophisticated image processing problems are associated to ever more complex fitness functions. Consider for example the problem of identifying incomplete contours of objects in an image. One of the most efficient method to solve it goes under the name of 'active snakes' (Kass et al. 1987). These algorithms are based on the idea of superimposing contours (snakes) on the image, and of subjecting these snake contours to some internal forces (minimizing the complexity of the shape of the contours) and some external forces (maximizing the fit of the contours on features of the image). Starting from some initial snake contour guess, these forces slowly change the shape of the snake on the image until it stabilizes on some significant feature within the image (usually the actual contour of an object).

The effect on the snake of internal and external forces is determined by a fitness function depending on several parameters that encode the relative strength of the forces, and to each choice of parameters (to adjust heuristically on the specific problem) corresponds a distinct fitness function. Effectively, we have an entire family of fitness functions, all suitable in principle to solve the problem of finding contours of objects in a given image. However, the dynamics of the snake on the image can be very complex, and different choices of fitness functions lead the snakes to stabilize on very different contours.

The tentative ways in which image processing problems translate into difficult optimization problems makes it clear that fitness functions are often ad hoc, and several parameters and fitness terms constraining the optimal solution need to be determined to make them plausible candidates to solve them. The significance of any specific fitness function is weakened, even though we need to have one such function to apply optimization methods. This is in line with the general fact that any application of forcing reduces the significance of our structural understanding of a phenomenon and a related problem. Which of these potentially distinct, but all equally plausible, fitness functions should we optimize? As Brandt notes in (Brandt 2001):

> this combination of penalty terms creates a monstrous minimization problem [...] It is extremely difficult to solve – *and unnecessarily so.*

The difficulty that Brandt highlights is due to the fact that for a general fitness function F there are no closed, analytical methods to solve the related optimization problem. The only way to look for a solution is often a local search, what he calls 'point by point minimization', an iterative method that, given a point x_i, gives as output a point x_{i+1} in a neighborhood of x_i such that $F(x_{i+1}) \leq F(x_i)$. The iterative process is started by selecting an initial guess x_0 for the potential solution and it is terminated when the distance $|x_{i+1} - x_i|$ between successive outputs of the process is smaller than some preset small $\delta > 0$, in which case we say that the point by point minimization has converged.

We have seen this local approach to optimization already in PSO methods, and indeed for all fitness functions used in the image processing problems above the solution is usually found with one of several recursive local optimization methods, such as gradient descent and Euler-Lagrange methods (Kass et al. 1987; Bresson et al. 2007), or interior point methods (Donoho et al. 2006; Boyd and Vandenberghe 2004). We suggest that for a sufficiently complex optimization problem and related fitness functions, arising from any empirical problem, a computational method that attempts to solve the problem is generally bound to be some type of point by point minimization. In particular, for a general fitness function, a point by point optimization process may not converge at all, may not converge quickly enough, or may converge to a point that does not minimize the fitness function on its domain; this limitation is intrinsic to these methods, rather than being incidental to some of the specific examples we have seen. Even more important, there is no guarantee that choosing slightly different fitness functions, with different parameters, will lead to similar solutions. In the next section we will see how "unnecessarily" difficult optimization problems and fitness functions can be reinterpreted and simplified by looking closely at the structure of point by point minimization processes.

3 Brandt's Principle

In this section we suggest that both PSO and the other methods developed by forcing optimization onto empirical problems are governed by one methodological principle, which we call 'Brandt's principle' (derived and generalized from considerations of Achi Brandt in 2001) which clarifies the role and appropriate usage of point by point minimization. We argue moreover that this principle shows a specific, operative way in which the link between structural understanding of a phenomenon and successful problem solving through data analysis can be broken, without apparent impact on the effectiveness of the problem solving itself. In particular, we show that the actual implementation strategies of Brandt's principle, while depending on partial, fragmented information on the phenomenon, are an expression of the microarray paradigm.

The starting point of our analysis is the strategy suggested in Brandt (2001), Brandt and Livne (2011) to solve image processing problems. Those reviews show how direct multiscale algorithms, that satisfy a whole set of contrasting criteria at different stages, by focusing on different scales at each of these stages, can perform as well as global regularization techniques in finding solutions of some ill posed problems (i.e. problems where the solution is not uniquely defined, and/or such that the solution is not robust under variations of the input of the problem). The relevance of multiscale methods is not surprising for image analysis, given what we have said in Sect. 2.1 concerning the way information about the relevant image are encoded differently at different scale of detail (contours versus texture for example). The main focus and interest in Brandt (2001), Brandt and Livne (2011) is to understand how to approach optimization problems pertaining to image processing and to partial differential equations with multiscale methods, and how to build, case by case, such methods. Here we would like to expand the scope of this insight to wider classes of problems and methods.

According to Brandt, in many cases, the search for optimization can be replaced by "a more universal and often far easier" approach, which just consists in admitting:

> a solution which is just the end product of a suitable *numerical process* not necessarily designed to satisfy, even approximately, any one governing criterion. (Brandt 2001, page 60)

The problem with this highly agnostic approach (in our sense) is how to identify suitable numerical processes. A criterion Brandt suggests for this purpose is this:

> the amount of computational work [in a numerical process] should be proportional to the amount of real physical changes in the computed system. Stalling numerical processes must be wrong. (Brandt and Livne 2011)

It would follow that what truly distinguishes a solution of a problem, or decides that the time has come to switch to a different solving algorithm, is a near-steady state of the algorithm output over successive iterations of the algorithm itself; stalling algorithms are wrong either because we have found a solution and we should terminate the algorithm (as it is the case for point by point optimization algorithms that have converged), or because we need to switch to a new algorithm. We state succinctly the radical idea that underlies the two quotations above in the following principle, which we name after Achi Brandt. And we claim that its domain of applicability is far more encompassing than the field of multiscale methods, and that it can be taken as the organizing principle of most data-driven computational methods:

Brandt's principle: An algorithm that approaches a steady state in its output has found a solution to a problem, or needs to be replaced.

This principle proposes a fundamental shift away from the search of best solutions to problems. This shift underscores the data-driven nature of the methods in data analysis, and complements it. Since Brandt's principle does not have a built-in criterion to establish when a solution has been found (instead of having reached a stage in which one has to switch from an algorithm to another), the output of the relevant numerical process is to be checked, by external validation means, for its usefulness: a fuzzy concept in general, which nevertheless does have quite precise renderings in many specific applications. For many empirical problems, a precise task has to be achieved, and any method that can achieve such a task would be suitable. Think, for example, of the problem of having an autonomous, driverless car travel from one location to another in rough terrain; achieving this task, in any way, is the really interesting problem, because of the dangers of the car getting overtopped, or otherwise incapacitated. Finding the very best route in terms of shortest distance, or fuel efficiency is secondary. In general, the distinction we make here is between optimality, which usually designate some singular points in the space of solutions, and usefulness, which rather designate entire regions in the space of solutions.

While Brandt's principle deemphasizes the search for optimal solutions to problems, it is satisfied by local, point by point optimization methods. We have already seen in Sect. 2 that wide classes of computational methods, including PSO and image processing methods, can ultimately be expressed in this way. Moreover, most data analysis methods attempt to fit a model on data by minimizing some suitable error function (Hastie et al. 2009). Because the fitness functions that correspond to such methods are usually both ad hoc and complex, local minimization techniques are preferred in this context. This is true in particular for artificial neural networks, currently some of the most powerful data analysis techniques (Izenman 2008, chapter 10). As we pointed out in Napoletani et al. (2013b), the success of artificial neural networks can be ascribed in great part to the effectiveness of a particular local recursive process to estimate their parameters, the backpropagation algorithm. In turn, this algorithm is essentially a form of gradient descent algorithm (Hastie et al. 2009, Sect. 11.4).

In light of the pervasiveness of local optimization techniques in data analysis, we can therefore claim that most of its methods satisfy Brandt's principle when implemented in an effective algorithmic form. Moreover, some of the most important types of classification methods, such as committee machines (Izenman 2008, chapter 14) can be shown to satisfy Brandt's principle without an explicit local optimization implementation. This is particularly significant when we consider that boosting algorithms, singled out in Napoletani et al. (2011) as one of the most significant embodiments of the microarray paradigm, are a type of committee machines (Izenman 2008, Sect. 14.3). In fact, committee machines are built starting from more basic, structurally similar classification algorithms (classifiers). A standard committee machine switches among these basic classifiers, each attempting to find a solution for a slightly different version of some initial classification problem, and it eventually combines them into a single, potentially more accurate classifier (we refer to Napoletani et al. (2011) for a general methodological analysis of classification problems and classifiers). Crucially, the decision of switching among the basic classifiers is made on the basis of the convergence to a stable tentative solution to the classification problem at each iteration of the process. A similar, sequential approach is also used in hybrid classification methods (Delimata and Suraj 2013). The difference with respect to committee machines is that the basic classifiers utilized by hybrid methods do not necessarily share a common structure.

We did not emphasize so far the aspects of an algorithm that are uniquely related to a specific problem. Brandt talks about "suitable" algorithms, but what does it mean in the context of a specific problem? To begin to answer this question, we recall that, in Sect. 2.2, the fitness functions for image processing problems are built piece by piece on the basis of partial assumptions on the images. While individually these assumptions do encode some understanding of the structure of images, the resulting fitness functions, with all their parameters to be chosen, do not offer a transparent way to interpret the solution to a given problem, and their complexity does not generally allow for a convergence to the optimum. Fitness functions are forced on the problem and they can, at best, be justified as rough quantitative a priori constraints on the form of a solution.

More generally, in image processing the a priori constraints are our assumptions on what constitutes a natural image or a contour, and these assumptions affect the choice of image dictionaries. It is possible however to make this choice, and establish our assumptions on images, automatically, for example by allowing sparsity to be the governing criterium of image representation, and using large databases of natural images to search for the best dictionary elements that, on average, optimize sparsity on a given image of the database (Olshausen and Field 1996; Field 1999). This process is computationally intensive, but it can be achieved in many cases, and the resulting choices of dictionaries can then be used for further image processing problems. Of course, the assumption of sparsity is not justified from the data per se, but is required by our limited resources.

We note also that the use of specific prior assumptions on a phenomenon to force optimization is similar to the basic approach of pattern theory (Grenander and Miller 2007) and Bayesian analysis (Berger 2010) (other very powerful theories to model and classify complex phenomena), where the solution to empirical problems is contingent on the identification of appropriate priors on the probability distributions of the stochastic processes used to model the underlying phenomena.

From this analysis, we can conclude that the building blocks of methods that respect Brandt's principle are based on the unavoidable inclusion of partial pieces of prior information (what, following a customary usage, we can call 'priors', *tout court*) on a phenomenon. This is an important point that we need to clarify, as apparently the presence of priors proper to a particular phenomenon may be seen to be in contradiction with the microarray paradigm, and the whole agnostic approach to data analysis. One could argue indeed that, if for each problem we should identify the exact prior information that is necessary to solve it, we would have no reason to say that the microarray paradigm is at work.

Still, as we have seen in Sect. 2.1, even the use of specific priors (fitness functions in that case) does not guarantee that we can understand the reason a specific output of an optimization method solves the relevant empirical problem. If this is not the case, and it happens very often, the problem solving is still agnostic in our sense, and the microarray paradigm applies. In other terms, the fact that some priors are proper to a certain problem does not mean that their identification depends on a structural understanding of the subjacent phenomenon.

In data analysis we assume a huge quantity of data about a phenomenon, and from these data we can extract many significant features of the phenomenon, in an entirely data driven way if necessary, to the effect that these features are generally not structurally related to each other. More than that, it is often quite hard to guess in advance which subgroup of the features so detected will be useful to solve a given problem. This large number of priors can be partially structured, for example in the form of plausible guesses on the mathematical nature of the data (as in Bayesian analysis, where we may have information on the shape of probability distributions about the phenomenon), or it can be simply in the form of large collection of raw data from the phenomenon. In both cases, it is the sheer size of priors that set us squarely within the microarray paradigm: we need as many priors as possible to solve a problem in a data driven way, and to have an excess of partially structured priors is as opaque and agnostic as to have none.

We finally note that effective hybrid classification methods differ significantly in the order and the type of the basic classifiers they use. Similarly, image processing methods based on optimization allow for significantly different, equally suitable fitness functions. This suggests that, when building a method based on Brandt's principle, the specific way the priors are put together is not essential: what counts is the ability to distinguish when an algorithm produces dynamically changing outputs, and when instead it gives nearly steady outputs. And the reason for the usefulness of the alternating process advocated by Brandt's principle is that it does not explore the solution space randomly, but according to priors that, no matter how they are combined, encode preferential ways to look at the problem itself. Therefore the expectations on algorithms governed by Brandt's principle are the following: they have to be easy to build, starting from a large number of weak assumptions on a phenomenon and a problem about it; their nearly steady outputs have to be exceptional and clearly identifiable; and switching among steady states outputs has to provide a fast exploration of the solution space of the problem at hand.

4 Data Analysis of Historical Phenomena

We have seen in Sect. 3 how the process of forcing optimality and its effectiveness can be reinterpreted in light of Brandt's principle, and we have argued that this principle governs most agnostic, data-driven methods. In this section we explore the appropriateness of using such methods, consistent with Brandt's principle, to solve problems about historical phenomena (introduced in Napoletani et al. 2014). We perform this analysis by looking at one of the organizing principles that have been suggested in recent years to make sense of biological processes, the so called principle of developmental inertia (Minelli 2011). We will show the generality of developmental inertia as an organizing principle of historical phenomena, subject to morphogenesis (a term used here in the general and original meaning of change of characteristics). Finally we will draw parallels between Brandt's principle and the principle of developmental inertia to find evidence of the relevance of the former (and of data analysis in general) for the solution of problems about historical phenomena.

In Napoletani et al. (2014) we suggested that biology and social sciences could be preferred domains of exploration of data analysis, as sciences concerned with "historical phenomena", i.e. phenomena significantly constrained by their past, historical development. We briefly argued moreover that it is possible to describe historical phenomena as "those phenomena whose development can only be constrained locally (in time and/or space) by (potentially multiple) optimization processes acting on subsets of variables, and in such a way that the functions to be optimized change over long periods of time" (Napoletani et al. 2014). This characterization was motivated by an analysis of fitness landscapes in evolutionary biology. There are important relations between these phenomena and Brandt's principle, in particular in the way the latter allows to reinterpret point by point minimization algorithmic processes. To understand these relations, we first explore the general structure of historical phenomena, starting with ideas from developmental biology.

In Minelli (2011) Alessandro Minelli, in order to conceptualize the way organisms structure themselves, suggests the *principle of developmental inertia*; generally stated, it asserts that we can see:

> [biological] developmental processes... as deviations from *local self-perpetuation of cell-level dynamics* (Minelli 2011, page 119).

Similarly to the concept of inertial conditions in physics, the main purpose of the notion of developmental inertia is to identify an appropriate singular state in the relevant processes, the null (or inertial) state of these processes, this is the state that the system would stay in, if no perturbations (external to this very state) had occurred. The intrinsic nature of such null state is less relevant than the nature of the perturbation acting on it, allowing the system to evolve along different lines. It is argued in Minelli (2011) that:

> [developmental] inertial conditions [...] *do not represent origin*, in the ordinary meaning of the word, but only a convenient "zero" term of comparison for the study of something that happens in time– a segment of history, without prejudice of what happened before it.

The identification of suitable inertial conditions in biological systems leads to fruitful reinterpretations of asymmetry, segmentation and regeneration in complex full-grown organisms, exactly by pointing out the appropriate, null state of biological development. A particularly simple example can be seen in the context of embryo development, where the appropriate inertial state of embryos is the reproduction, symmetrical in space and indefinite in time, of the same basic tissue structure. However, in a real system, developmental inertia is constantly perturbed (Minelli 2011, page 123), which allows complex, inhomogeneous and asymmetrical organisms to form.

Developmental inertia may be a key principle in conceptualizing biological developmental processes, but it would need to have much wider applications to be an organizing principle of general historical phenomena. Since social phenomena are arguably, together with biological ones, the most important type of historical phenomena, the principle of developmental inertia should apply to these phenomena as well. Showing case by case the usefulness and validity of this principle in this new context would require a large survey in itself (which cannot be performed here). However, it would appear that, if we look at a social system as a collection of interacting individual agents with their own decision rules for behavior (Epstein 2006), collective agreement among such individuals can be assumed to be an inertial state. For example, social norms have been shown to have a tendency to become entrenched and to be self-enforced (Epstein 2001), to the effect that their entrenchment and their spreading across individuals could be taken, locally in time and space, as an inertial state in a society.

A better, more general route to justify the principle of developmental inertia for social phenomena is to rely on the broad, strong homology of social and biological systems. This homology is made explicit in accounts such as the "living systems theory" detailed in Miller (1978), that emphasize the notion of process in all living systems, whether biological or social (Bailey 2006). Assuming its truth, it is likely that general principles from biology transfer to social phenomena, and we can conclude that the principle of developmental inertia applies to a wide range of historical phenomena, both biological and social in nature.

We also note that, at any given time, there is accumulation of structures in the states of historical phenomena; for example, accumulation of tissues differentiation in the embryo, and accumulation of legacy social norms in societies. This accumulation of structures gives clues about the inertial states that occurred along the history of a phenomenon: we can identify a structure as distinct from the whole (for example a specific tissue in an embryo) only because some inertial state (the repeated proliferation of that tissue) is replaced by some other inertial state (the differentiation into a new tissue). This suggests focusing on those characteristics of an historical phenomenon that can be ascribed to some unperturbed developmental inertial state that occurred in its past. We shall refer to them as the developmental structures of such a phenomenon.

In light of the broad applicability of the principle of developmental inertia to historical phenomena, it is striking how similar this principle is to Brandt's principle. Both principles identify appropriate null states: self-perpetuating cell dynamics for developmental inertia; near-steady state outputs of algorithms for Brandt's principle. Both principles see the breakdown and deviation from a null state as essential: to morphogenesis processes in complex organisms in the case of developmental inertia; to computational processes that solve problems in the case of Brandt's principle. Moreover, to have a near-steady state output of an algorithm implies that its (dynamical) computational processes are nearly repeating themselves, so that we can speak of self-perpetuation of dynamical processes also in the context of Brandt's principle. We believe this strong similarity is fundamental to the understanding of the effectiveness of data analysis, and we highlight it in the following proposition:

The principle of developmental inertia, as applied to historical processes, and Brandt's principle, as applied to computational processes, are homologous.

For historical phenomena, there is no compact description of the totality of developmental structures, so that it is necessary to identify and collect as many individual structures as we can to solve problems, which is in line with the microarray paradigm. Indeed, each developmental structure of an inertial state essentially differs from, and cannot be reduced to, the developmental structure due to another, contemporary or subsequent, inertial state. If it is necessary to focus on distinct developmental structures to solve a problem, these structures have to be identified, and harnessed, independently of each other. The author of Minelli (2011) talks of distinct centers of developmental dynamics that are "everything everywhere", so that the proliferation of developmental structure does not reduce itself to a simple, unifying description. However, the collection of all such developmental structures does encode, to some extent, the state of the historical phenomenon.

Computational methods that respect Brandt' principle, or 'Brandt's methods', are uniquely suitable to take advantage of the incoherent proliferation of developmental structures, since each of these structures can be used as a prior for a corresponding algorithm, a building block for Brandt's methods. Alternating among these algorithms, when their output reaches a steady state, may be the preferred way to successfully explore the solution space of a problem. For general problems about historical phenomena, a suitable Brandt's method may be as efficient a way to search for solutions as any other method. The lack of global finality or optimality of developmental structures frustrates any attempt at faster exploration of the solution space.

As we have seen in Sect. 3, Brandt's principle offers a fruitful, theoretical scaffolding for forcing optimization and, more generally, for data analysis. The homology of Brandt's principle with the principle of developmental inertia suggests something more: data analysis, rather than being simply an heuristic, preliminary set of tools, could actually be the privileged way to approach historical phenomena and their problems with quantitative, theoretical tools.

References

Arfken, G. B., & Weber, H. J. (2005). *Mathematical methods for physicists*. Boston: Elsevier Academic Press.

Bailey, K. D. (2006). Living systems theory and social entropy theory. *Systems Research and Behavioral Science, 23*, 291–300.

Berger, J. O. (2010). *Statistical decision theory and Bayesian analysis*. New York: Springer.

Bhutada, G. G., Anand, R. S., & Saxena, S. C. (2012). PSO-based learning of sub-band adaptive thresholding function for image denoising. *Signal, Image and Video Processing, 6*(1), 1–7.

Boyd, S., & Lieven, V. (2004). *Convex optimization*. Cambridge/New York: Cambridge University Press.

Brandt, A. (2001). Multiscale scientific computation: Review 2001. In J. B. Timothy, T. F. Chan & R. Haimes (Eds.), *Multiscale and multiresolution methods: Theory and applications*. Berlin/New York: Springer.

Brandt, A., & Livne, O. E. (2011). *Multigrid techniques: 1984 guide with applications to fluid dynamics*. Philadelphia: Society for Industrial and Applied Mathematics.

Bresson, X., Esedoḡlu, S., Vandergheynst, P., Thiran, J.-P., & Osher, S. (2007). Fast global minimization of the active contour/snake model. *Journal of Mathematical Imaging and Vision, 28*(2), 151–167.

Chen, S. S., Donoho, D. L., & Saunders, M. A. (2001). Atomic decomposition by basis pursuit. *SIAM Review, 43*(1), 129–159.

Delimata, P., & Suraj, Z. (2013). Hybrid methods in data classification and reduction. In A. Skowron & Z. Suraj (Eds.), *Rough sets and intelligent systems – Professor Zdzisław Pawlak in memoriam. Intelligent Systems Reference Library, 43*, 263–291.

Donoho, D. L., Elad, M., & Temlyakov, V. N. (2006). Stable recovery of sparse overcomplete representations in the presence of noise. *IEEE Transactions on Information Theory, 52*(1), 6–18.

Donoho, D. L., & Johnstone, I. M. (1994). Ideal spatial adaptation by wavelet shrinkage. *Biometrika, 81*(3), 425–455.

Donoho, D. L., & Johnstone, I. M. (1995). Adapting to unknown smoothness via wavelet shrinkage. *Journal of the American Statistical Association, 90*(432), 1200–1224.

Dorigo, M., et al. (2008). Particle swarm optimization. *Scholarpedia, 3*(11), 1486.

Epstein, J. M. (2001). Learning to be thoughtless: Social norms and individual computation. *Computational Economics, 18*, 9–24.

Epstein, J. M. (2006). *Generative social science: Studies in agent-based computational modeling*. Princeton: Princeton University Press.

Field, D. J. (1999). Wavelets, vision and the statistics of natural scenes. *Philosophical Transactions of the Royal Society A, 357*, 2527–2542.

Grenander, U., & Miller, M. (2007). *Pattern theory: From representation to inference*. Oxford: Oxford University Press.

Hastie, T., Tibshirani, R., & Friedman, J. (2009). *The Elements of statistical learning*. New York: Springer.

Izenman, A. J. (2008). *Modern multivariate statistical techniques: Regression, classification, and manifold learning*. New York: Springer.

Kass, M., Witkin, A., & Terzopoulos, D. (1987). Snakes: Active contour models. *International Journal of Computer Vision, 1*(4), 321–331.

Kennedy, J., & Eberhart, R. C. (1995). Particle swarm optimization. In *Proceedings of the IEEE 978 International Conference on Neural Networks, Perth, WA* (Vol. 4, pp. 1942–1948). New York: IEEE.

Mallat, S. (2008). *A wavelet tour of signal processing*. San Diego: Academic Press.

Miller, J. G. (1978). *Living systems*. New York: McGraw Hill.

Minelli, A. (2011). A principle of Developmental Inertia. In B. Hallgrimsson & B. K. Hall (Eds.), *Epigenetics: Linking genotype and phenotype in development and evolution*. Berkeley, CA: University of California Press.

Napoletani, D., Panza, M., & Struppa, D. C. (2011). Agnostic science. Towards a philosophy of data analysis. *Foundations of Science, 16*(19), 1–20.

Napoletani, D., Panza, M., & Struppa, D. C. (2013). Artificial diamonds are still diamonds. *Foundations of Science, 18*(3), 591–594.

Napoletani, D., Panza, M. & Struppa, D. C. (2013). Processes rather than descriptions? *Foundations of Science, 18*(3), 587–590.

Napoletani, D., Panza, M. & Struppa, D. C. (2014). Is big data enough? A reflection on the changing role of mathematics in applications. *Notices of the American Mathematical Society, 61*(5), 485–490.

Nasri, M., & Pour, H. N. (2009). Image denoising in the wavelet domain using a new adaptive thresholding function. *Journal of Neurocomputing, 72*, 1012–1025.

Olshausen, B. A., & Field, D. J. (1996). Emergence of simple-cell receptive field properties by learning a sparse code for natural images. *Nature, 381*, 607–609.

Pedersen, M. E. H., & Chipperfield, A. J. (2010). Simplifying particle swarm optimization. *Applied Soft Computing, 10*(2), 618–628.

Poli, R. (2008). Analysis of the publications on the applications of particle swarm optimisation. *Journal of Artificial Evolution and Applications*. doi:10.1155/2008/685175.

Reynolds, C. W. (1987). Flocks, herds, and schools: A distributed behavioral model. *Computer Graphics, 21*(4), 25–34.

Wegener, I. (2005). *Complexity theory: Exploring the limits of efficient algorithms*. Berlin; New York: Springer.

Object Oriented Models vs. Data Analysis – Is This the Right Alternative?

Jürgen Jost

1 Introduction: The Basic Issue

Traditionally, there has been the distinction between pure and applied mathematics. Pure mathematics – so the story goes – discovers, creates and investigates abstract structures (see Jost 2015) for their own sake, while applied mathematics applies existing mathematical tools and develops new ones for specific problems arising in other sciences. The interaction between pure and applied mathematics takes place in both directions. Applied mathematics utilizes concepts and methods developed in pure mathematics, and problems from diverse applications in turn stimulate the development of new mathematical theories. And traditionally, those problems arose within a clear conceptual framework of a particular science, most notably physics. In this essay, I want to argue that this distinction between pure and applied mathematics is no longer useful – if it ever was –, and that the challenge of large and typically rather diverse data sets, typically arising from new technologies instead of theoretically understood and experimentally testable concepts, not only calls for new mathematical tools, but also necessitates a rethinking of the role of mathematics itself.

In fact, already in the past, what was called applied mathematics often developed domain independent methods. And this domain independence typically led to a gain in generality and mathematical depth. Statistics, for instance, has become so powerful precisely because it developed methods and concepts that apply to essentially any field, from particle physics to the social sciences. And the error estimates and convergence rates provided by numerical analysis are valid for any application of a particular numerical method in the engineering sciences or elsewhere. Nevertheless,

J. Jost (✉)
Max Planck Institute for Mathematics in the Sciences, Inselstr.22, 04103, Leipzig, Germany

Santa Fe Institute, NM 87501, USA
e-mail: jost@mis.mpg.de

J. Lenhard, M. Carrier (eds.), *Mathematics as a Tool*, Boston Studies in the Philosophy and History of Science 327, DOI 10.1007/978-3-319-54469-4_14

such applications usually took place within a particular field with a well developed theoretical framework that provided interpretations for the statistical or computational results obtained. And in other cases, like modeling with differential equations, the specific properties of a concrete scientific theory yielded key ingredients for the mathematical structures. The development of the underlying scientific theory and the mathematical model often went hand in hand, and they deeply depended upon each other. Mathematical modeling was an essential ingredient of the scientific strategy, and in that sense, mathematics was more than a tool. In contrast, nowadays mathematicians are often confronted with data sets of obscure quality, perhaps even of dubious origin, and without any firm theoretical foundation. Even the distinction between meaningful data and meaningless noise may not be clear at all.

Therefore, as data collection these days is typically ahead of theoretical understanding, mathematics should radically face the lack of theory and look at what there is, the data, and see what it can do with them. And what I want to call for are not ad hoc methods for every concrete data set, but rather an abstract analysis of the deeper structural challenges. Of course, modern mathematics is developed and sophisticated enough to provide appropriate and helpful tools for basically any data set, but this by itself is too narrow a scientific perspective. For me as a mathematician, mathematics is more than data analysis. We need a conceptual rethinking.

Let us take a look at the situation from the perspective of the sciences. It seems to me that in the presence of data, there are two different, but interwoven issues:

1. The epistemic question, or the role of theory: Can we have data without a theory? And if not, does the theory have to be specific for the domain from which the data are collected?
2. The ontological question, or the role of models: Do we need, or have to postulate, specific objects underlying the data? What is it that the data tell us something about, and how can we, or should we model that?

This distinction may sound a little like issues debated between scientific realists, positivists, and empiricists. But even if we adopt the latter stance and use, for instance, van Fraassen's criterion that a "theory is empirically adequate if it has some model such that all appearances are isomorphic to empirical substructures of that model" where "appearances" are "structures which can be described in experimental and measurement reports" (van Fraassen 1980, p. 64), the problem of data without such an empirically adequate theory remains.

Of course, there are some easy answers that seem rather obvious:

- Data that cannot be interpreted within some theoretical framework are meaningless. Much of current data collection is solely driven by particular technologies rather than by scientific questions. As Sir Peter Medawar put it, "No new principle will declare itself from below a heap of facts".[1] And for "facts", we might want to substitute "data".

- If we simply analyze data with intrinsic formal tools whose precise functioning we may not even understand, then science becomes agnostic (Napolitani et al. 2011). We need to know first about what the data are telling us something. That is, we need an underlying structure from which the phenomena revealed by the data are derived. The meaning of data depends on the context from which they are collected.

These concerns are clearly valid. Nevertheless, I think that the issues are somewhat more subtle.

1. To what extent does theory have to be domain specific? Or more precisely, what is the relationship between general theoretical issues – to be elaborated below – and domain specific ones? Or more positively, will the current situation of "big data" lead to new types of theories that start from a data intrinsic rather than a domain specific perspective, and could that possibly lead to theoretical insights at a higher level of abstraction?
2. What is the right balance between a purely phenomenological approach and one that starts with a model involving underlying objects? Such objects could be either the carriers of the properties revealed by the data or at least offer formal structures to which quantities measured on the data are isomorphic.

A traditional perspective would consider mathematics only as a tool when – possibly very large – data sets are to be analyzed. One of the theses of this essay is that this process itself, data analysis, can and already has become an object of mathematical research. Thus, mathematics not only serves as a powerful tool, but by reflecting this role, gains a new perspective and lifts itself to a higher level of abstraction. Then, the domain of such mathematical inquiry no longer is a specific field, like physics, but mathematical technique itself. Mathematics then is no longer, if it ever was, a mere formal tool for science, but becomes the science of formal tools.

Another thesis of this essay is more concerned with the traditional role of mathematics as the formal analysis of models arising from specific domains. I shall argue here against the principle of reductionism. The thesis will be that in the empirical domains where regularities have been identified that could be cast into mathematical structures, the relations between the different levels at which such structures emerge are very intricate. A simple formal structure at one level typically can neither be derived from an even simpler structure at a lower level, nor can be used to identify the relevant structure at a higher level. The structures at both lower and higher levels can be substantially more complex. Again, it is a general task for mathematical research to identify and more deeply understand the principles of such transitions between different scales and levels. That is, what is at issue are not so

[1] I learned this quote from Peter Schuster.

much the phenomena at the different levels, but rather the transition laws between levels. Much is known here already, both from the side of mathematics and that of physics, but a more abstract and fundamental mathematical theory is still missing.

There might even be unexpected connections between those two aspects, a mathematics of and not only for data analysis on one side, and the principles of transitions between different levels on the other side. At least, there is the question of how such transitions can be inferred from the data. I regard this as a fundamental question for any abstract theory of data.

Before addressing these issues in detail, I should point out that there is another, perhaps even more fundamental and important role for mathematics in the sciences, although that role will not be discussed in this essay. This consists in providing a framework for conceptual thinking and formal reasoning. This is valid across all disciplines, and actually badly needed in many of them. I shall address this aspect elsewhere.

2 The Role Model of Physics

In some way or another, there is a particular role model behind many discussions, that of classical physics. According to that perspective, every science should strive towards such a model. We should have objects whose properties and interactions can be formally described by mathematical relations, in particular, differential equations. And we should be able to conduct experiments whose outcomes can be understood within an encompassing theoretical framework. Deviations from that role model are regarded as deficits.

Newton's theory combined the terrestrial dynamics as developed by Galilei and others and Kepler's laws of celestial motions in a unified theory of gravity. The basic objects were pointlike masses. Such a concept would have appeared meaningless to Descartes who considered extension as the basic quality of matter. In Newtonian dynamics, however, the essential feature is the mass of an object, and this allowed him to work with extensionless points carrying a mass as idealized objects. Newton's theory allows for an exact solution of the two-body problem.[2] Besides points, there are other fundamental constituents of physical theories. These are the fields, like the electromagnetic one. In physical theories, fields are considered to be no less real than material objects. In particular, they also carry forces which make physical interactions possible. Also, while the two-body problem admits an exact solution in Newton's theory, this is no longer true for the three-body

[2]The one-body problem, however, was eventually understood as being ill posed. As Leibniz first saw, any physical theory has to be concerned with relations between objects, and an irreducible point mass can only entertain relations with other such objects, but not with itself, because there is no fixed external frame of reference independent of objects. The latter aspect is fundamental in Einstein's theory of general relativity. (See for instance the exposition by the author in Riemann 2016.)

problem. Although that problem is perfectly well posed within Newton's theory of gravity, it can in general no longer be solved in closed form, unless particular symmetries pertain. This was first realized by Poincaré. Nevertheless, one can develop approximation schemes or compute numerical solutions of the differential equations to any desired degree of accuracy with sufficiently large computer power. At a more theoretical level, the stability against perturbations becomes a subtle issue, as developed in KAM theory (after Kolmogorov, Arnol'd, and Moser), see Siegel and Moser (1971), Treschev and Zubelevich (2010). Also, there are mathematical examples (Xia 1992; Saari and Xia 1995) where internal fluctuations in such a system can grow without bounds until one of the bodies is ejected to infinity.

The paradigmatic objects of this theory are the celestial bodies in the solar system, and the theory is concerned with their orbits. Even when the theory is restricted to the basic configuration of the sun, the earth, and the moon, there are more than two such objects. And these objects are by no means pointlike, but rather extended and possessing a non-homogeneous internal structure. If one models them in more detail, the resulting theory becomes much more complicated, and at best one may hope for numerical approximations, but these will inevitably be of limited scope and validity.

But this then raises the question why such a simple theory as Newton's is applicable at all to such a complicated configuration, and in particular, why numerical solutions of Newton's equations lead to such accurate descriptions and predictions of the movements of the planets. (For the sake of the argument, we ignore here the corrections necessitated by Einstein's theory.) When we change the scale, either going down to the internal structure of the celestial bodies, or up to configurations of many gravitating objects, no such simple theory applies.

Is Newton's theory then a completely singular instance, and would we therefore be ill-advised to consider it as a role model of a scientific theory? Well, it is not completely singular, and in order to understand the issues involved better, let us consider another example.

The fundamental equation of quantum mechanics is Schrödinger's equation

$$\sqrt{-1}\hbar\frac{\partial\phi(x,t)}{\partial t} = -\frac{\hbar^2}{2m}\Delta\phi(x,t) + V(x)\phi(x,t) \tag{1}$$

for the quantum mechanical state $\phi(x,t)$ at position x and time t, where Δ is the Laplace operator, a second order partial differential operator, $V(x)$ is the potential at x, and $\hbar$, m are physical constants. The Schrödinger equation no longer describes the deterministic behavior of pointlike objects, but rather the evolution of probabilities. The dynamics of these probabilities are still deterministic. And in fact, for the hydrogen atom, the Schrödinger equation is exactly solvable. For more complicated atoms, we encounter the difficult numerical problems of quantum chemistry. The reason why it applies so well to the hydrogen atom is that its nucleus consists of a single proton which under normal conditions essentially behaves like an elementary, indivisible ("atomic" in the literal sense) particle. In fact, however, the proton is not elementary, but is composed of more elementary particles, the quarks (see e.g. Close 2004). That is, the fact that the hydrogen atom can be described by the Schrödinger equation does not have a theoretical, but an empirical reason, the so-called quark confinement. Of course, one may then try to derive this from the standard model of particle physics, but that is a different issue. (For the theoretical background, see e.g. Weinberg 2000.) What remains is the empirical validity of Schrödinger's equation. The standard model, in contrast, is a model whose deeper justification still is a matter of intense debate. There are approaches like string theory (see e.g. Green et al. 1995; Polchinski 1998; Jost 2001, 2009; Zwieback 2009) which, however, are beyond the range of experimental testing for the time being. Why Schrödinger's equation applies so well (at least to the hydrogen atom) then remains mysterious at a deeper level. And it even applies to other atoms, although it is no longer exactly solvable already for the Helium atom. This is analogous to Newton's equation and the three-body problem.

The equations of Newton and Schrödinger may possess certain universal properties, and some aspects might be considered with the modern tools of renormalization groups, see for instance Batterman (2002), but the question why such equations appear at particular scales and not at others remains. Thus, my question is not why a mathematical description of nature is possible, or why mathematics is so effective as a tool for describing physical reality, but rather why it works so well at particular scales or under certain circumstances, but not in other situations. Historically, when the mechanical natural philosophy of the seventeenth century could capture so well the movements of celestial bodies, explain the acceleration of falling bodies or the parabolic trajectories of canon balls or identify the conservation laws of inelastic collisions, it was expected that this success could easily be extended to other domains and achieve, for instance, a mechanical explanation of animals, and, as some, like La Mettrie, believed, even of humans. Of course, this scientific program turned out to be a miserable failure, and biology as a modern science got off the ground only in the nineteenth century, on the basis of completely different principles than those of the mechanical natural philosophy of the seventeenth century.

I shall return in Sect. 3 to the difference between the theoretical validity of an equation and its exact solvability, as this is important for our topic. Before that, however, I'll proceed to larger physical scales. In the area of the physics of nano- and microstructures, on the one hand, there are atomic or molecular models, and on the other hand, there are continuum models. Either type of model can yield accurate descriptions within its range of validity and make precise predictions. One might then argue that the discrete (atomic or molecular) models are the basic ones that capture the underlying objects whereas the continuum models are purely phenomenological. This, however, will miss the fundamental mathematical question, the relationship between these two types of models and the transition between them when changing the scale. Again, this is an aspect to which I need to return. For instance, it seems that in many respects, the Navier-Stokes equations correctly – whatever that means – or at least adequately describe the behavior of fluids. These are continuum equations. The numerical solution, that is, an approximation by a numerical scheme, is one of the most important topics of scientific computation, and by now, there exist numerical schemes that can solve them to a very high degree of accuracy. Nevertheless, within the mathematical theory of partial differential equations, the general existence of a solution to date has not been established, and in fact, this is considered to be a challenging and difficult problem. The issue of turbulence which is still not understood reveals that there are some deep problems lurking here. Interestingly, while turbulent flow seems to be rather chaotic, there might exist some statistical regularities. In any case, the Navier-Stokes equation serve as the most important model in fluid dynamics.

When we move further up with regard to the physical scale, we come to the example already discussed, the dynamics of the orbits of the planets in the solar system which (except for relativistic corrections) are so well described by Newtonian mechanics. Planets, however, are no elementary bodies, but composite objects. So, let us repeat the question why such complicated objects as planets, just think of the geology of the earth, follow such elementary laws. If we go to a smaller scale, everything breaks down, or at least becomes substantially more complicated. For instance, which mathematical or physical theory can explain the rings of Saturn?

And if we go to configurations of large numbers of gravitating bodies, Newton's theory, although applicable in principle (modulo relativistic corrections, again), becomes useless. We should need to change to a statistical description à la (Boltzmann 2000). Such a statistical description would still be mathematical; it would derive its power from ignoring the details of the lower level structure.

Physics then loses some of its special role. It seems to be a general phenomenon that in many domains particular scales exist at which a rather simple mathematical model can capture the essential aspects, sometimes even accurately in quantitative terms, while that need no longer hold for smaller or larger scales. It might seem that this occurs less often in domains other than physics, but there do exist positive examples. In chemistry, we have the reaction kinetics. For instance, reaction kinetics of Michaelis-Menten type are important in biochemistry. They depend on important simplifications. In particular, they assume that the chemical substances participating in the process in question are uniformly distributed across the cell. The internal spatial structure of the cell is completely ignored. As another example, the differential equations of Hodgkin-Huxley type model (Hodgkin and Huxley 1952) the generation of the spike of a neuron in a quantitatively exact manner, see e.g. Murray (2008) and Koch (1999). The model operates at the cellular level and describes the dynamics there in a deterministic manner. If one goes down to the lower, molecular, scale, things get much more complicated, and the deterministic quantities in the Hodgkin-Huxley equations become stochastic, that is, probabilities for the opening and closing of certain ion channels. Similarly, if one moves up and considers interacting systems of neurons, things become quite complicated, and completely different types of models are called for. Thus, we have a relatively simple, formally closed, and quantitatively accurate model at the cellular level, but no such model at the smaller molecular or the larger tissue scale. I shall present the mathematical structure of biochemical reaction kinetics and the biophysics of the Hodgkin-Huxley model in detail below – and readers not interested in those mathematical or physical details can skip those –, but let me first emphasize again the general point. Not only in physical, but also in biological systems, there may exist one or more particular scales at which a gross simplification can lead to simple, but nevertheless quantitatively accurate models, and these scales then correspond to particularly useful levels of description. Typically, however, this is no longer so when we move to either smaller or larger scales. This leads to the question why and how such particular levels emerge, at which such simple quantitative models work. And also, why are such levels apparently so rare?

Let us now describe the mechanism of biological reaction kinetics in some detail to see what is involved. General references for biochemical kinetics are Murray (2008) and Klipp et al. (2005), and we follow here the presentation in Jost (2014). The basis is the law of mass action which states that the reaction rate of a chemical reaction is proportional to the concentrations of the reactants raised to the number in which they enter the reaction. That expression is proportional to the probability that the reactants encounter and react with each other. Let us consider the simple reaction

$$S_1 + S_2 \leftrightharpoons 2P \tag{2}$$

that converts $S_1 + S_2$ into $2P$ with forward rate k_+ and backward rate k_-. That is, when a molecule of substance S_1 encounters one of S_2, they react and form two copies of P with rate k_+. Conversely, two P-molecules together can decay into a copy of S_1 and a copy of S_2 at the rate k_-. Thus, the chemical reaction can take place in either direction, but at possibly different rates. If we denote the respective concentrations by s_1, s_2, p and consider them as functions of time, then (2) leads to the differential equations

$$\begin{aligned} \frac{ds_1}{dt} &= -k_+ s_1 s_2 + k_- p^2 \\ \frac{ds_2}{dt} &= -k_+ s_1 s_2 + k_- p^2 \\ \frac{dp}{dt} &= 2(k_+ s_1 s_2 - k_- p^2) \end{aligned} \tag{3}$$

whose solutions $s_1(t), s_2(t), p(t)$ then give the respective concentrations at time t as functions of their initial values. In enzymatic reactions, there also is the complex ES of the enzyme E and the substrate S. The Michaelis-Menten theory makes the simplifying assumption of a quasi-steady state for the complex ES, that is, its concentration is not changing in time, and this concentration can then be simply computed. This assumption, that the concentration of an intermediate product remains constant, reduces a multidimensional process to a single equation. The resulting simple systems of ordinary differential equations capture the concentration level of substances involved in biochemical reactions in the cell well enough for many purposes. In contrast to the Schrödinger equation which considers the state

$\phi(x, t)$ as a function of the position x and time t, the chemical concentrations s_1, s_2, p in (3) are considered as functions of t only. That is, the model assumes that they are homogeneously distributed in the cell. This means that the concentrations are assumed to be the same across the cell at each fixed time t. This is, of course, a gross simplification, and the model can be refined by allowing for varying concentrations and diffusion effects. The point I want to make here, however, is that even under this (and the further Michaelis-Menten type) simplification, the solutions of systems like (3) can effectively describe the values of concentrations of chemical reactants in cells.

We next turn to the other example discussed above, the Hodgkin-Huxley model (Hodgkin and Huxley 1952) for the generation of spikes in neurons,[3] and give a brief description of the model and its dynamics, see e.g. Jost (2014) and Jost (to appear) for more details. Formally, the model consists of four coupled (partial) differential equations, but they are more difficult than those for biochemical reactions, and so, we refrain from writing them down here. Instead, we only discuss the qualitative features of the model. The basic dynamic variable is the membrane potential V of the neuron. The potential is caused by the different densities of charged ions in- and outside the cell. The boundary of the cell, the cell membrane, is impermeable to most charged ions, except for channels that are selectively permeable for certain specific ions. That permeability in turn depends on the membrane potential as well as on the concentration of certain intracellular and extracellular substances. The differential equations of Hodgkin and Huxley then link the temporal changes of these quantities, that is, the membrane potential and three gating variables that control the permeability of the cell membrane for specific electrically charged ions. Concerning the evolution of the membrane potential, the idea is that the time derivative of this potential is proportional to the derivative of the electrical charge, hence to the current flowing. That current in turn is the sum of an internal membrane current and an external current. The latter represents the external input to the cell. The internal membrane current is a sum of specific terms, each proportional to the difference between the potential and some specific rest term. These proportionality factors then depend on the corresponding gating variables. Thus, the dynamical interplay between the membrane current and the gating variables is the key point. When the system is near its resting value and some positive current is injected that lifts the membrane potential V above some threshold, then a positive feedback between V and the fast one, m, among the gating variables sets in. That is, the potential V rises, and positively charged sodium ions flow in, rising the potential further. Soon, the neuron emits a spike. But then, the two slower gating variables, h and n, take over, causing an inactivation of the inflow of the sodium ions and reversing the potential by an outflow of potassium ions. The potential then drops even below its resting value, but after some time, during which no further spike is possible, recovers to that latter value, and the neuron is receptive again to repeat

[3]The original model was developed only for a particular type of neuron, the giant squid axon, but similarly models have subsequently been developed for other classes of neurons as well.

the process in response to some new external current. The interaction of two time scales is important for these dynamical properties. The positive feedback between the fast variables V and m triggers the spike, whereas the slow variable h ultimately stops the inflow of positive ions, and n causes other positive ions to flow outwards, to reverse the effect and bring the potential back to (or, more precisely, below) its resting value. As already mentioned, this particular model was conceived for the giant squid axon. For this neuron, quantitative measurements were easier than for other neurons, and therefore, Hodgkin and Huxley could carefully fit the parameters of their model. Following the pioneering work of Hodgkin and Huxley, then also models for other classes of neurons were developed. While the details are different, the essential principles are the same. Thus, we have quantitative accurate models of the biophysics of neurons that operate at the cellular level, even though the details at the smaller, molecular level are much more intricate. No such simple models exist at that scale.

3 Validity vs. Computability

We return to Poincaré's insight into the impossibility of an exact solution (a solution in closed form, that is, in the form of an explicit formula for the positions of the bodies involved at every instance of time) of the three-body problem. Even if a mathematical theory of nature were exactly valid in principle – let us ignore the quarks and the internal structure of the planets for the sake of the argument here and assume that for instance, Newton's or Schrödinger's equations were exactly valid –, it could not offer an exactly solvable description of the dynamical behavior of its objects. It is a mathematical question when and under which particular conditions an exact solution by an explicit formula is possible. This has nothing to do with the range of validity of the theory in question. Almost all differential equations cannot be solved by an explicit formula (they do not constitute completely integrable systems in the language of classical mathematics). The search for closed solutions, that is, to demonstrate the complete integrability of a dynamical systems, was one of the highlights of nineteenth century mathematics. In the twentieth century, in the wake of Hilbert, mathematics turned to a different approach. Instead of attempting to construct an explicit solution of a specific differential equation, the problem was converted into showing the existence of solutions for large classes of differential equations. This was the guiding theme in particular for partial differential equations, see for instance Jost (2013). A key point was to separate the abstract question of the existence from the explicit representation of a solution. Once the existence of a solution had been demonstrated from abstract principles, it then became the task of mathematics to understand the properties of such abstract solutions, and in particular to find out under which conditions singularities can be avoided, and then to derive approximation schemes for such solutions and to convert them into algorithms for their numerical construction with a precise error control. This is basic for all the modern applications of partial differential equations in the engineering sciences and elsewhere. This does not mean, however, that mathematics becomes purely instrumental.

A similar issue arises for the weather forecast. Again, it was Poincaré who first realized the specific aspects of chaotic dynamics. Some meteorological models like the Lorenz equations indeed exhibit chaotic behavior. Therefore, we cannot accurately predict next week's weather, because of the amplifications of tiny fluctuations by chaotic dynamics. It is claimed, however, that the global climate 50 years from now can be predicted within certain bounds, for any scenario of greenhouse gas emissions. The reason why this is feasible is that on a longer time scale, the daily weather fluctuations average out, and only the long term trend remains, and that is precisely what the climate models try to capture. In any case, the peculiarities of chaotic dynamics as exhibited by weather models are not an artefact caused by any deficiency of a model, but are – according to deep mathematical insights – rather grounded in specific structural features of the pertinent equations. The mathematics of chaos goes much beyond the rather simple exponential divergence of individual trajectories, and gains positive insight via such concepts as invariant measures (for details, see e.g. Jost 2005). It can even elucidate the universal nature of many aspects of chaotic dynamics, thereby going much beyond any specific object domain.

In this section, I have described two different trends that distinguish twentieth century from nineteenth century mathematics. On the one hand, the ideal of writing down an explicit solution has been abandoned for the less ambitious aim of approximating a solution to any specified degree of accuracy, or in practice at least as accurately as the computing facilities permit. When one has a good theoretical control of the numerical scheme, then higher accuracy simply requires more computer power. In that regard, mathematics has become more instrumental, concentrating on the computational tools rather than on the explicit formulae. On the other hand, general structural insights into chaotic dynamics teach us that gains in accuracy may require an exponential increase in computational effort and therefore quickly become unrealistic. Thus, in most problems we cannot hope to be able to write down an explicit solution of the mathematical model, and even a very accurate approximation of a solution may not be computationally feasible.

4 What Are the Objects?

As described at length, the objects of Newton's theory are modelled as points that possess masses, but no extension. In solid state physics, physicists are working with objects that have more internal properties and structure and therefore might appear more realistic. Electromagnetic and other fields are objects of physical theories with less material substrate. The nature of subatomic particles is not so clear, even though one might grant them some reality because they produce discernible effects (Hacking 1983; Falkenburg 2007). When we go to string theory, particles are nothing but excitation modes of completely virtual "objects", the strings (see e.g. Green et al. 1995; Polchinski 1998; Jost 2001, 2009; Zwieback 2009). Such an excitation mode is similar to a Fourier coefficient. There is nothing material about it, and it is a good question in which sense that should be considered as a physical object. Clearly, however, it is a theoretical object, even though it might not possess any independent reality.

Many biological structures are modelled as discrete entities, even though they might be composed of atoms, molecules, cells, or individuals in a complex manner. Mendel's laws are a good example. Mendel conceived of genes as abstract entities, without any clue about their physical implementation or realization. Although Mendel's laws are not exactly valid, for many purposes this does not matter. This leads to the question how such discrete entities can emerge from some continuum at a lower scale. Conversely, large ensembles of such discrete objects are often best modelled by continuum models, as in population genetics, see e.g. Hofrichter et al. (2017).

In any case, even though it has been discovered in molecular biology that the material substrates of genes are nucleotide sequences, the modern biological conception of a gene is more abstract than such a nucleotide sequence. We refer to Scherrer and Jost (2007) and the discussion in the journal Theory in Biosciences about this issue (Scherrer and Jost 2009). Similarly, the other fundamental biological concept, the species, is more abstract than the collection of individuals composing a population (see for instance Breidbach and Jost 2004). Thus, "objects" in modern biology are abstracta like genes or species that do not directly correspond to physical entities.

This is not fundamentally different in the social sciences, even though there are fewer examples of successful mathematical models. At present, the validity of the mathematical models of macroeconomic theory is somewhat controversial. Likewise, it is an issue of debate to what extent game theory can adequately capture human behavior (see for instance the discussion in Kabalak et al. 2015). On the other hand, in recent years there has been definite progress in modelling traffic dynamics with quasi-physical theories inspired by microscopic particle physics, gas kinetics or fluid dynamics (see Helbing 2001), notwithstanding the fact that individual drivers can behave very differently from each other.

Another issue that has been debated already 200 years ago by social scientists and mathematicians is to what extent the law of large numbers provides not only an empirical, but also a conceptual basis for mathematical modelling of social phenomena and dynamics, see Hacking (1990). When the samples are large enough, statistical laws can lead to arbitrarily precise predictions. This is the foundation of such domains as demography or demoscopy. When we recall Boltzmann's description of statistical ensembles or the above scale transitions, this appears no longer fundamentally different from what has been discussed above for physics and biology.

5 Scales and Levels

In the preceding, I have already discussed the fact that the difficulty and feasibility of a description or a model of a complex system can be rather different at different levels. In this section, I want to analyze this issue more systematically, drawing upon the collaboration with Nihat Ay, Nils Bertschinger, Robin Lamarche-Perrin, Eckehard Olbrich and Oliver Pfante within the EU Project MatheMACS (Mathematics of Multilevel Anticipatory Complex Systems). Let me start with the terminological distinction between *levels* and *scales*. According to the definitions that we propose

- Scales refer to observables. Scales are determined by the measurement process, the data and their representations at different resolutions. For instance, concerning the dimension of length, we might consider the scales of micrometers, millimeters, meters, and kilometers, or other, smaller or larger ones.[4]
- Levels refer to descriptions or models. Levels arise in the modeling process through the identification of entities that lend themselves to useful analysis. Thus, we could model a biological system at the molecular, cellular, tissue or organismal level.

Clearly, the two concepts are interdependent. Levels rely on measurements taken at characteristic scales. They also rely on a choice of observables that are measured. Scales can be chosen relatively arbitrarily, as long as the relevant measurements can be performed, but a useful choice of scale should depend on the identification of a level. Levels, however, should be distinguished by characteristic properties of the model they allow. That is, at a given level, particular regularities should arise that do not pertain at other levels.

First of all, there are the systems that do not possess any characteristic scale. Such systems have been objects of research in both mathematics and physics. One of the concepts proposed here is that of self-similarity. A self-similar structure looks the same at any scale. Such structures were popularized by Mandelbrot (1982) under the name of fractals. The relevant mathematical theory had already been created earlier, by Hausdorff in 1918. Hausdorff had developed a general concept of dimension, and many fractals possess a non-integer Hausdorff dimension. In particular, fractals are in some sense highly irregular. Nevertheless, there also exist such fractals with an integer dimension, and so, mathematically, a more subtle mathematical characterization is necessary, see Steffen (1996) and the references therein. The concept of a fractal also comes up in the theory of chaotic dynamics where the attractors and the boundaries of their domains of attraction could have such a fractal

[4]In different fields, it may be different what is considered as a larger or a smaller scale. In geography, for instance, larger scale means higher resolution, that is, a smaller reduction factor. In other areas, like physics, a larger scale means the opposite. We shall follow the latter terminology. Thus, at a larger scale, many details from a smaller scale may disappear whereas larger structures might become visible.

structure. The corresponding discoveries about iterations of polynomial maps were also made in 1918/1919, by Julia and Fatou, see Devaney (1986) for a more recent description. In the physics literature, scalefree structures are usually characterized by a power-law behavior, as opposed to an exponential decay of correlations. That is, we essentially find correlations at any scale. It turns out that such power-law behavior is rather ubiquitous, from the financial data analyzed by econophysicists to the degree sequences of empirical networks in different domains. From a more theoretical perspective, such systems can even serve as some kind of universal models, when arising as limits of renormalization group flows (see e.g. Cardy 1996), and they are basic constituents of conformal field theory and the theory of critical phenomena, see Di Francesco et al. (1997) and Zinn-Justin (1993).

Of course, no real system can be self-similar at all scales or possess correlations of all orders. Such properties can only hold within a certain range of scales. Nevertheless, as asymptotic idealizations, systems with such properties can be quite useful as theoretical tools.[5] Scalefreeness, however, is not an aspect I want to focus upon here. Let me rather return to a system with well-defined objects at some particular scales. For instance, take a couple of Newtonian particles, at some scale, for instance gas molecules or celestial bodies. When there are only a few of them, their kinetic or dynamical interactions can be described by a mechanical model. For the celestial bodies, that is, when gravity is the active physical force, we have already seen above that for more than two particles, an explicit solution of the equations of motion in general cannot be achieved. But even worse, when the numbers of particles becomes large, like the stars in a galaxy or the gas molecules in a container, the description in terms of classical mechanics is no longer feasible, and we need to turn to a less explicit statistical description à la Boltzmann. Here, a particular configuration of, say, gas molecules, given in terms of their positions and momenta, is a microstate, but at the macrolevel, one only has collective observables like entropy or temperature. These observables are not meaningful at the level of the individual particles, because they represent statistical averages. More precisely, any macrostate can be the result of many different microstates, and the more microstates underlie a given macrostate, the higher the latter's entropy, that is, the more likely it is to occur. (There are some subtle issues here concerning the interpretation of the probabilities involved, see Jaynes (2003), but we do not enter into those here.) In summary, here a deterministic description at the level of individual particles becomes unfeasible and useless due to their large numbers and the impossibility of accurately measuring all their positions and momenta, and instead a statistical description at a higher level is used. It can also be the other way around, that a statistical description at a lower level yields to a deterministic description at a higher level in terms of statistical averages. As fluctuations may average out, these averages themselves may well obey deterministic laws. We see this, of course, in the transition from the quantum level to that of micro- or mesoscopic physics, but

[5] Batterman (2002) and Lenhard (2013) emphasize the fact that such idealized systems that arise as asymptotic limits in some theory are employed as models for empirical systems.

for instance also in cells when we go from the molecular to the cellular level. Recall our discussion of the Hodgkin-Huxley equations in Sect. 2.

Besides the dichotomy between deterministic and stochastic models, there also is the dichotomy between discrete and continuous models. When we move up or down the scales, or better, from one level to the next, again the relation can go either way. Discrete particles (atoms, molecules, stars, cells, individuals, ...) can give rise to continuum models at a higher scale. However, out of underlying continua, also discrete structures can emerge. Partly, these are simply physical processes, for instance when interstellar dust condenses into a star by the force of gravity. Partly, the discrete entities are crucial constructs of models. The gene as a concept in molecular or evolutionary biology emerges from some underlying level of molecular interactions or reproductions of individuals in a population, see Jost and Scherrer. The continuous dynamics of the Hodgkin-Huxley model gives rise to the discrete event of a spike. Or putting it somewhat differently, the neuron as a biological system transforms an analogous input into a binary output, that is, it chooses between the alternatives of spike vs. no spike. Thus, models of information transmission in neural systems can operate with discrete events as the carriers of discrete bits of information. From a formal point of view, such phenomena are genuinely nonlinear, and there exist theoretical concepts, like those of critical threshold, bifurcation, or phase transition, to analyze them, see for instance (Jost 2005, 2014). Here, however, rather than analyzing the dynamical mechanisms that give rise to such discrete events in a continuous setting, I want to utilize this as an example of a higher level description in discrete terms with an underlying continuous dynamics at a lower level. Interestingly, at the level of discrete events, the spiking pattern of a neuron is often modelled as a Poisson process (see for instance Dayan and Abbott 2001; Gerstner and Kistler 2002; Jost to appear), that is, as a stochastic process, instead of a deterministic one. Here, the reason why one switches from a deterministic to a stochastic description is somewhat different from the case of the Boltzmann gas. In the latter case, a statistical description is called for because of the larger number of particles involved and the infeasibility or impossibility to measure all of them. Here, in contrast, we have a single event, the generation of a spike by a neuron that at one level is modelled by a deterministic dynamical system.[6] Also, this system does not exhibit chaotic behavior that would make long term predictions impossible. Of course, near the critical threshold above which the neuron spikes and below which it returns to rest, the dynamical behavior naturally is unstable in the sense that arbitrarily small perturbations or fluctuations can induce a switch from one state to the other. At another, more abstract, level it is instead modelled as a stochastic process. This is not only simpler, because it dispenses us of having to deal with the subtle nonlinear behavior of the Hodgkin-Huxley system, but it also offers the advantage that we can now look at the relation between the strength of the neuron's input and the single parameter that characterizes the

[6]Interestingly, stochastic perturbations of this deterministic system produce genuinely nonlinear effects, see Gutkin et al. (2009) and Tuckwell and Jost (2012).

stochastic process, the firing rate of the neuron. Put simply, the hypothesis would be that the neuron fires more frequently on average when it receives a stronger input. (Whether, or perhaps more precisely, to what extent neuronal systems really employ a rate coding as the preceding might suggest is an unresolved and intensely debated issue in the neurosciences, but here, I do not enter that discussion, and rather refer to Grün and Rotter (2010) and Wibral et al. (2014) for the information theoretical analysis of neuronal spike trains.)

I now turn to the formal analysis of the relation between different levels, following (Pfante et al. 2014a). (Let me also mention earlier results in Shalizi and Moore (2003) and Görnerup and Jacobi (2010) and many other papers, and the case study in Pfante et al. 2014b.)

We consider a process

$$\phi : X \to X'. \tag{4}$$

This could, for instance, be the transition from the state of a system at time t to its state at time $t+1$. We assume that this is a Markov process, in the sense that knowledge of the state at prior times $t-1, t-2, \ldots$ does not contain any information beyond that contained in the state at time t relevant for the state at time $t+1$. That is, the future is conditionally independent of the past given the present. We also have an operator

$$\pi : X \to \hat{X} \tag{5}$$

that is considered as a projection, coarse graining, averaging, or lumping. This simply means that the $\hat{}$ indicates a description at a higher level, with less detail and resolution. The question then is whether we can transform the transition directly at the higher level, that is, whether there exists a process

$$\psi : \hat{X} \to \hat{X}'. \tag{6}$$

that is self-contained in the sense that it does not depend on anything in X that is not already present in the upper level $\hat{X}$. In other words, we ask whether we can close up the following diagram (make it commutative in mathematical terminology Jost 2015).[7]

[7]A perhaps somewhat technical point concerning the representation by this diagram: Often, one thinks of a projection as going down, instead of up, and one would then represent X in the top and $\hat{X}$ in the bottom row. Since, however, we think of $\hat{X}$ as a higher, more abstract, level, we rather represent that higher level in the top row of our diagram.

$$\begin{array}{ccc} \hat{X} & \xrightarrow{\Psi} & \hat{X}' \\ \uparrow \pi & & \uparrow \pi \\ X & \xrightarrow[\phi]{} & X' \end{array} \tag{7}$$

Expressed as a formula, commutativity means that

$$\psi(\pi(x)) = \pi(\phi(x)) \tag{8}$$

for all $x \in X$.

There exist several criteria to make this precise.

I **Informational closure**: All we need to know to determine the state $\hat{X}'$ or to predict its statistical properties is already contained in the state $\hat{X}$. The higher process is informationally closed, i.e. there is no information flow from the lower to the higher level. Knowledge of the microstate will not improve predictions of the macrostate. The upper level process is self-consistent in the sense that it does not need to perpetually draw information from or about the lower-level states.

II **Observational commutativity**: It makes no difference whether we perform the aggregation first, and then observe the upper process, or we observe the process on the microstate level, and then lump together the states. In that sense, the upper level process seems autonomous, as once initialized, it appears to unfold on its own, without needing any further updating by details from the lower level process.

III **Commutativity** in the sense of (8): There exists a transition kernel ψ such that the diagram (7) commutes.

IV **Markovianity**: $\hat{X}, \hat{X}'$ forms again a Markov process. We recall that we assume that X, X' yield a Markov process, that is, the current state of lower level process contains everything needed to compute its next state, and it does not need to draw upon past states any further, as these have transmitted all relevant information to the current state. But it does not follow in general that the upper process is Markovian as well, and so, IV indeed is a nontrivial condition. It could compensate a lack of information about the current lower level state by structural constraints in order to also utilize information from past states. In particular, at the upper level, information about past states can improve its prediction about the next state, whereas such information is not useful at the lower level. See the example in Pfante et al. (2014b), and also the discussion in Sect. 9.

We show in Pfante et al. (2014a) that I implies both II and IV, and II implies III, whereas IV does not imply III in general. Also, for a deterministic process, but not necessarily for a stochastic one, conversely III implies II which in turn implies I. III of course is a formalization of I and II, and as we show, in the deterministic case, they are equivalent. In the stochastic case, the notions turn out to be somewhat different, as information concerns statistical properties, and these are different from the results of particular observations. IV, of course, is a condition that is meaningful only for stochastic dynamics.

6 Data Without Underlying Objects?

When we start with data, then an object behind these data is first a hypothesis, a construct. A well discussed example are the gigantic data sets created by modern particle accelerators where the elementary particles that are supposed to produce those data are postulated by theory, but where the only evidence for them has to be extracted from those data (for a conceptual discussion, see Falkenburg 2007). More generally, physics studies phenomena and data, like the scattering matrix in quantum mechanics that records the input and output of a quantum mechanical system. Physicists do not think in terms of objects, in contrast to the picture of Newtonian physics depicted above. In fact, precisely these scientists are those that drive the data analysis approach also in other domains. For instance, neoclassical economic theory has developed elaborate models for describing the functioning of an (idealized) economy, but then came the so-called econophysicists that simply took the data from financial and other markets and searched for statistical regularities (Mantegna and Stanley 1999). They used methods from nonlinear time series analysis, for instance. Those methods are not sensitive to the origin of the data, nor do they depend on any object oriented models. Rather, they look for intrinsic regularities in possibly chaotic dynamics (Kantz and Schreiber 1997). For instance, they identify (chaotic) attractors and determine their (possibly fractal) dimension. These, however, are not ad hoc methods; they are based on deep insights from the theory of dynamical systems. For example, the center manifold principle, see e.g. Jost (2005), (which, for instance, is the basis of Haken's slaving principle Haken 1983) tells us that in a dynamical system, typically most directions are quickly relaxing to their equilibrium, and the essential aspects of the dynamics are determined by very few slow variables.

In fact, it had already argued by Koopmans (1947) in his review of a book by Burns and Mitchell (1946) that measurements of economic quantities without underlying economic concepts are useless; and that time, the criticism was directed against a use of econometric techniques without a theoretical framework grounded in economics, but the same type of criticism would also, and perhaps even more forcefully, apply to the approach taken by econophysics.

We should also remember that Newton's theory that can derive and predict the motion of celestial objects is a rather exceptional feat even in the history of astronomy. Astronomy started with data collection, and these data then were typically organized by models with an ontological status that was dubious at best. The old Babylonians simply interpolated between observational data by assuming that the positions of the celestial bodies varied linearly between the extrema and then suddenly changed direction, see the systematic studies of Neugebauer (1967, 1969), who speaks of zigzag functions to describe this interpolation scheme. The Ptolemaic model as described in the Almagest employed more and more epicycles without really accounting for their ontological status. The Aristotelian system of the celestial spheres was more explicit in that regard, but more difficult to reconcile with the astronomical data. (In fact, Ptolemy also not only tried to represent data as in the Almagest, but also developed a planetary hypothesis in which Aristotle's celestial spheres were fattened to provide space for the epicycles needed in Ptolemy's system to account for the astronomical data.) And Kepler had to labor painstakingly for many years through the data compiled by Brahe. He first tried a preconceived scheme, an organization of the solar system in terms of Platonic solids, which he then abandoned because it did not fit the data. In particular, he was concerned with the orbit of Mars. He tried many variants before he finally arrived at fitting the orbit as an ellipse. Thus, by trying to account for the data, in the end he succeeded in discovering his laws of planetary motion. As already discussed, the laws that Kepler had empirically discovered were then derived by Newton from his law of gravity, that is, within an explicit physical model. Of course, without knowing Kepler's laws, Newton might not have found his theory of gravity.

I should concede, however, that this picture of large data sets that are first collected without an adequate theory and only subsequently inspire deep physical theories is not without exceptions. At least one exception springs to mind, Einstein's theory of general relativity. This theory was developed on the basis of an abstract principle, general covariance, and was only subsequently tested against and confirmed by empirical data.

Also, the traditional view of physics is that it does not collect more or less arbitrary data, but conducts specific experiments whose results acquire their meaning within an established theoretical framework.

On the other hand, however, we should also discuss the approach of Alexander von Humboldt. The novelty of his expeditions rests in the systematic and comprehensive collection of all kinds of data with all the measuring devices available at his time, at a time when there was still very little theoretical understanding of the processes shaping the ecology of the earth. His expedition in South America took some years, but the evaluation of the data collected during that expedition took him several decades. His work then launched the scientific discipline of physical geography and helped and inspired several other scientific fields, even though no single coherent and encompassing theory emerged from his data. Thus, Humboldt had gathered huge amounts of data, but these data did reveal only very few coherent patterns, like the dependence of climatic zones on altitude and latitude. In other words, not only was the model missing, but also the data analysis largely failed in discovering general structures.

7 Towards an Abstract Theory

In fact, it is a quite general finding that there exist profound analogies between models in physically very different domains. It is actually one of the main driving forces of mathematics to consider the corresponding structures and relations abstractly and independently of any particular instantiation and to work out general theories. This will then make it possible, in turn, to apply these mathematical theories to new domains. It is then irrelevant whether such a domain constitutes some independent reality or whether one simply has a collection of data. For instance, similar statistical phenomena show up in quantum mechanics, in the analysis of biological high-throughput data, or in the description of social phenomena.

Understanding the laws and regularities of the transition between scales (see e.g. Pavliotis and Stuart (2008) for the mathematical background), or in a more ambitious formulation, a mathematical approach towards the issue of emergence (for instance, Jost et al. 2010), is a theoretical challenge that transcends individual domains. Nevertheless, it should lead to fundamental insight into the structure of reality, at an abstract level.

In particular, it is a fundamental question in the theory of complex systems to what extent general laws apply across different domains and disciplines, and where the applicability of a general theory ends and a more concrete modelling of the details of the specific system is required. For instance, which analytical concepts and mathematical tools apply simultaneously to cells, neural systems, psychological systems, and societies, or at least to several of them, and where do the specific peculiarities of each of these fields enter?

In a different direction, we may ask for a theory of structure in high-dimensional spaces. As will be explained below, even though the data may possess a large number of degrees of freedom, these degrees of freedom typically do not vary completely independently in large data sets, but rather obey some nonlinear constraints. Thus, we are dealing with intrinsically lower dimensional geometric structures in high dimensional spaces. Since the particular structure will usually not be known before having analyzed the data, we need to consider spaces of such structures. This leads to new challenges for the mathematical field of geometry. These spaces will then carry a probability density, telling us about the a-priori likelihood of finding them in a particular data set. In statistical terms, we then have some prior hypothesis about the data set, and such a prior then has to be turned into a posterior by Bayes' rule on the basis of the data observed. Of course, this then should be done in an iterative manner. In a different direction, we may employ methods from algebraic topology or metric geometry in order to discover robust qualitative features of specific data sets, see Carlsson (2009), Bačák et al. (2015).

8 The Challenge of Big Data

Both the abundance of huge data sets in almost all disciplines and the availability of the computational power to formally analyze them are quite recent developments. We have the high-throughput data in molecular and cellular biology generated by gene sequencing, microarrays or various spectroscopic techniques, the imaging data in the neurosciences, the email or movement data of mobile phone users, the data of transactions in financial markets at millisecond resolution, the linking patterns of the world wide web, and so on.

And almost uniformly across disciplines, we see a major transition from model driven to data driven approaches. The latter approaches typically depend on the statistics of large data sets. These statistics are utilized automatically, and statistical regularities within large data sets are only used in an implicit fashion, without making them explicit. Examples abound. In molecular biology, the problem of protein folding had for a long time been approached by explicit physical models. The problem consists in predicting the three-dimensional folding pattern from the linear sequence of the amino acids constituting a polypeptide on the basis of molecular attractions and repulsions. As the energy landscape is quite complicated, there are typically many metastable states, and finding the configuration of minimal energy, i.e., the configuration supposedly assumed by the polypeptide in the cell, therefore is computationally quite difficult, even though powerful Monte Carlo type schemes have been developed. Recently, however, it turns out that the best predictions are achieved by data bank searches without any physical model in the background. One simply compares the sequence at hand with those sequences where the folding pattern is already known, for instance by X-ray crystallography, and then makes a statistical prediction based on sequence similarities.

Perhaps the preceding becomes clearer when we consider a hypothetical approach to weather forecast. A model driven approach develops a detailed dynamical model of cloud formation and movement, ocean currents, droplet formation in clouds, and so on, and measures the many parameters of such a model, or perhaps also tries to fit some of them on the basis of observations. Such a model would be given by a system of coupled partial differential equations (PDEs), for which one has to solve an initial value problem. The initial values are determined by current measurements on a grid of measurement stations that is dense as possible, naturally with a higher density on land than on the oceans. Since the dynamics described by the PDE model tends to have chaotic aspects (see also the discussion in Sect. 3), the precision of the measurements of the initial values is of utmost importance in order to have somewhat accurate predictions for a few days. Likewise, the details of the model and its parameters are crucial. In contrast, a data driven approach would like depend on accurate measurements, but it would then try to identify those constellations in the past whose values are closest to those presently recorded, and then use the known weather dynamics from the past for those constellations to predict the weather derived from the current values. This would simply require large data bases of the past weather recordings, and perhaps some scheme of taking weighted averages over similar constellations in the past, but it would not need any model of weather dynamics. Such an approach should then naturally be expected to improve as the data base grows over time.

In the geosciences, the standard scheme consists now in estimating 2-point correlators from noise correlations at different locations (see e.g. Ritzwoller et al. 2011; Garnier and Papanicolaou 2012), instead of using physical models of, for instance, wave propagation in geological media. The noise sources can be physical or caused by human activities. While the differences matter at a technical level because of different noise characteristics, this does not affect the principle of the method.

Or to present another example that plays a prominent role in Napolitani et al. (2011), microarrays record the simultaneous expression of many different genes in a particular cell condition by exposing the corresponding RNAs in the cell simultaneously to an array of pieces of complementary DNA sequences and then simply recording which of those DNA sequence pieces find RNA partners. The underlying biological rationale is the following. The DNA pieces are part of the genome that is selectively transcribed into RNA which may then be further translated into polypeptides, the building blocks of proteins. Proteins carry out most of the essential operations in a cell. Therefore, mechanisms of gene regulation should ensure that precisely those proteins are manufactured that are needed by the cell in a particular situation. Therefore, precisely those pieces of DNA should be transcribed into RNA that encode the right proteins (there are some problems with this oversimplified account, see Scherrer and Jost (2007), but the microarray technology happily ignores them). Thus, a microarray tests the expression patterns and intensities of many DNA segments – which stand for genes in this simplified model – simultaneously, because the transcribed RNAs bind to the complementary DNA pieces offered by the microarray.

In computer linguistics, for the purposes of automatic translation, models grounded in syntactic and semantic theory are replaced by statistical techniques that simply utilize cooccurence patterns of words in large corpora. In particular, Google uses so-called n-gram models for the purpose of automatic translation. This simply means that one derives the relative frequencies of strings of n words from databases containing trillions of entries. $n = 5$ is a typical value. That is, the meaning of a word – insofar as one should still speak about meaning here, in the absence of any semantic concepts – is determined by the environment, that is, a few words preceding and following it, in which it occurs. We shall analyze the conceptual shifts that this implies in more detail in Sect. 9.

In economics, as already mentioned, techniques from statistical data analysis, like nonlinear time series analysis, are applied to financial data sets in order to find subtle patterns that do not follow from theoretical models. This is the new field called econophysics. Recently, econophysicists also started to develop economic models. They thereby depart from a purely statistical approach that solely analyzes economic or financial data and now build models of economies or financial markets themselves. Their models, however, are in contrast to the classical economic models that employ a so-called representative agent. The latter stands for models with many identical typical agents that are ideally analytically tractable. The agent based models of econophysicists instead utilize very simple, but possible diverse and heterogeneous agents that fit well to large scale computer simulations. That is, instead of a single type of agent that might be rather sophisticated and grounded in economic theory, here many simple agents are employed. These agents depend on a couple of parameters, and the parameter values can and typically do differ between the agents. The representative agents of economics are usually assumed to be fully rational – the argument being that non-optimal agents are quickly exploited by their more clever competitors and thereby driven out of the market. The diverse agents of agent based models are not at all assumed to be rational. They rather follow relatively simple empirical rules, and the purpose of the models is to uncover the collective effects of such behavior of many agents through systematic computer simulations. The validity of the simulation results usually remains unclear, however. An important issue is that the models of (neo)classical economic theory strive for the ideal of exactly solvable equations, or at least for describing the economy by a small and carefully specified set of explicit equations. On the basis of these equations and their solutions, they want to achieve analytical insights into the working of the economy. The agent based models of econophysicists, in contrast, employ much less rigid models, possibly with many parameters, and large numbers of equations that can only be numerically solved. Analytical insight no longer is the foremost aim, and one rather wants to identify unexpected nonlinear effects and critical transitions between different regimes triggered by small variations of crucial parameters.

Even in elementary particle physics, after the confirmation of the Higgs boson, that is, a prediction made by the standard model, at the LHC, in the future one will probably move towards the automatic analysis of large scale scattering data in order to find unpredicted events that may not fit into any current theoretical model. In any case, already for finding evidence for the Higgs boson, to a large extent automatic data analysis methods have been employed. Such methods may find patterns of correlations or untypical events in huge data sets by completely implicit methods. As described and analyzed for instance in Falkenburg (2007), the empirical evidence is gathered in four steps. At the basis, there are position measurements. Adjacent position measurements are then combined into tracks. Adjacent tracks in turn are combined into events. Finally, statistical ensembles of scattering events contain resonances. All this is done in a completely automated way. The tools have no special affinity to particle physics, even though particle physicists are among the people making the most advanced use of them. For instance, so-called neural networks encode the patterns of their training sets in a rather indirect and completely implicit manner in synaptic connection weights (see e.g. Bishop 1995). When they are subsequently applied to the real data, they produce corresponding associations which may then be interpreted as patterns in the data set. Neural networks and other such schemes find applications in a wide range of data domains. In fact, the methods themselves also change, and for instance, neural networks are sometimes replaced by other methods like support vector machines.[8]

This phenomenon, the transition from model to data driven approaches, occurs not only in individual and specific domains, but also at the formal level. The field of scientific computing currently undergoes a similar transition. The field of machine learning is concerned with the development of techniques for the automatic analysis of large data sets. Statistics, the science of finding structure in data, is moving into a similar direction, and in fact a profound convergence between machine learning and statistics seems to take place. In particular, the computer intensive Bayesian methods take over much of more traditional parametric statistics. Often, such methods are combined with stochastic search algorithms like Monte Carlo methods or Markov type models. Lenhard (2013) also points out the interdependence between data dynamics and computational modelling.

Big data are said to be characterized by large or even huge values of the three **V**s, that is,

- **V**olume: often petabytes ($1000^5 = 10^{15}$ bytes)/day, possibly more
- **V**ariety: heterogeneity of data types, representation, and semantic interpretation
- **V**elocity: arrival rate and reaction time

Processing big data requires adapted strategies and methods, and it can be decomposed into five phases

1. Acquisition and recording: filtering, compression, metadata generation
2. Information extraction, cleaning, and annotation
3. Integration, aggregation, and representation; data base design
4. Analysis and modeling; querying and mining the data
5. Interpretation; visualization; possibilities of interaction of human observers with machine processing

For our purposes, phase 4 is the most relevant. Although computer power is rapidly growing, and cloud computing or even access to supercomputers becomes ever more available, large data sets still may give rise to the "curse of dimensionality". This means that computer time will increase exponentially with the number of degrees of freedom and therefore quickly exceed even the capacities of supercomputers, unless clever use of specific structures within a data set is made. Therefore, corresponding data analysis techniques are being developed. An example is compressed sensing, see Donoho (2006), Candès et al. (2006), Fornasier and Rauhut (2010) and Foucard and Rauhut (2013). Here, the idea is to use a specific sparsity assumption, for instance that a large and complicated acoustic data set might be generated by a small number of sound sources only. For instance, the sound sources might be humans carrying on conversations in a crowd. Formally, sparsity

[8]Support vector machines are efficient classifiers that use a high-dimensional linear feature space (see Christianini and Shawe-Taylor 2000; Schölkopf and Smola 2002; Steinwart and Christmann 2008).

means that a high-dimensional vector or matrix possesses only few entries that are different from 0. One does not know beforehand, however, which are the nontrivial ones and how many of them there really are. Thus, there is a not very explicit, but still rather powerful and constraining structural assumption. Or, in a similar vein, one might assume that the data points one has in some high-dimensional space are in fact constrained to some low-dimensional manifold (Belkin and Niyogi 2003). This low-dimensional manifold, however, need not be a linear space, and therefore, even though it possesses only few intrinsic degrees of freedom, it may still stretch into many of the ambient dimensions (for the background, see Jost 2011). Thus, the data are supposed to possess some rather constraining type of regularity, but again, this regularity is not explicit, and it is the task of a machine learning scheme to use such an abstract assumption in an efficient manner. When one incorporates such an assumption, one may drastically reduce the number of computations needed to identify the underlying structures from the data set, for instance the sound sources in the above example. The mathematical techniques required are typically nonlinear and therefore on the one hand more flexible in discovering and utilizing hidden regularities and patterns in the data, but then, on the other hand, require more specific, but ideally still automatic, adaptations to the data set at hand.

In many respects, human cognition is still superior to automatic data analysis techniques. In order to utilize the power of human cognition, however, it is necessary to convert the data into a format that is familiar to humans. This leads us into the domain of visualization, a newly emerging scientific discipline between computer science, mathematics, and psychology. Thus, the aim is to develop formal methods that can convert a data set into a form in which humans can easily discern patterns. At a more advanced level, the combination of machine learning tools and human perception may become interactive, which was the goal of the EU funded research project CEEDs (The Collective Experience of Empathic Data Systems).

9 Big Data and Automatic Translation, or a Paradigm Shift From Linguistic Theory to Language Processing

Modern linguistic theory and philosophy is founded upon basic oppositions, for instance between

- *langue* (the abstract system of a language) vs. *parole* (the concrete utterance) (de Saussure 1995)
- *diachronous* (across time) vs. *synchronous* (simultaneous) (de Saussure 1995)
- *competence* (the ability for the correct syntax of one's native language) vs. *performance* (the actual production of utterances) (Chomsky 1965)
- *deep structure* vs. *surface structure*, i.e., a language independent representation in an abstract structure is transformed into a sentence according to the syntax of a specific language (Chomsky 1965), and this transformation obeys the general

rules of abstract grammar; the latter have become more general and abstract themselves in the course of Chomsky's work (Chomsky 1981, 1995)

- *spoken* vs *written* language, again from de Saussure (1995) and emphasized more recently by Derrida (1967).

Perhaps the most important opposition is that between de Saussure's (1995)

- *paradigmatic* alternatives and *syntagmatic* series; this means that a word at a given position in a sentence can be chosen paradigmatically from a list of words that could grammatically occupy this position, but has to obey the syntactic rules of the sentence. This is also an opposition between *absence* and *presence*, or between *selection* and *constraint*; the alternatives that have not been chosen are absent, but the syntactic constraints are present through the other words in the sentence.

These oppositions are (more or less) formal, but not directly mathematical. They gave rise to a formal theory of language, and for some time, one attempted to utilize that theory for purposes of automatic translation. The idea was essentially to automatically infer the grammatical relationships within a given sentence, that is, the dependencies between the different words and grammatical tokens and the internal references, like the referents of pronouns and anaphora. That grammatical structure could then be transformed into the corresponding structure of another language, and the meanings of the individual words could be correlated with the help of good lexica. For instance, for some time (Pollard and Sag 1994) was popular as a suitable grammatical theory for such purposes. All such attempts, however, have more recently been brushed aside by the big data approach to automatic translation, as developed and pushed in particular by Google. That approach completely ignores, and often even ridicules, linguistic theory, and rather draws upon correlations in huge linguistic corpora. My purpose here is to analyze the underlying conceptual shift and to describe what mathematical structures are behind this approach. Those mathematical structures are completely different from those developed from the context of formal linguistics. And the people working on automatic translation within this paradigm express little interest, if at all, in the conceptual basis of their endeavor. Nevertheless, that will be important for this essay.

Thus, the oppositions sketched above no longer play a role. Instead, we see the *corpus*, that is, a data base of texts in the language(s) in question, as the single basic element. The corpus scales by size. Thus, a small corpus might be seen as corresponding to *parole*, whereas big corpora can approach the *langue* side of de Saussure's dichotomy, and the same applies to Chomsky's dichotomy between *competence* and *performance*. More precisely, there no longer is any such abstract thing as *competence*. The possibly huge collections of linguistic data are all there is. Corpora do not care much whether they have been assembled from contemporary, hence essentially simultaneous texts or whether they result from scanning texts over longer periods of time. The only relevant criteria are of a practical nature, digital availability and computational capacity. Also, corpora can as well be based on automatically recorded spoken language as on written texts. The difference is again largely irrelevant, or at least plays no basic conceptual role. (More precisely, what remains is the technical difference between off-line and on-line processing and translation.)

More importantly and interestingly, the opposition between de Saussure's *paradigmata* and *syntagmata* is also resolved. This best explained through the mathematical concepts of stochastic processes and Shannon information (see e.g. Cover and Thomas 1981; Shannon and Weaver 1998; MacKay 2003; Jost 2005). De Saussure talked about alternative words at a given position in a sentence. This is qualitative, but not quantitative. Information theory, in contrast, would quantify the probabilities of different words to occur at such a position. These probabilities then will not only depend on the abstract properties of that position, but also on the concrete words before and behind it in the sequential string forming the sentence or the text. That is, we do not just have probabilities for the occurrences of words at a given position, but we rather have transition probabilities from a word, or more precisely, a segment of a few words, to the next. In fact, automatic translation today mostly works with pentagrams, that is, overlapping strings of five words.[9] Thus, we no longer have de Saussure's qualitative opposition between selection and constraints, but both aspects are combined and made quantitative within the concept of a stochastic process. (In fact, this is a slight oversimplification. A stochastic process would occur if one has to guess or produce the next word on the basis of those preceding it. Thus, a process describes a temporal sequence as in actual speech. A corpus, however, is not a process unfolding in time, but is simultaneously given as a whole. Therefore, the probabilities are determined not only by the preceding words, but also by the following ones. The underlying mathematical principle, however, is not fundamentally different.) Thus, while the big data approach to automatic translation does not care about an underlying conceptual structure, its analysis nevertheless sheds some light on the limitations of formal language theory, and it points towards mathematical structures that replace the qualitative oppositions by quantitative probabilities. These probabilities are nothing but relative frequencies of pentagrams, i.e., certain word constellations, and they can be automatically computed from the corpora at hand. The larger the corpus, the more accurately such probabilities can be determined, in the sense that further additions to the corpus will likely change them only very little. In a certain sense, this approach asks more precise questions than formal linguistics. When analyzing a temporal sequence, like a spoken text, it would not simply ask at each instant "what could come next?", but rather "how well can I guess what comes next?", and the latter is quantified by Shannon's information (Shannon and Weaver 1998).

Of course, this approach to automatic translation has its limitations. It cannot capture long range dependencies. In a nutshell, the basic assumption underlying the

[9]In the terminology of statistical physics, the correlation length of word sequences in texts becomes relatively small after five words. In fact, one would expect that it never becomes exactly zero, that is, there do exist correlations of arbitrary length, whatever small. Thus, in technical terms, when moving from a word to subsequent ones in a text, we have a stochastic process that does not satisfy a Markov property for strings of words of any finite length.

approach is that the transition between overlapping pentagrams satisfies a Markov property, that is, no information from more than five words back in the sequence is needed for the probability of the next word. From the syntactic perspective, this for instance does not adequately capture the structure of the German language where the different components of the verb can be separated by many words in a sentence. From the semantic perspective, the meaning of some sentence in a text may refer to other, much earlier parts of that text, or even to a context outside the text itself. In formal terms, let us recall also our discussion about Markovianity at different levels in Condition IV in Sect. 5.

One of the central, but to a large extent still unresolved, issues of linguistics, that between syntax and semantics, between structure and meaning, is simply bypassed by automatic translation. A human translator would first try to extract the meaning of a text and then express that meaning as well as she can in the other language. Automatic translation, in contrast, simply transforms one structure into another one.

10 The Issue, Again and Hopefully Clearer

We are faced with the following alternative to which no general answer can be given, but which needs to be evaluated in each individual situation.

1. The significance of data collected depends on the specific context from which they are taken, and they cannot be fully understood without that context.
2. Data sets typically possess internal structure, and much of such structure generalizes across different disciplines, domains and contexts, or is at least accessible to context independent methods. Identifying and understanding that structure with formal tools will then in turn enable us to use the data to learn something new and insightful about their context.

This is nothing but the old alternative between specificity and generality, as, for instance, already described by Kant. It is easy to deplore too much of an emphasis on either side and support this by supposedly misguided scientific case studies, and to declare a one-sided practice as agnostic or unscientific. The real challenge consists in finding the appropriate balance between the two aspects in each case. Currently, big data sets offer new opportunities for formal methods and computational models, and this may shift the balance for a while.

Concerning the role of mathematics, this might transcend the alternative between a mathematics of content that establishes and analyzes relations between the concepts of a theory and a purely auxiliary mathematics that is relegated to the role of data handling and preprocessing. What should emerge rather is a mathematics that develops new abstract concepts for analyzing and representing data spaces. For instance, a partial merging of the domains of statistics, high dimensional geometry, information theory and machine learning might take place. Such a mathematics would help to detect structures, as opposed to either formalizing structures proposed and developed by other scientific disciplines or to offering tools for fitting given data into such a structure. This process can only enhance the role and the importance of mathematics, as it becomes liberated from being instrumentalized by conceptualizations that are externally imposed. Instead of accepting models from particular domains, mathematics would itself propose abstract metamodels, like sparsity, smoothness, or symmetry.

11 Some Consequences

The preceding has important implications for the role of models and hypotheses in the scientific process. From the perspective of Bayesian statistics, we begin the process of scientific inquiry with prior hypotheses constructed by us, in whatever way we may find reasonable or on the basis of whatever prior experience we may have. During the process, such a prior hypothesis gets transformed into a posterior one on the basis of the observations made or the data acquired. This is achieved by applying a fixed rule, that discovered by Bayes. The role of models then is relegated to provide sets of parameters that have to be fitted to the data. This is in stark contrast to the ideal of a model in physics. Such a model should contain only very few, and ideally no free parameters at all that are not theoretically determined but need to be measured. Even for such a nearly ideal model, one may question whether empirical adequacy should constitute a proof of the correctness of the model. In a model with many free parameters that need to be fitted to the data, certainly empirical adequacy can no longer count as a proof of the correctness of the models, and indeed, in complex situations, the term "correct model" may lose its meaning entirely. **There no longer is such a thing as a correct model.** There may only be a superior fit of the parameters to the data collected according to unknown probabilities. See Hasse and Lenhard (2017) for a more detailed analysis of this issue. Alternatively, we might be seeking regularities in data on the basis of certain structural hypotheses, as described in Sect. 8. Again, such structural hypotheses like sparsity do not come from the data at hand, but are applied by us in order to get some handle on those data. In a certain sense, they constitute a structural prior, although that prior is usually not updated in a Bayesian manner.

While all this may sound rather agnostic, in practice one might get quite far with those schemes of Bayesian updates, parameter fitting, and structural hypotheses. It is challenge for mathematics to analyze this issue theoretically.

Putting it somewhat differently: Data analysis depends on prior structural assumptions. That could be the prior of a Bayesian approach, or it could be the choice of a family of models as in parametric statistics. It could be a general structural hypothesis like sparsity, a low-dimensional manifold (for instance arising as the center manifold of a dynamical system), certain invariances or symmetries etc. This is the epistemic side. At the ontological side, is there anything underlying the data matching those assumptions? One may argue that these assumptions are our constructions, and that therefore there is no guarantee of any correspondence with the data source. However, a simple theory can sometimes enable us to discover specific phenomena that would otherwise not emerge from beneath the heap of data. One may also argue, as I have tentatively done in this essay, that in some situations, such assumptions could be justified by the structures from which the data are derived, but there is no guarantee for that. The question then is whether those instances where it holds – we have discussed Newton's theory of planetary motion, the Schrödinger equation, biochemical reaction kinetics, the Hodgkin-Huxley equations or traffic dynamics – are just lucky coincidences, or whether there are more systematic structural aspects of reality – whatever that is – that make our speculations sometimes so successful.

12 The Dream of Mathematics

Does this bring mathematics closer to its dream of a science that is purely structural, abstract, and independent of specific content? We have argued already above that mathematics seeks regularities. Perhaps these regularities are discovered in specific domains, but hopefully, they should apply also in other domains, and ideally, they should be universal. For instance, while Lie's theory of symmetry groups was originally motivated by the symmetries of the systems of classical mechanics, the theory as such is abstract. In particular, through the work of W.Killing and E.Cartan, it lead to the classification of all possible continuous symmetries (see Hawkins (2000) for the history). This turned out to be of fundamental importance for quantum mechanics, and even further for quantum field theory. It permeates much of pure mathematics. It is also relevant in all applications where continuous symmetries arise. Even more generally, the notion of a group (Lie groups are a special type of continuous groups), as developed by Gauss and Galois, first arose from studying solutions of algebraic equations, that is, a specific mathematical problem, but it then become something of a paradigm of a mathematical structure as such, and it is now important in almost all fields of mathematics, as well as in many areas of physics and other disciplines.

Perhaps data science enables further steps in this direction. Here, even the original problems are no longer domain specific, as the symmetries and invariances of the systems of classical mechanics, but by their very nature already general and abstract when they apply to all kinds of large data sets. It is then natural that the mathematical tools developed to investigate these problems are at least as abstract as those problems themselves. Thus, data science may lead to an abstract theory of structures, that is, the purest form of mathematics. Incidentally, this brings fields of mathematics into focus that hitherto have been considered as applied and not pure mathematics. Statistics is an example. On the one hand, it can be mathematically treated in terms of the geometry of families of probability distributions, see e.g. Amari et al. (2007), Ay et al. (2017), and on the other hand, high-dimensional statistics and machine learning might also become a science of distributions of geometric objects in high-dimensional spaces, again independent of any specific content. But there should be mathematical structures even more abstract and general than that.

Acknowledgements I am very grateful to my partners in the ZIF project, Philippe Blanchard, Martin Carrier, Andreas Dress, Johannes Lenhard and Michael Röckner, for their insightful comments and helpful suggestions which have contributed to shaping and sharpening the ideas presented here. Martin Carrier and Johannes Lenhard also provided very useful comments on an earlier version of my text. Some of my own work discussed here was supported by the ERC Advanced Grant FP7-267087 and the EU Strep "MatheMACS".

References

Amari, S. I., Nagaoka, H., & Harada, D. (2007). *Methods of information geometry*. Oxford: American Mathematical Society.

Ay, N., Jost, J., Lê, H. V., & Schwachhöfer, L. (2017). *Information geometry*. Heidelberg: Springer.

Bačák, M., Hua, B. B., Jost, J., Kell, M., & Schikorra, A. (2015). A notion of nonpositive curvature for general metric spaces. *Differential Geometry and its Applications, 38*, 22–32.

Batterman, R. (2002). *The Devil in the details: Asymptotic reasoning in explanation, reduction, and emergence* (Oxford studies in philosophy of science). Oxford/New York: Oxford University Press.

Belkin, M., & Niyogi, P. (2003). Laplacian eigenmaps for dimensionality reduction and data representation. *Neural Computation, 15*, 1373–1396.

Bishop, C. (1995). *Neural networks for pattern recognition*. Oxford/New York: Oxford University Press.

Boltzmann, L. (2000). Entropie und Wahrscheinlichkeit, Ostwalds Klassiker 286, Harri Deutsch.

Breidbach, O., & Jost, J. (2004). Working in a multitude of trends. Species balancing populations. *Journal of Zoological Systematics and Evolutionary Research, 42*, 202–207.

Burns, A., & Mitchell, W. (1946). *Measuring business cycles* (Studies of business cycles, Vol. 2). New York: National Bureau of Economic Research.

Candès, E., Romberg, J., & Tao, T. (2006). Stable signal recovery from incomplete and inaccurate measurements. *Communications on Pure and Applied Mathematics, 59*, 1207–1223.

Cardy, J. (1996). *Scaling and renormalization in statistical physics*. Cambridge/New York: Cambridge University Press.

Carlsson, G. (2009). Topology and data. *Bulletin of the American Mathematical Society, 46*, 255–308.

Chomsky, N. (1965). *Aspects of the theory of syntax*. Cambridge: MIT Press.

Chomsky, N. (1981). *Lectures on government and binding*. Dordrecht: Foris.

Chomsky, N. (1995). *The minimalist program*. Cambridge: MIT Press.

Close, F. (2004). *Particle physics: A very short introduction*. Oxford/New York: Oxford University Press.

Cover, T., & Thomas, J. (1981). *Elements of information theory*. Hoboken: Wiley.

Christianini, N., & Shawe-Taylor, J. (2000). *An introduction to support vector machines and other kernel-based learning methods*. Cambridge/New York: Cambridge University Press.

Dayan, P., & Abbott, L. F. (2001). *Theoretical neuroscience*. Cambridge: MIT Press.

Derrida, J. (1967). L'écriture et la différence. Seuil: Paris.

Devaney, R. L. (1986). *An introduction to chaotic dynamical systems*. Benjamin/Cummings: Menlo Park.

Di Francesco, P., Mathieu, P., & Sénéchal, D. (1997). *Conformal field theory*. New York: Springer.

Donoho, D. (2006). Compressed sensing. *IEEE Transactions on Information Theory, 52*, 1289–1306.

Falkenburg, B. (2007). *Particle metaphysics. A critical account of subatomic reality*. Berlin/New York: Springer.

Fornasier, M., & Rauhut, H. (2010). *Theoretical foundations and numerical methods for sparse recovery*. Berlin/New York: de Gruyter.

Foucard, S., & Rauhut, H. (2013). *A mathematical introduction to compressive sensing*. New York: Birkhäuser.
Garnier, J., & Papanicolaou, G. (2012). Correlation based virtual source imaging in strongly scattering random media. *Inverse Problems, 28*, 075002.
Gerstner, W., & Kistler, W. (2002). *Spiking neuron models*. Cambridge: Cambridge University Press.
Görnerup, O., & Nilsson Jacobi, M. (2010). A method for finding aggregated representations of linear dynamical systems. *Advances in Complex Systems, 13*, 199–215.
Green, M., Schwarz, J., & Witten, E. (1995). *Superstring theory* (2 Vols.). Cambridge/New York: Cambridge University Press.
Grün, S., & Rotter, S. (2010). *Analysis of parallel spike trains*. New York: Springer.
Gutkin, B., Tuckwell, H., & Jost, J. (2009). Inhibition of rhythmic neural spiking by noise: The occurrence of a minimum in activity with increasing noise. *Naturwissenschaften, 96*, 1091–1097.
Hacking, I. (1990). *The taming of chance*. Cambridge: Cambridge University Press.
Hacking, I. (1983). *Representing and intervening*. Cambridge/New York: Cambridge University Press.
Haken, H. (1983). *Synergetics* (3rd ed.). Berlin/New York: Springer.
Hasse, H., & Lenhard, J. (2017). Boon and bane: On the role of adjustable parameters in simulation models. In *Mathematics as a tool: Tracing new roles of mathematics in the sciences*. Cham: Springer. This volume.
Hausdorff, F. (1918). Dimension und äußeres Maß. *Mathematische Annalen, 79*, 157–179.
Hawkins, T. (2000). *Emergence of the theory of Lie groups: An essay in the history of mathematics* (pp. 1869–1926). New York: Springer.
Helbing, D. (2001). Traffic and related self-driven many-particle systems. *Reviews of Modern Physics, 73*, 1067–1141.
Hodgkin, A., & Huxley, A. (1952). A quantitative description of ion currents and its application to conductance and excitation in nerve membranes. *The Journal of Protozoology, 117*, 500–544.
Hofrichter, J., Jost, J., & Tran, T. D. (2017). *Information geometry and population genetics*. Heidelberg: Springer.
Jaynes, E. T. (2003). *Probability theory: The logic of science*. Cambridge: Cambridge University Press.
Jost, J. (2001). *Bosonic strings*. Providence: American Mathematical Society/International Press.
Jost, J. (2005). *Dynamical systems*. Berlin/New York: Springer.
Jost, J. (2009). *Geometry and physics*. Berlin/London: Springer.
Jost, J. (2011). *Riemannian geometry and geometric analysis* (6th ed.). Heidelberg/New York: Springer.
Jost, J. (2013). *Partial differential equations* (3rd ed.). New York: Springer.
Jost, J. (2014). *Mathematical methods for biology and neurobiology*. London: Springer.
Jost, J. (2015). *Mathematical concepts*. Cham: Springer.
Jost, J. Mathematical neurobiology. Concepts, tools, and questions. Monograph, to appear.
Jost, J., Bertschinger, N., & Olbrich, E. (2010). Emergence. A dynamical systems approach. *New Ideas in Psychology, 28*, 265–273.
Jost, J., & Scherrer, K. A conceptual analysis of genetic coding and regulation in the light of modern molecular biology. Manuscript to be submitted.
Kabalak, A., Smirnova, E., & Jost, J. (2015). Non-cooperative game theory in biology and cooperative reasoning in humans. *Theory in Biosciences, 134*, 17–46.
Kantz, H., & Schreiber, T. (1997). *Nonlinear time series analysis*. Cambridge: Cambridge University Press.
Klipp, E., Herwig, R., Kowald, A., Wierling, C., & Lehrach, H. (2005). *Systems biology in practice*. Weinheim: Wiley-VCH.
Koch, C. (1999). *Biophysics of computation*. New York: Oxford University Press.

Koopmans, T., (1947). Measurement without theory. *The Review of Economics and Statistics, 29*, 161–172.

Lenhard, J., (2013). Coal to diamonds. *Foundations of Science, 18*, 583–586.

MacKay, D. (2003). *Information theory, inference, and learning algorithms*. Cambridge: Cambridge University Press.

Mandelbrot, B. B. (1982). *The fractal geometry of nature*. San Francisco: Freeman.

Mantegna, R., & Stanley, H. E. (1999). *An introduction to econophysics: Correlations and complexity in finance*. Cambridge: Cambridge University Pres.

Murray, J. (2008). *Mathematical biology* (2 Vols., 3rd ed.). New York: Springer.

Napolitani, D., Panza, M., & Struppa, D. (2011). Agnostic science. Towards a philosophy of data analysis. *Foundations of Science, 16*, 1–20.

Neugebauer, O. (1967). Problems and methods in Babylonian mathematical astronomy. Henri Norris Russell Lecture. *The Astronomical Journal, 72*, 964–972.

Neugebauer, O. (1969). *The exact sciences in antiquity* (2nd ed.). Dover.

Pavliotis, G., & Stuart, A. (2008). *Multiscale methods. Averaging and homogenization*. New York: Springer.

Pfante, O., Bertschinger, N., Olbrich, E., Ay, N., & Jost, J. (2014). Comparison between different methods of level identification. *Advances in Complex Systems, 17*, 1450007.

Pfante, O., Olbrich, E., Bertschinger, N., Ay, N., & Jost, J. (2014). Closure measures and the tent map. *Chaos, 24*, 013136.

Polchinski, J. (1998). *String theory* (2 Vols.). Cambridge: Cambridge University Press.

Pollard, C., & Sag, I. (1994). *Head-Driven phrase structure grammar*. Chicago: University of Chicago Press.

Riemann, B. (2016). Über die Hypothesen, welche der Geometrie zu Grunde liegen, edited with a historical and mathematical commentary by Jost, J., Springer, 2013; English translation: On the hypotheses which lie at the bases of geometry, Birkhäuser.

Ritzwoller, M., Lin, F. C., & Shen, W. (2011). Ambient noise tomography with a large seismic array, *C. R. Geoscience*. doi:10.1016.

Saari, D., & Xia, Z. (1995). Off to infinity in finite time. *Notices AMS, 42*, 538–546.

de Saussure, F. (1995). Cours de linguistique générale, published by C. Bally, A. Sechehaye and A. Riedlinger, critical edition by T. de Mauro, Payot.

Scherrer, K., & Jost, J. (2007). Gene and genon concept: Coding vs. regulation. *Theory in Biosciences, 126*, 65–113.

Scherrer, K., & Jost, J., (2009). Response to commentaries on our paper Gene and genon concept: Coding vs. regulation. *Theory in Biosciences, 128*, 171–177.

Schölkopf, B., & Smola, A. (2002). *Learning with kernels*. Cambridge: MIT Press.

Shalizi, C., & Moore, C. (2003). What is a macrostate? subjective observations and objective dynamics, Technical report. arXiv:cond-mat/0303625v1.

Shannon, C., & Weaver, W. (1998). In R. Blahut & B. Hajek (Eds.), *The mathematical theory of communication*. Urbana: University of Illinois Press.

Siegel, C., & Moser, J. (1971). *Lectures on celestial mechanics*. Berlin/New York: Springer.

Steffen, K., (1996). Hausdorff-Dimension, reguläre Mengen und total irreguläre Mengen (pp. 185–227). In E. Brieskorn (ed.), Felix Hausdorff zum Gedächtnis, Vieweg.

Steinwart, I., & Christmann, A. (2008). *Support vector machines*. New York: Springer.

Treschev, D., & Zubelevich, O. (2010). *Introduction to the perturbation theory of Hamiltonian systems*. Heidelberg/New York: Springer.

Tuckwell, H., & Jost, J. (2012). Analysis of inverse stochastic resonance and the long-term firing of Hodgkin-Huxley neurons with Gaussian white noise. *Physica A, 391*, 5311–5325.

van Fraassen, B. (1980). *The scientific image*. Oxford: Clarendon Press.

Weinberg, S. (1995/1996/2000). *The quantum theory of fields* (3 Vols.). Cambridge/New York: Cambridge University Press.

Wibral, M., Vincente, R., & Lizier, J. (Eds.). (2014). *Directed information measures in neuroscience*. Heidelberg: Springer.

Xia, Z. (1992). The existence of non-collision singularities in Newtonian systems. *Annals of Mathematics, 135*, 411–468.

Zinn-Justin, J.-J. (1993). *Quantum field theory and critical phenomena* (2nd ed.). Oxford/New York: Oxford University Press.

Zwieback, B. (2009). *A first course in string theory* (2nd ed.). New York: Cambridge University Press.

The manufacturer's authorised representative in the EU is Springer Nature Customer Service Centre GmbH, Europaplatz 3, 69115 Heidelberg, Germany. If you have any concerns regarding our products, please contact ProductSafety@springernature.com

Printed and bound by CPI Group (UK) Ltd, Croydon, CR0 4YY
01/10/2026
02225860-0003